Professors, Physicians and Practices in the History of Medicine

Archimedes

NEW STUDIES IN THE HISTORY AND PHILOSOPHY OF SCIENCE AND TECHNOLOGY

VOLUME 50

Archimedes has three fundamental goals; to further the integration of the histories of science and technology with one another: to investigate the technical, social and practical histories of specific developments in science and technology; and fi nally, where possible and desirable, to bring the histories of science and technology into closer contact with the philosophy of science. To these ends, each volume will have its own theme and title and will be planned by one or more members of the Advisory Board in consultation with the editor. Although the volumes have specifi c themes, the series itself will not be limited to one or even to a few particular areas. Its subjects include any of the sciences, ranging from biology through physics, all aspects of technology, broadly construed, as well as historically-engaged philosophy of science or technology. Taken as a whole, *Archimedes* will be of interest to historians, philosophers, and scientists, as well as to those in business and industry who seek to understand how science and industry have come to be so strongly linked.

More information about this series at http://www.springer.com/series/5644

Gideon Manning • Cynthia Klestinec
Editors

Professors, Physicians and Practices in the History of Medicine

Essays in Honor of Nancy Siraisi

Springer

Editors
Gideon Manning
Claremont Graduate University
Claremont, California
USA

Cynthia Klestinec
Miami University
Oxford, Ohio
USA

ISSN 1385-0180 ISSN 2215-0064 (electronic)
Archimedes
ISBN 978-3-319-56513-2 ISBN 978-3-319-56514-9 (eBook)
DOI 10.1007/978-3-319-56514-9

Library of Congress Control Number: 2017939877

Printed on acid-free paper

This Springer imprint is published by Springer Nature
The registered company is Springer International Publishing AG
The registered company address is: Gewerbestrasse 11, 6330 Cham, Switzerland

Acknowledgments

It was with admiration and goodwill toward Professor Nancy Siraisi that we (the editors) began to conceive this *Festschrift*. It was, however, the energy and dedication of the contributors, as well as Jed Buchwald, the general editor of the *Archimedes* series from Springer, that made this *Festschrift* possible. We wish to thank Jed and the contributors for their encouragement and patience. Manning also would like to acknowledge the sage advice of Noel Swerdlow and Mordechai Feingold throughout the production process; Steve Hindle and his staff at the Huntington Library, especially Juan Gomez, for creating an environment so conducive to work and research during the final preparations; and, most of all, the unwavering support of Melissa Pastrana, without whom Manning's role in this volume would never have been realized.

September 2016

Cynthia Klestinec
Gideon Manning

Contents

Contributors

Domenico Bertoloni Meli teaches the history of science and medicine at Indiana University, Bloomington. He is the author of *Equivalence and Priority: Newton Versus Leibniz* (1993, 1997), *Thinking with Objects* (2006), and *Mechanism, Experiment, Disease* (2011). He is currently working on a project on the visualization of disease from the Renaissance to the nineteenth century.

Ann Blair teaches early modern European history and book history at Harvard University. She is the author of *The Theater of Nature: Jean Bodin and Renaissance Science* (1997) and *Too Much to Know: Managing Scholarly Information Before the Modern Age* (2010) and articles in the history of science and of learned practices. She is currently examining the role of amanuenses in early modern scholarship.

Chiara Crisciani is professor of history of medieval philosophy at Pavia University in Italy. Her research focuses on medieval scientific thought, with special emphasis on scholastic medicine, the development of alchemy in the West, and the relationship between these two disciplines through the early modern period. Among her recent publications include the "Introduction to Arnold of Villanova" in the modern edition of his *Tractatus de humido radicali* (2010) and "Medicine and philosophy in the Middle Ages: sisters, companions, rivals" (2012).

Paula Findlen is Ubaldo Pierotti professor of Italian history and department chair at Stanford University. Her publications on early modern science and medicine include *Possessing Nature: Museums, Collecting, and Scientific Culture in Early Modern Italy* (1994); *The Italian Renaissance: Essential Readings* (2002); *Merchants and Marvels* (with Pamela Smith, 2002); and *Early Modern Things: Objects and Their Histories* (2013). Most recently, she has been enjoying one of Nancy Siraisi's fabulous footnotes for a project on Alessandra Giliani and other medieval women invented by their early modern Italian forgers.

Anthony Grafton teaches European history at Princeton University. His books include *Joseph Scaliger* (1983–93), *The Footnote: A Curious History* (1995), and (with Joanna Weinberg) *I Have Always Loved the Holy Tongue* (2011). He is currently studying early modern visions of the origins of Christianity.

Hiro Hirai has a Ph.D. (1999) in history of science and philosophy, University of Lille 3 (France) and is a research fellow at Radboud University (the Netherlands). He has published widely in early modern natural philosophy, medicine, and chemistry, including *Le concept de semence dans les théories de la matière à la Renaissance* (2005) and *Medical Humanism and Natural Philosophy* (2011).

Danielle Jacquart is directeur d'études EPHE, section des sciences historiques et philologiques. Her many books and essays include *Sexualité et Savoir médical au Moyen Âge* (1985), *La Science médicale occidentale entre deux renaissances* (1997), and, most recently, *À la recherche du corps médiéval: études sur la repré sentation médicale et son écriture* (2012).

Cynthia Klestinec is associate professor of English at Miami University of Ohio. In addition to *Theaters of Anatomy: Students, Teachers, and Traditions of Dissection in Renaissance Venice* (2011), her recent publications include *Renaissance Surgeons: Anatomy, Manual Skill and the Visual Arts* (2016) and "Translating learning surgery" (forthcoming in a special issue, coedited with D. Bertoloni Meli, of the *Journal of the History of Medicine and Allied Sciences*). Her current project explores the solid, architectural body in late Renaissance Italy.

Ian Maclean is professor of Renaissance studies (emeritus) at the University of Oxford and emeritus fellow of All Souls College. Among his recent publications are *Logic, Signs and Nature in the Renaissance: The Case of Learned Medicine* (2001); *Le monde et les hommes selon les médecins de la Renaissance* (2006); *Learning and the Market Place* (2009); and *Scholarship, Commerce, Religion: The Learned Book in the Age of Confessions* (2012). He is currently working on the learned book in the early enlightenment.

Gideon Manning is a visiting scholar at Claremont Graduate University and has taught the history and philosophy of science and medicine at the University of Pittsburgh, the College of William and Mary, and the California Institute of Technology. Among his recent publications include "The history of 'hylomorphism'" (2014), "Descartes and the Bologna affair" (2015), and "The Vitality and Importance of Early Modern Galenism" (forthcoming). He is currently editing a volume on the history of the concept of death as well as working to understand the role of medicine in early modern philosophy through a study of Rene Descartes and his reception.

Michael McVaugh is Wells professor of history (emeritus) at the University of North Carolina. He has written *Medicine Before the Plague* (1993) and *The Rational Surgery of the Middle Ages* (2006) and is one of the coeditors of the medical works of Arnau of Villanova (1975–). At present, he is engaged in a number of projects involving medieval medical translation from Arabic into Latin.

Vivian Nutton is an F.B.A. and professor of the history of medicine (emeritus) at University College London, writing particularly on medicine from the Greeks to the Renaissance. His publications include *Girolamo Mercuriale, De arte gymnastica* (2008); *Galen: On Problematical Movements* (2011); and *Ancient Medicine* (ed. 2, 2013). He is at present preparing a translation of Vesalius' *Institutiones anatomicae*, which will incorporate Vesalius' own notes for a never completed revision.

List of Figures

List of Illustrations

Introduction

The German word *Festschrift* is a compound of *fest*, which derives from the Latin *festum*, meaning "festival," and *schrift*, an abstractum of *schreiben*, which means "writing." It literally means "festival-writing," though in keeping with an academic tradition extending back to the nineteenth century, a more appropriate translation is "celebratory volume." And that is precisely what this is. The essays collected here celebrate Nancy Siraisi, distinguished professor emerita in history at Hunter College and the City University of New York (CUNY). Most *Festschriften* are tied to a momentous occasion, such as a seventieth birthday or a retirement, but this volume arose sui generis, from a collective desire to celebrate Siraisi's scholarly achievements and to acknowledge the professional and personal impact she has had on each of us, on the discipline of history, and on the history of medicine and science in particular. Siraisi's work is marked by lucid prose, profound originality, authoritative analysis of texts, and an accessibility that invites attention and response: she has been an invaluable guide through the periods spanning the Middle Ages to the seventeenth century, always with an eye toward a better understanding of medicine's connections to the broader intellectual, social, and cultural context in which it was taught, studied, and practiced. Her work embodies that rare combination of impeccable scholarship and deep insight guaranteeing its endurance, not only on its own terms but also through its influence on the rest of us.

The contributors to this volume came to know Siraisi and her work at different moments in their careers. Some have known her for more than 40 years, others have been collaborators, and still others were first her students, if only students of her work, before they came to know her personally. Some are primarily historians of medicine, but others historians of science, Europe, or the book. Some are from the United States, some from Europe, and one from Asia; in the spirit of the *Festschrift* genre, this is a volume with international provenance. Yet for all our differences of origin and expertise, we testify to the many positive influences Siraisi has had on our professional lives. This experience, shared by the contributors, led each of them to accept the invitation to participate in this volume. The essays, we hope, give expression to the highest regard in which Siraisi is held by so many.

***

As one would expect, Siraisi's career is marked by many distinctions. Since her earliest publication, "The *Expositio Problematum Aristotelis* of Peter of Abano" in 1970, she has led the way in connecting medicine to the broader intellectual world of the medieval, Renaissance, and early modern periods. As Siraisi herself has emphasized, "medieval and Renaissance learned physicians participated fully in the intellectual movements as well as the social and cultural environment of their age," and, true to her word, she has pursued that intellectual history in context, carefully reading texts and probing the daily practices of physicians.

During Siraisi's 40-year career, the history of medicine has changed dramatically, gaining broad acceptance among other more specialized historians. Siraisi's inclusive approach has been an important part of this transformation. In the announcement the History of Science Society prepared on the occasion of awarding Siraisi its highest honor—the George Sarton Medal for lifetime achievement—she was praised for her "richly nuanced and detailed studies that attend to texts and textual traditions, to individual lives and daily practices, to institutional settings and social relations, to disciplinary distinctions and literary genres." To cover such a range of topics, Siraisi has become an expert in many fields. Her expertise and engagement in these fields has been acknowledged and furthered by her service to diverse scholarly societies, including the History of Science Society, the American Association of the History of Medicine, and the Renaissance Society of America, of which she was president between 1994 and 1996.

Of course, Siraisi's preparations and interests began earlier, as she details in her intellectual autobiography, which is included in this volume. She was born in prewar England and received a B.A. (1953) and an M.A. (1958) from the University of Oxford, where she studied with the medievalist Beryl Smalley, whose study of medieval Bible commentary utterly transformed medieval studies. After a 15-year break from the academy, Siraisi returned to pursue a graduate degree in New York City, where she was encouraged not only by New York's public school system but by the distinguished medievalist and historian of science and the universities, Pearl Kibre, who would become Siraisi's dissertation advisor and, later, coauthor. Siraisi received her Ph.D. (1970) from The Graduate Center, CUNY, and that same year began as assistant professor at Hunter College, CUNY, where she taught general medieval, Renaissance, and early modern European history in the Department of History until her retirement as distinguished professor in 2003.

In the year of her retirement from teaching, Siraisi received the History of Science Society's George Sarton Medal. The following year, in 2004, she received the Paul Oskar Kristellar Award for Lifetime Achievement of the Renaissance Society of America, and in 2005, she was awarded the American Historical Association Award for Scholarly Distinction. Subsequently, in 2008, she was awarded a John D. and Catherine T. MacArthur Foundation fellowship, and, in 2010, Siraisi was honored by the ACLS as their Haskins Prize Lecturer, the honor that prompted her to write her intellectual autobiography. In recounting the process that led to Siraisi's nomination as the Haskins Prize Lecturer, the president of the ACLS, Pauline Yu, noted:

> No candidate to our memory has received a nomination from so many societies at the same time. The Renaissance Society of America, the Medieval Academy of America, the American Association for the History of Medicine, and the History of Science Society all agreed when nominating her for this honor that Nancy Siraisi's 'life of learning' has been distinguished, varied, international, and collegial. She has had a profound influence not only on the scholarship but also on the scholars in her field.

***

Directly or indirectly, Siraisi's many honors derive from her scholarly publications. Yet given the number and range of Siraisi's published works, it would be foolish to attempt to convey the richness of her research or to synthesize her many insights. Instead, we wish to highlight some of the major shifts, trends, and discoveries that have marked Siraisi's scholarly career in order to provide a glimpse of her work's appeal. To do this, we will not discuss the dozens and dozens of refereed articles and book chapters included in her bibliography (reproduced at the end of this volume); we limit ourselves to the main lines of inquiry and innovations that emerge in her several monographs and volumes of collected essays.

These begin with her first book, which grew from her dissertation: *Arts and Sciences at Padua: The* Stadium *of Padua before 1350* (published in 1973). Not only is the title a clear announcement of the book's content, but for those historians active at the time, it marked one of the first efforts to look beyond Padua's role in the emergence of modern science during the seventeenth century and toward the Paduan schools in the medieval period, between 1222 and 1350 to be exact. Focused on medicine, astrology, and the *libri naturales* of Aristotle, this is a work of intellectual history, which explores its subject primarily through the texts known to have been used at Padua or written by those associated with its schools. In its published form, Siraisi's dissertation was expanded to include her subsequent research on manuscripts and printed sources in England, France, and Italy related to Padua's curriculum. *Arts and Sciences at Padua* signaled the beginning of a productive scholarly career centered primarily, though not exclusively, on northern Italy and on the professors, physicians, and practices of medicine.

Next came Siraisi's *Taddeo Alderotti and His Pupils: Two Generations of Italian Medical Learning* (1981), which won the American Association for the History of Medicine William H. Welch Medal. Still focusing on the medieval period—Alderotti died in 1295—the content of this book, like *Arts and Sciences at Padua* before it, is clearly indicated by its title, although, if anything, the content of *Taddeo Alderotti and His Pupils* goes beyond what is advertised on the title page. For Siraisi demonstrates that medical learning, somewhat paradoxically to the modern expectation, was not just about medicine. In the medieval period, medicine's practitioners were involved with the whole of medieval intellectual life. This important insight would go on to inform much of Siraisi's later scholarship and its focus on the broad significance of the medical world of the medieval and Renaissance periods.

But in this work, Siraisi focuses on Alderotti, whom Dante called "the Hippocratist," and his students, who came to him with diverse backgrounds and went on to have equally diverse careers, not all of them in medicine. Siraisi illuminates the activities of Alderotti's circle as well as its version of scholasticism. She describes not just their efforts to integrate Avicenna's *Canon* into the curriculum, or their interest in the *de anima* tradition in natural philosophy when the topic turned to mental illness, or even the connection they saw between medicine and morality evident in their commentaries on Aristotle's *libri morales*, but also their lasting impact on European medicine, which included new attention to human dissection and to the literary forms of medical expression. Especially important for the last is the emphasis she places on disputation and the *questio* method and on the genre of *consilia*. Indeed, Siraisi gives Alderotti and the other Bolognese masters credit for the introduction of the genre of the *consilium*, which she discusses at length in a chapter all its own. A type of medical reportage focused on individual diseases and specific recommendations for treatment, Siraisi points us to the underappreciated role of Alderotti's *consilia*—185 in all—which were actively copied while remaining entirely in manuscript for centuries, and she puts to rest the erroneous view that the *consilium* genre was, at its inception, identified with Hippocratic case histories, such as those in the Hippocratic *Epidemics*.

Following her work on Alderotti, his students, and the vibrant culture of scholasticism, Siraisi shifted the focus of her research to the Renaissance. With her full command of the medieval period, this push forward transformed Siraisi into our most distinguished guide to the history of medicine ranging from the medieval period through the seventeenth century. Her own words explain why her research developed in this way:

> A good case can... be made for treating medicine from about 1300 to about 1600 as a distinct entity, whatever that entity may be termed. Although medicine at the beginning of this period represented no sharp break with, but on the contrary drew heavily upon, the classical, Islamic, and earlier medieval past, thirteenth-century innovations of lasting significance included the teaching of medicine in a university context, the emergence of a secular learned medical profession, the rise of Montpellier and Bologna (and other north Italian cities) as major European centers of medical education, the occasional practices of writing medical consilia for individuals and dissecting the human cadaver, and the widespread application of scholastic methodology in medical exposition.

Siraisi began her career by examining universities at a time when their main inspiration came from an earlier period, from the Ancients, and from ancient texts, and her first publications drew attention to significant and lasting changes taking place in the medieval medical world. As she traced the impact of those changes, she continued to explore the relevance of medical learning and links to its social and cultural context, extending her research as far as the seventeenth century.

Avicenna in Renaissance Italy: The Canon and Medical Teaching in Italian Universities After 1500 (1987) was Siraisi's next monograph. Its focus is Avicenna's *Canon* and, in particular, sections 1.1 and 1.2 of the *Canon*, which relate to medical theory, as well as the later commentaries written about these sections. She reveals that even in the case of Avicenna and his commentators, "it is often far from easy to disentangle tradition and innovation." Take the example of Santorio Santorio, the late sixteenth- and early seventeenth-century physician credited with introducing systematic controlled experimentation and quantification into the study of physiology. Siraisi's masterful discussion of Santorio shows the extent to which he, like other sixteenth-century medical humanists, expresses allegiance to ancient authorities and, at the same time, advances his efforts at innovation. Not only did Santorio measure and record daily fluctuations in his body weight over several decades, his interests in quantitative measurement found their way into his commentary on Avicenna's *Canon*, which included illustrations of a thermoscope and pulsilogium. The point, of course, is that even in a traditional commentary tied to a text with a long history, there was room for innovation. Now, 30 years after it was published, no finer work has been written about Avicenna's reception and the Latin commentary tradition he inspired. Its broader implications are also clear, for the *fortuna* of Avicenna's *Canon* shows that, in Siraisi's words, "it has a place not only in the history of medicine more broadly considered, but also in other histories: those of humanist and baroque learning, and of science education in universities during the so-called Scientific Revolution."

Although there is no obvious or direct inference to the quality of a scholar's teaching from her written work, an exception may be made for the introductory textbook, and Siraisi has written one of the most effective specialized introductions to the history of medicine. Since the publication of *Medieval and Early Renaissance Medicine: An Introduction to Knowledge and Practice* in 1990, which won the Watson Davis and Helen Miles Davis Prize from the History of Science Society, this text has been required reading for every student studying medieval and Renaissance medicine. It persists not only because of its clarity but also because of its comprehensive approach to the topic, treating everything from the formation of medical institutions and their curricula, to routine and innovative practices among physicians and surgeons, even to approaches to disease. Such a valuable introduction could not have been written by a scholar who was not also a careful and committed teacher. It is our best evidence of what Siraisi offered her students for more than 30 years.

Next for Siraisi came a work focused on a single individual, the highly idiosyncratic and prolific Girolamo Cardano. *The Clock and the Mirror: Girolamo Cardano and Renaissance Medicine* (1997) has the only title from among Siraisi's monographs that does not immediately announce her subject. It is about Cardano, yes, and Renaissance medicine too. But the "the clock and the mirror?" The phrase comes from Cardano himself, who maintained that the "studious man should always have at hand a clock and a mirror." The clock was needed to keep track of time, especially important "if [one] is a professor, teaches, or writes," and the mirror so that the studious man can observe his changing body. These references appear in one of Cardano's medical works, and Siraisi directs us to Cardano's mirror and the extent of his medical interests, revealing the depth and self-possession of this fully Renaissance man, who rose from poverty and his illegitimate birth to become a significant force in Renaissance medicine. In discussing her own strategies for studying "a compulsive reader, writer, and would-be masterer of disciplines" like Cardano, Siraisi explicitly draws on methods from "the history of the book and reading, the uses of narrative and self-expression, and the formation of disciplines," but in her introduction, she again reminds us that "the understanding of a system of knowledge and practice as highly textual as Renaissance medicine" demands that "the content of texts remains of primary importance. The task of struggling with the texts and endeavoring as far as possible to explain the ideas they transmit remains inescapable."

Although Cardano had long been noted as a reformist and innovator, not least by Cardano himself, Siraisi's scholarly methods allow her to convey the extent to which Cardano both drew upon and measured his achievements in relation to the learned scholarship of his day. Exploring this apparent tension, common among his contemporaries, she shows Cardano to be someone for whom "medicine encompassed every aspect of human knowledge, power and technique." Siraisi also gives extensive attention to Cardano's daily medical practice, connecting it to his abiding interest in astrology and the occult and even to the reception history of the influential Renaissance anatomist, Andreas Vesalius. For, in spite of Cardano's great admiration for Vesalius and his enthusiasm for the newly minted visual epistemology generated by Vesalius's anatomical illustrations, Cardano recognized that Vesalius's work gave physicians a professional and rhetorical advantage over other practitioners in the sixteenth-century medical marketplace. Concluding with a discussion of "medical narratives," Siraisi draws extensively on Cardano's medical and autobiographical writings. With an enviable command of the texts, she shows that the many narrative forms employed by Cardano come together to provide a "Renaissance prehistory of the early modern case history," a topic that she would go on to explore in greater depth in her next monograph.

By the late 1990s, many of Siraisi's essays published between 1975 and 1998 had become standard reading in the history of medicine, and these, along with some of her more recent work, were collected in *Medicine and the Italian Universities, 1250–1600*. This volume includes Siraisi's reflections on her own work and the changes she had witnessed in the history of medicine to this point in her career. She even shares her belief that "more still needs to be done fully to integrate the world of medical learning into our general historical picture of medieval, Renaissance and early modern intellectual life." Siraisi's efforts to this end, as well as the depth of her scholarship, are evident on every page of this collection. While each paper stands alone, many of the essays have gone on to inform her larger projects and the projects of other scholars. Picking favorites from among the essays would be difficult, but for us (the editors), "Vesalius and the Reading of Galen's Teleology" (Ch. 13) and "Vesalius and Human Diversity" (Ch. 14) are fundamental, as is "Signs and Evidence: Autopsy and Sanctity in Late Sixteenth-Century Italy" (Ch. 15). Focusing just on her work on Vesalius, Siraisi's reading of Galen shows evident mastery of the Hellenistic physician's view of teleology, but, more than this, it links this highly technical aspect of Galen with Vesalius's own efforts to reveal the design and perfection in the fabric of the human body. The issues this paper raises—about reliance on an ancient source and rejection of the authority of the past—have continued to inform scholarship on Vesalius and his transitional role away from and simultaneously toward ancient authorities. Equally foundational, Siraisi's study of diversity in the first and second editions of Vesalius's *De humani corporis fabrica* (1543, 1555) brought clear focus to the issue of normative anatomy and to the manner in which Renaissance figures differentiated between mere variation and genuine abnormality. Here too, Siraisi extended the reach of scholarship on Vesalius by showing that, by the second edition of 1555, he had begun to study diversity in anatomy in a systematic way.

Siraisi's next book considered the practice of writing history and its relationship to medicine, which she initially treated through the figure of Cardano in *The Clock and the Mirror*. The inaugural monograph of the series, *Cultures of Knowledge in the Early Modern World*, with the University of Michigan Press, *History, Medicine and the Traditions of Renaissance Learning* brings together two aspects of Renaissance learning: the physician's reinvigorated interest in historical aspects of medicine and the Renaissance enthusiasm for the writing of history. Not only did physicians bring historical material into their medical writings (Part One), but physicians active in different locales—e.g., Milan, Rome, Vienna—produced works of history (Part Two). Siraisi presses the point that physicians and the domain of medicine were very much a part of the dominant intellectual traditions of the Renaissance, traditions of inquiry and writing as well as antiquarianism. Moreover, physicians were qualified to write history not only because of their humanistic training, something they shared with their educated peers, but also because of the content of medicine itself, which during the Renaissance included "increased emphasis on the recording of particulars, the construction of narrative, and the analysis of past events." There were, to put it simply, epistemic genres or styles of thinking created through a humanist medical education that played a significant role in the development of historical writing. A remarkable twofold consequence of Siraisi's conclusion is that we should abandon accounts of "medical humanism" tied only to the application of classical philological skills. Instead, medical humanism should be recognized, first, as a robust combination of classical philology, antiquarianism, biography, history, and rhetoric, and, second, we should expect that humanists who were not physicians, nevertheless, would have read medical texts. Accordingly, any scholar interested in Renaissance humanism is well advised to look anew at physicians and the full range of their interests and publications.

Siraisi's latest monograph, *Communities of Learned Experience: Epistolary Medicine in the Renaissance* (2013), derives from three lectures inaugurating the Singleton Distinguished Lecture Series for the Charles Singleton Center for the Study of Pre-modern Europe at Johns Hopkins University. The lectures, slightly expanded for publication, focus on the genre of *epistolae medicae*, i.e., letters of learned physicians, including university professors as well as court and town physicians, which were written as part of an effort to establish networks and share information and descriptions of rare cases. Patterned on ancient models, letters were considered the appropriate vehicle for reciprocal and open exchange among learned physicians. In fact, they were a means to establishing contacts in both international and regional centers. According to Siraisi, they were even seen as a companion to medical *consilia* (written advice for individual patients and a topic Siraisi explored in her 1981 book on Alderotti), since medical letters incorporated dialogue and expressions suited to conversation between like-minded individuals, as well as disputation and debate on medical issues. From her three case studies, she shows how these letters served professional interests by advertising expertise, instantiating one's membership in a medical community, and excluding from membership other medical persons deemed unworthy. Drawing on letters from the court physician Johann Lange, for example, Siraisi describes how his letters police professional boundaries, where, in one instance, he denounces the users of astrology, which he considered "trashy finery." Encouraging more research and a methodology that incorporates the history of the book and engages with the genres of medical writing, Siraisi introduces us to the ways that learned medical practitioners engaged the epistolary networks and epistolary practices of their time.

Between publishing these monographs and the volume containing a selection of her essays, Siraisi also published significant edited collections that have included collaborations with other leading scholars working in Renaissance history, history of medicine, and the history of science. Most recently, this includes *Historia: Empiricism and Erudition in Early Modern Europe* with Gianna Pomata and, before that, *Natural Particulars: Nature and the Disciplines in Renaissance Europe* with Anthony Grafton (the latter is another instance in which her work inaugurated a series—this time in connection with the Dibner Institute conferences). Finally, Siraisi has continued the beneficent work she began in her earliest publications of making texts available to the scholarly community, with the forthcoming collection *Medical Humanism* in the *I Tatti Renaissance Library* series (with translations by Craig Martin). The latter collection promises to deliver access to canonical texts never before translated into English, including works by Giambattista Da Monte, Giovanni Argenterio, G. M. Eustachio, and many others.

One might have the impression from our survey that Siraisi's achievement has been to bridge the gap between medieval and Renaissance medicine, and in a sense this is entirely correct, but it is more appropriate to credit her with the awareness that there are no fixed chronological boundaries and that the historian's task is to explore where the texts and other source material takes her, using the very best research tools at hand. In this effort, Siraisi is both a model and, as we noted at the outset, a guide. Her published work shows how the core elements of social and cultural history are deeply connected to the reality that medicine's learned practitioners perceived their calling as an intellectual one, even as they promoted themselves in different ways in an ever more complex medical marketplace. Never a conventional intellectual historian, Siraisi has crossed boundaries, not just of time, but of disciplines and techniques, and her unique combination of skills has led to new insights about gender, the body, the history of the book and reading, the reception of ideas, and the production of knowledge, to name just a few of the areas of research that have felt her influence. In short, through her research, Siraisi has elevated the history of medicine.

***

The idea for this volume came to life when its contributors agreed to participate. They were asked to choose topics that "relate to medicine and the wider world of learning and knowledge, as Siraisi's own work does," and they were given the freedom to pick from among her varied interests and the periods in which she has worked. Danielle Jacquart and Michael McVaugh choose to focus on the medieval period, with essays ranging from the thirteenth to the early fourteenth century. The remaining essays, by Chiara Crisciani, Ian Maclean, Vivian Nutton, Anthony Grafton, Hiro Hirai, Ann Blair, Paula Findlen, and Domenico Bertoloni Meli, take us from the fifteenth through the seventeenth centuries. Each of the chapters draws inspiration from the research methods that Siraisi has championed, and there is a strong focus on issues she has discussed at length, including curricular growth and humanist's innovation, the formation of disciplines, and the collection and transmission of medical and scientific ideas, sustained by the Republic of Letters and the text-based traditions of learned medicine. Each essay, in addition, reinforces one of the central themes of Siraisi's scholarship, namely, that physicians were more than simply healers. Rather, they were learned men whose activities were drawn from and, in turn, influenced the intellectual traditions of their age.

Following Siraisi's autobiography, we have organized the essays into three parts. The first part includes the essays from Jacquart and McVaugh and emphasizes the connections between medicine and late medieval scholasticism. In particular, Jacquart's contribution explores the relationship between medicine and theology through commentaries on both the Old and the New Testament. Situating the late medieval commentaries in relation to Ambrose of Milan and Augustine and then within the context of thirteenth-century developments in university teaching, Jacquart shows several instances in which biblical exegesis is tied to current debates about medical topics, such as the link between the mind and body and the physiological explanation of nutrition. Just as Siraisi has made a career of revealing the presence of medical ideas in a wider field of textual debate—extending the canon of the history of medicine—Jacquart shows how a closer look at biblical commentaries can sometimes deliver "the personal opinions or doubts of a medieval author" on medical subjects "out of their usual context."

In his essay, "In a Montpellier Classroom," McVaugh invites us to turn our gaze away from the extraordinary names for which Montpellier is known—e.g., Arnau of Villanova and Bernard of Gordon—and to look, with him, at a set of scattered manuscript sources, deriving from the teachings of "ordinary" masters, including one Cardinalis, Pierre de Capestang, and Bernard of Angarra. From these sources, McVaugh shows how the texts of the old *ars medica* were situated alongside the "new Galen" and how critical, debated questions were treated in what was initially (presumably) an exchange between master and student. Thus, when Pierre responded to the question of whether *ptisana* was truly *viscosa*, his response was much longer than Cardinalis' treatment had been and perhaps the earliest example recorded of a master's authoritative resolution of a formal dispute. We are witnessing, according to McVaugh, a transition: school texts were becoming familiar enough that masters, instead of producing fine-grained commentaries, focused their attention on *questiones*, formerly a "subsidiary feature" of classroom and text-based learning. McVaugh uses these sources to reconsider the medical curriculum: for example, he notes, "Bernard de Angarra's *dubitata* on *De complexionibus* and *De malitia complexionis diverse*. . .reveal that those works were indeed being explained to Montpellier students, probably well before a formal requirement came into place." McVaugh's insights leave us with a more robust understanding of the classroom experience and the ways it shaped the teaching of medicine at Montpellier. His essay also includes two appendices—translations of Bernard de Angarra, *questiones* on Hippocrates, *Aphorisms* I.2, and a student's notes on Bernard de Angarra's questions on Galen, *De malitia complexionis diverse*—which, by their content, emphasize the exchange between teacher and master and, by their form, join Siraisi's work in making texts available to a wider scholarly community.

The second part includes the contributions by Crisciani, Nutton, Grafton, and Maclean, all of which examine either the Renaissance physician's choice of genre for the expression of medical ideas or the physician's place in the Republic of Letters. The third essay, Crisciani's "Medicine as Queen: The *Consilia* of Bartolomeo da Montagnana," develops a topic explored by Siraisi in her early work: the genre of *consilia*. As one would expect, over the centuries the genre changed, and Crisciani, whose scholarship elsewhere is our best guide to these developments, here elaborates them through the extended example of Montagnana. Among her many insights, she concludes that it was this "literary genre in which the concept of medicine as a *scientia operativa* is best expressed in the fifteenth century." Moreover, she goes on to suggest that specialization and the growth of universities drove many of the changes in the form and content of *consilia*. But she cautions that the *consilia* were not only representative of scholastic medicine; when read carefully, they show that learned physicians also used the experiences of other practicing physicians at the patient's bedside, "as an extension of [their] senses and as intermediaries between [their] text and the actual [and particular] body of each individual patient."

The next two contributions are by Nutton and Grafton, respectively. Both examine the historical writing of the English medical humanist John Caius, who was twice president of the London College of Physicians, roommate to Vesalius in Padua, founder of Caius College, Cambridge, and a prolific, if erasable, historian. They also share the belief that Caius is a figure Siraisi might easily have studied had she turned her attention to sixteenth-century England. In "John Caius, Historian," Nutton discusses nearly all of Caius' historical works, from the early *De antiquitate—On the antiquity of Cambridge University*—to Caius' *Annals of the College of Physicians*. Nutton's primary focus, however, is on the late *History of Cambridge University*, which he considers very different from the *De antiquitate*. For one thing, the *History* includes a detailed survey of mostly accurate information, at least beginning in the thirteenth century, "of the university and town based on personal observation and familiarity with the archives in greater depth and at greater length than anything previously undertaken." And for another, the *History* is less overtly polemical than the *De antiquitate*. Yet Caius, the curmudgeon, is very present in the later history, in which he laments both Cambridge's 1542 adoption of new pronunciation for Greek and Latin and the behavior of modern students, with their "fine clothes, ruffled shirts, slashed saffron breeches, fancy hats, dashing beards." Nutton's close examination of the *History* also identifies "a more subtle attack on contemporary Cambridge, and, more specifically, on the religious changes there in [Caius'] lifetime." If the consensus nevertheless remains that the *History* is a flawed work, something Nutton does not altogether disagree with, he maintains that adopting a "dismissive verdict is to adopt a perspective that takes little notice of the context in which Caius was writing." Indeed, an important element of Renaissance humanism includes a return to an earlier age, an element expressed by Caius in the *History* as a yearning. This element, as Nutton points out, reveals that Caius, like other humanists, has "a much greater kinship with [his] medieval predecessors" than with modernity.

Reaching a similar conclusion in "A Medical Man Among Ecclesiastical Historians: John Caius, Matthew Parker and the History of Cambridge University," Grafton focuses exclusively on Caius' early *De antiquitate*. He begins with the observation that Caius' early work looks to be "a particular kind of history—one normally focused on the history of the church and produced more often by groups than by individuals." With his usual wit and insight, Grafton builds a persuasive case for believing that, in fact, Caius' 1568 work was produced with the help of others and modeled after the histories of the church. Specifically, Grafton maintains that *De antiquitate* was "the result of intensive collaboration and consultation" with Archbishop Matthew Parker and the team of scholars who worked with him. The effects of the collaboration, furthermore, did not only benefit Caius. Grafton notes that Parker would subsequently adopt several argumentative strategies that likely derived from Caius and that Caius helped Parker "bring a new and more sympathetic perspective to bear on the Anglo-Norman historians whom he and his secretaries published." Generalizing from his study of Caius the historian, if Grafton is right about *De antiquitate*, the image of the solitary humanist is one of many "generalizations [that are] as untrue as they are widespread," and he recommends "it is well past time" that this particular generalization be "laid to rest."

In the next essay, "A Medical Collection Anatomized: The *Catalogus bibliothecae Hieremiae Martii* (1572)," Maclean takes as his subject what is in all likelihood the first printed library catalogue, one announcing the sale of the physician Hieremiae Martius' library. Aligned with Siraisi's research on the "wider scholarly interests of doctors," Maclean describes the range of Martius' collection, along with its gaps and apparent omissions (what evidence suggests it could have included but did not). Avoiding generalization and conscious that Martius' library may have contained more than the *Catalogue* indicates, Maclean evaluates Martius' claim in the *Catalogue* that his library is "comprehensive, well-chosen, up-to-date and includes rare items." Maclean compares what we know of Martius' library to those of other collectors, including Casper Peucer, whose library is "marked by its holdings in chemistry and divination," and Girolamo Mercuriale, whose library reflects a "broader sweep of interests and geographical origins." Against these other collections, Martius' collection emphasizes "holdings in *practica* and botany." He was a practitioner of "somewhat limited means with a professional commitment to nosology, pharmacology and therapeutics, a strong interest in botanical study, both for its own sake and as a source of *materia medica*," and so, in a sense, this is unsurprising. The surprise comes when we learn that it was his open-mindedness that led him to collect "the writings of Paracelsus and empirics outside the learned profession." But as Maclean also points out, even our surprise must be balanced against the possibility that Martius' collection habits were closely coordinated with the market. For some of the libraries, Maclean examines "contain bibliographical guides, and evidence that the catalogues of the Frankfurt Book Fair were used as vehicles for purchasing, even in faraway England and in Catholic Italy which officially was hostile to the circulation of such publications as they contained references to much heretical material and to the works of authors who had been placed on the Index." Martius' collecting habits, while personalized, should be understood in this context.

The third and final part includes the essays by Findlen, Blair, Hirai, and Bertoloni Meli, which range from the collecting practices of sixteenth-century natural history, to the history of the book in the same period, to the mysteries of embryological development as discussed in the early seventeenth century, to the representation of pathology and the genre of *observationes*. In "The Death of a Naturalist," Findlen uses the life and death of Luca Ghini to show the extraordinary development of botany during the mid-sixteenth century. She continues her groundbreaking work on the period in which dependence on ancient sources diminished and naturalists began to cultivate diverse forms of engagement with the world, such as observation and collecting. Ghini was the first physician to have taught medicinal simples at the University of Bologna. He was engaged all of his life in the field of natural history, in debates about where and how it should be studied, taught, and practiced, and in a vision of collaborative research that involved collecting, the establishment of gardens, and a wide network of correspondence. Findlen uses these aspects of Ghini's life to explain how Ghini "embodied a new breed of physician who advocated learning from experience, initially as an antidote to problems and lacunae in ancient natural history and ultimately as a technique in its own right, capable of generating new knowledge." Just as Siraisi's work has emphasized the role of reception, Findlen shows us the full impact of Ghini as felt by subsequent naturalists and made evident in letters and books, in collecting practices, and in the formation of research interests that organized natural history in the wake of his death.

Turning from the garden to the book, Ann Blair focuses on the many and varied dedications of Conrad Gessner, a physician, humanist, and avid *dedicator* in "The Dedication Strategies of Conrad Gessner." She makes extended use of the paratext and demonstrates that Gessner broadened his portfolio, in a sense, by using the dedication to do much more than land a single wealthy patron. According to Blair, Gessner used dedications to navigate "local hierarchies and international networks," addressing collective entities as well as important and wealthy individuals and even printers. For the last, he praised them in a manner more consistent with socially prominent individuals in order "to incite" the printers to continue publishing learned texts. Drawing on the methodologies associated with the history of the book—especially the politics of print—Blair continues her meticulous work on the strategies associated with book use. She shifts from the perspective of readers and their annotating practices to the perspective of writers, like Gessner, who explored the form of the dedication and multiplied its functions. From her analysis—a parallel to Siraisi's on epistolary networks—we see not only how the Republic of Letters was sustained and strengthened but also how books were carefully coordinated with plans for circulation and with the intellectual and economic agendas of humanists and printers.

The next essay is Hiro Hirai's "Imagination, Maternal Desire and Embryology in Thomas Fienus." The topic is one that garnered ever more attention in the early modern period: the process of embryological development and the physical effects of the mother's emotions and desires on the growing fetus. As Hirai notes, Renaissance philosophers, such as Marsilio Ficino and Pietro Pomponazzi, believed that the mother's imagination could cause various physical effects in the fetus, including birthmarks and even deformities. Hirai then shifts to do something of which Siraisi would approve, namely, consider what physicians had to say about this widely held popular belief. He specifically examines the work of Thomas Feyens or Fienus of Antwerp, a learned physician who wrote a monograph about the power of the imagination. With sources ranging from Aristotle, Augustine, and Aquinas to Galen, Avicenna, and the aforementioned Renaissance philosophers, Fienus's *De viribus imaginationis tractatus* addresses nearly every conceivable issue related to his topic. These include the soul-body relation, the faculty of the imagination itself, and, of special interest, the "working of the mother's imagination in fetal formation." Fienus' defense of the power of the mother's imagination is extensive. It includes evidence drawn from the testimony of experience as well as references to ancient, medieval, and early modern sources. Hirai is careful to show that Fienus was selective in the evidence he choose and the positions he advocated. For example, Fienus avoids Platonic accounts, like those in Ficino and the physician Jean Fernel. Instead, he promotes the notion of formative powers, which were emphasized many generations earlier, by the physicians Niccolo Leoniceno and Jakob Schegk. As the heir to these earlier physicians, Fienus links himself to the originators of medical humanism, and, as Hirai notes, this motivation may help explain his defense of the power of the mother's imagination.

The last of the essays explores the development of the *observationes* genre in the seventeenth century and, in particular, the significance of illustrations of rare and pathological cases. In "Gerardus Blasius and the Illustrated Amsterdam *Observationes* from Nicolaas Tulp to Frederik Ruysch," Bertoloni Meli locates Blasius, a Dutch anatomist and physician, between two celebrated figures—Tulp and Ruysch—and provides a richly detailed account of the ways that images serve collections of *observationes*, addressing the subject matter of the image, the styles of representation, and the relationship between the image and the surrounding text. Although illustrations in anatomical texts have received careful study by scholars in several disciplines, the illustrations in the epistemic genre of medical, surgical, and anatomical *observationes* have not. Bertoloni Meli explains that the collections of *observationes* published in Amsterdam in the second half of the seventeenth century reflect the interest that medical practitioners took in isolating anomalies—rarities and pathologies—and sharing them with a community of learned practitioners. The illustrations of these collections are often embedded in narratives that document the presence of surgeons, physicians, and students at autopsies and clinical events. Tulp, for example, sought to describe the spine and its pathological condition as much as to defend the surgeon, whose name is not mentioned. He also says that he had seen a similar tumor, presumably with the same spine condition as well, six times, thus offering valuable information on the frequency of the condition. Because the conventions of this genre were in the process of acquiring stability, the collections do not emphasize clear or standard classifications, which remained, according to Bertoloni Meli, quite fluid in the period. Instead, he emphasizes a range of cross-referencing systems, indicating that these *observationes* were a closer kin of anatomical, surgical, and natural historical cabinets of curiosity than of the systematic approaches associated with the later clinic.

***

It has been an honor for us and for the contributors to celebrate Nancy Siraisi. We are all profoundly grateful to her for her scholarship.

A Life of Learning

When I consider the distinction and varied accomplishments of previous speakers, I am especially sensible of the honor of being asked to give the Haskins Prize Lecture. Unlike some of my predecessors, I can offer neither significant achievements in the public or institutional sphere nor a striking or particularly unusual life history. What I do share with a number of previous speakers is the feeling that if I have been fortunate enough to have enjoyed a life of learning, that privilege has owed much to chance or, more bluntly, been due more to good luck than good management. But the good luck surely included encountering teachers—and later in my career, colleagues—who inspired and helped me; I will have occasion to mention a number of them.

The England of my earliest childhood is doubtless now gone beyond recall from collective memory as well as from my own. All I can say is that mine was a conventional middle-class family with many connections among the Anglican parish clergy. Ours was not an intellectual home, but I owe to my mother's example the most important of all qualifications for a historian, a lifelong pleasure in reading. My memories of early schooling are vague, largely because of constant moves caused by my father's service in the British Air Force. In just 3 years when he was stationed in various parts of Canada, my educational experience ranged across public schools in small towns in Alberta and Saskatchewan, a one-room schoolhouse in rural British Columbia, and a semester in a private school in Montreal. Not surprisingly, I recall more what I learned about adjusting to change than about the curriculum in any of these institutions.

Our family's return to England when I was 11 evidently meant to my parents that it was time to get serious about my education. I spent the next 6 years in a girls' boarding school and hated every minute of its collective living arrangements and enforced togetherness. Since the school was small and sent few students on to universities, any possibility of success in what was then called Higher School Certificate or in university entrance examinations meant intensive coaching in subjects in which the student was likely to do well and the elimination of most else. The result in my case was an intensive focus on history, which by the last

2 years of high school meant almost exclusively nineteenth-century English political history. Deplorable as this educational approach may have been in terms of broad culture, it was in its own terms fairly effective. At any rate, I succeeded in obtaining admission to Oxford; what made it possible for me to go there was a "state studentship" (i.e., scholarship) provided under the terms of the Butler Education Act of 1944, which paid entirely for all costs associated with my university education. (State support for British university students was subsequently scaled back considerably as demand for access to higher education broadened, but a fortunate few of us in the 1950s and 1960s benefited from a level of public generosity unequaled before or since.)

I went to Oxford—my college was Saint Hilda's—to read history in 1950. I have to admit I was a frivolous undergraduate. In my defense I can only offer that I had, after all, just emerged from 6 years of being confined in the hated boarding school. By contrast, the university seemed to offer limitless freedom and autonomy, even given the considerable restrictions surrounding students in the women's colleges in the early 1950s. In retrospect, it is, alas, clear that I wasted many of the intellectual opportunities that Oxford could have given me.

Nevertheless, it was there that I encountered the first of the teachers who inspired me and first made the acquaintance of the Middle Ages as an object of serious study. That teacher was my tutor Beryl Smalley, a distinguished medievalist whose publications include books on important and fascinating topics in intellectual history: *The Study of the Bible in the Middle Ages* and, later, *Historians in the Middle Ages*. I cannot now remember if I actually read the first of those books as an undergraduate or a little later; if I did read it as an undergraduate, it can only have been for private relaxation. For it was a time when medieval studies in Britain were powerfully influenced by T. F. Tout's work on administrative history, while the requirements of the then Oxford history curriculum in general were strongly focused on the history of England. Consequently, the medieval history I actually studied with Beryl Smalley was, once again, English—and I do mean English, not British—political and administrative history. I do not know that the household administration of Edward I is an entirely desirable way to introduce 18-year-olds to the Middle Ages. Yet I also glimpsed a somewhat different and more engaging view of the emergence of medieval and indeed of Christian Europe by selecting the earliest chronological period offered—the third to the seventh centuries—for the required study of a span of history of some part of the world outside the British Isles.

At the end of the undergraduate curriculum, I emerged, as I deserved, with a second-class degree. Naturally, it occurred to none of my teachers to suggest that I go on to pursue graduate work in history with an academic career as a possible ultimate goal, any more than it occurred to me to imagine such a future for myself. The picture of university education that I have just sketched seems no doubt as remote as that of the medieval and Renaissance universities with which I subsequently concerned myself. A few years ago, Peter Brown provided the audience for the Haskins Lecture with a view of the 1950s Oxford as it appeared to a brilliant graduate student and junior fellow of All Souls College; alongside those recollections, you may now place my memories of a much more ordinary undergraduate.[1] No thoughts of an academic career—or indeed a serious profession of any kind—entered my mind for about the next 15 years. A brief first marriage ended with the untimely death of my first husband. And after setting aside various well-meant suggestions for my future that did not appeal to me—especially that of applying for a position teaching history in a girls' boarding school—I followed a random collection of decidedly nonprofessional occupations. At different times and successively, I was, in London, an editorial assistant on a trade magazine published by an association of cement manufacturers and a research assistant on a company history; in Rome, a teacher of English in a commercial language school; and, in New York, a personal secretary to an administrator at the original Museum of Modern Art and, later, an editorial assistant on a children's encyclopedia. The move to New York was dictated by a desire to see a little more of the world—in Rome, I had for the first time worked alongside American colleagues and liked them—but not made with any intention of staying in the United States permanently. But in New York, I met my husband, then also a recent arrival in the United States, and so we stayed.

By the mid-1960s, with two small children, I began to reconsider my options. In particular, I began to wonder whether some kind of teaching might not be preferable to a return to full-time editorial work. A first step seemed to be to accumulate some graduate credits, and to that end, I enrolled as a part-time non-matriculated student in the M.A. program in history at Hunter College of the City University of New York (CUNY). All I had in mind was a pragmatic, tentative move toward a modest, probably part-time, teaching job. I did not, of course, know that the fates were waiting for me in the shape of the instructor in the first class I took at Hunter College. Professor Pearl Kibre was a widely respected medievalist, whose research interests lay in the history of universities and the history of science and medicine; she was a longtime collaborator of Lynn Thorndike, with whom she produced the monumental *Catalogue of Incipits of Mediaeval Scientific Writings in Latin.* She was also a powerful personality. Brushing aside my own reservations, she urged me to enroll full time in the Ph.D. program in history (then recently established) of the CUNY Graduate Center, where she would be my mentor.

My memory of graduate school, spent commuting between home responsibilities and a properly demanding mentor, is, I must admit, one of permanent exhaustion; I'm sure it bore hardly on my family, good though they were about it. Nevertheless, in professional terms, the results were worth the effort. My completion of the Ph.D. in 1970 coincided with Pearl Kibre's retirement, and she recommended me for a position at Hunter College. I taught general medieval, Renaissance, and early modern European history in the undergraduate and M.A. programs at Hunter

[1]Peter Brown's 2003 Charles Homer Haskins Lecture was published as ACLS Occasional Paper No. 55. It is available on the ACLS website at www.acls.org/publications/0P/Haskins/2003_PeterBrown.pdf

College—and from 1976 also in the Ph.D. program at the Graduate Center—until my own retirement in 2003.

I am in no doubt whatsoever that just two things—both encountered after my arrival in this country—made my life of learning possible and that without either I would not have had a professional career at all, let alone the opportunity to address this distinguished audience. The first was Pearl Kibre's very determined encouragement and guidance; opportunities offered by public higher education in New York were the second. Without the CUNY Graduate Center and a generous Lehman fellowship from New York State, I would have been very unlikely—for both financial and family reasons—to have had any other opportunity for graduate education. Thereafter, Hunter College, to be sure, imposed a heavy teaching load (I taught three courses a semester for 20 years, until a promotion in 1990 reduced the number), but successive department chairs were consistently very supportive of my research, and research leave policies were relatively generous. (In this and other respects, CUNY faculty benefit from the existence of a faculty union.) And of course in New York, I was excellently placed for local and regional library resources, as well as for summer travel to European libraries (which became easier as my children grew older).

If there is a moral to my talk, it is perhaps a plea for support of public higher educational institutions, even in hard economic times. I might add that after I was fortunate enough to be granted a MacArthur fellowship in 2008, I received a letter from a slight acquaintance of many years earlier, in effect assuming that a scholarly career in a public institution must have been pursued in the face of many difficulties and disadvantages. His sympathy was well intentioned, but in my view misplaced. From graduate school to defined benefit pension, the City University of New York treated me very well, and I have no complaints, so much for the external circumstances that enabled me to pursue the pleasures of historical research and writing.

But social, historiographical, and personal intellectual developments all had a part in shaping the work that resulted and indeed in reshaping my ideas and interests as the years went by. Fresh from writing a dissertation on the early history of the University of Padua, I began as a medievalist with a very narrow view of intellectual history, albeit one that was, thanks to Pearl Kibre's influence, already directed toward science and medicine. In the early 1970s, the by then 50-year-old "revolt of the medievalists" was still in some quarters a living cause. From the 1920s, Charles Homer Haskins, for whom this lecture is named, and other scholars of the European Middle Ages had stressed the originality and lasting importance of medieval culture as a corrective to the Burckhardtian emphasis—the medievalists would have said overemphasis—on the special significance of the fifteenth-century Renaissance. Lynn Thorndike of Columbia University, the author of the massive, multivolume *History of Magic and Experimental Science*, went a good deal further; he denied not just the significance but the very existence of the Renaissance as a distinct historical moment and viewed humanism as having been a negative influence for the development of science.

Thorndike had a strong intellectual influence on Pearl Kibre, who had, as I have already noted, been his collaborator, an influence that did not end with his death in

1965. As a result, during my graduate student years, I became very familiar with Thorndike's historical views. It would be hard to find any historian of science now who would endorse that outlook in its entirety; yet in its own day, Thorndike's approach, like the views espoused even earlier by Pierre Duhem and George Sarton, helped to inspire research into significant but hitherto little-explored aspects of the Middle Ages. The call for attention to the sciences in the Middle Ages bore fruit in the detailed investigations by Marshall Clagett and others that made the 1960s and 1970s something of a golden age for the history of medieval physical and mathematical knowledge.

In the meantime, of course, Renaissance studies had not stood still, and a new field of early modern studies was in the process of developing. In particular, one of the most distinguished scholars who left Nazi Germany first for Italy and then for the United States, the historian of Renaissance humanism Paul Oskar Kristeller, taught at Columbia University from 1939 until 1976 and remained in New York and active in scholarship long after his retirement.[2] For me as for many others, Kristeller's work was foundational in building an understanding of the significance of Renaissance humanism, and Kristeller himself was extraordinarily generous with time and advice for a tyro medievalist at a neighboring public institution. If at the time I occasionally found myself torn between two very different views of European intellectual history, in the long run I found it helpful to have been exposed to both perspectives and to hold them in tension.

Meanwhile, I continued to develop an interest in the history of medicine in several different aspects—in relation to the history of science, as professional and social practice, and as a branch of medieval university learning. But the 1970s and 1980s were years in which the historiography of science and medicine was also changing. Historians of science—or some of them, for the field was going through its own culture wars at the time—were beginning to give more attention than before to life sciences and to the social context of scientific knowledge. And much more than previously, medical history was beginning to benefit from a rich body of research into the history of practices, practitioners, and patients in their social and cultural context.

All this was encouraging from the standpoint of someone whose initial training was in history rather than in medicine or science. Consequently, when I embarked on a study of a group of medical professors in late thirteenth- and early fourteenth-century Bologna, Taddeo Alderotti, whose fame was noted by Dante, and some of Taddeo's pupils—I took some trouble to portray not only their Latin medical learning but also their role as citizens and men of letters and as practitioners who counseled patients. It was in those years of the late 1970s and early 1980s, too, that I first spent several summers in research in Italy and made the acquaintance of younger Italian historians of medieval medicine and related disciplines, among

[2]Paul Oskar Kristeller's 1990 Charles Homer Haskins Lecture was published as ACLS Occasional Paper No. 12. It is available on the ACLS website at www.acls.org/publications/OP/Haskins/1990_PaulOskarKristeller.pdf

them Chiara Crisciani, Daniela Mugnai Carrara, and the late Jole Agrimi (my dissertation had, by contrast, been completed under serious time constraints, with a minimal period of research abroad). In 1985–1986, a blessed year at the Institute for Advanced Study in Princeton enabled me to bring together my thoughts on medieval medicine in the compass of a short survey of the field that summarized my work up to the mid-1980s.

Of the new scholarship on the history of medieval medicine in the United States, the work of Michael McVaugh, which combined an intensive analysis of Latin texts with a keen awareness of the social context and history of both learned and popular medical practice, greatly appealed to me. Accordingly, I was very pleased when the opportunity came to edit—jointly with McVaugh—a volume of essays on the history of medicine for the History of Science Society's annual *Osiris*. One theme that I think will emerge from this talk is how much I have benefited from—and enjoyed—collaborative projects and the scholarly contacts made in the course of them. In the volume for *Osiris*, which also appeared in 1990, we sought to bridge the late Middle Ages and the sixteenth century. Hence, preparation of the volume brought contact with scholars working on the history of Renaissance medicine, most notably Vivian Nutton, then of the former Wellcome Institute for the History of Medicine in London (subsequently the Wellcome Trust Centre for the History of Medicine at University College London), whose work on the many varieties of Renaissance Galenism opened up new areas of medical history to me. Thus, this project also contributed to my intellectual trajectory: since the mid-1980s, my own interests came more and more to center not on the Middle Ages or even the Quattrocento, but on the sixteenth century.

I was, of course, well aware that in many respects, sixteenth-century medicine was already a very well-studied topic. Its technical and scientific accomplishments and innovations, particularly in the areas of anatomy and botany, attracted the early attention of internalist historians of science and medicine; similarly, the Renaissance history of ancient medical texts—their transmission, reception, and influence—has been the subject of much scholarly investigation. And over the last generation, historians of medicine have devoted much attention to the social or cultural milieu of health, disease, and healing in the early modern world. Nevertheless, it seemed to me that there was still more to be said, especially on the subject of continuities between the worlds of medieval and Renaissance medical learning and education.

It was in pursuit of this idea that I embarked on a study of the usage in sixteenth-century universities of that most central of medieval medical texts, the Arabo-Latin *Canon* of Avicenna (Ibn Sina). In addition to giving me the opportunity to spend many happy hours of research in the Vatican Library and in various Italian libraries, the project produced abundant evidence, in the shape of commentaries and lectures on Avicenna's text, of the persistence in sixteenth-century medical education of traditional texts and teaching methods alongside innovations. Yet if this remained true even in some Renaissance universities most famed for innovation in medicine, it was also the case that the traditional forms were often penetrated by new views and new material. Accordingly, once the project on the Renaissance use of

Avicenna's *Canon* was concluded, it seemed time to turn to a different methodology and to a new subject matter; it was indeed high time to set aside my search for evidence of continuity in favor of trying to get a better understanding of the nature and limits of innovation in Renaissance medicine—that is, medicine in Europe between approximately the fifteenth and the early seventeenth centuries. It will by now be obvious that I do not mean to play down the continuing significance for that period of many aspects of the medical system born in the high and late Middle Ages; I hope it is equally clear that I believe that the Renaissance centuries are in important ways also appropriately termed "early modern."

I turned first to the paradigm of innovation in the Renaissance medical world: the *De humani corporis fabrica* of Andreas Vesalius and the remaking of anatomy. It seemed that I owed it to myself not to bypass what had been for generations of historians a central feature, a monument, of the field I was trying to make my own. I learned a great deal from devoting many hours to reading through the text of the *Fabrica* and comparing passages between the first edition of 1543 and the second, with attention to identifying revisions by the author. (A practical note: what made this possible was the availability for purchase of copies of microfilms of rare books in the collection of the National Library of Medicine in Bethesda, Maryland, a godsend for US-based scholars in the age before digitization.)

But although I published a couple of articles on aspects of the *Fabrica*, the idea of writing a book on Vesalius and his great book came to nothing. This was largely because I learned of two separate ongoing projects for an English translation of the complete *Fabrica*, one of which proposed to compare the first and second editions (as it happened, the first of the two proposed translations, by William Richardson and John Carman, was not completed until 2009; the second, with comparative treatment of the two editions, is still far from completion). Not only did the existence of these enterprises seem to make any account I could give seem superfluous, but both my own reading of the *Fabrica* and the composition of the two translating teams, each of which consisted of a classicist and an anatomist, reminded me that Vesalius and his response to Galen called for more expertise in ancient Greek medicine and more technical anatomical knowledge than I could lay claim to or was likely to be able to acquire. My encounter with Vesalius was enormously instructive in more ways than one, but I still think the decision not to proceed further was the right one.

Instead, I decided to focus my exploration of the nature and limits of innovation in Renaissance medicine on someone who seems to stand at the opposite pole of Renaissance medical knowledge from Vesalius: the polymath-physician, eclectic philosopher, astrologer, and mathematician—Girolamo Cardano. Of course, after studies of a group of medical teachers and practitioners and of the *fortuna* of a textbook, a quasi-biographical approach was in itself a departure, but it was not so much that I planned a methodological shift as that I was drawn by growing interest in Cardano. My introduction to Cardano, too, came about partially through chance and, in this instance, mortality. A fortuitous encounter in a library—where all the most serendipitous meetings take place—had introduced me to the historian of Renaissance philosophy Charles Schmitt, whose untimely death too soon thereafter was a sad loss to scholarship. Schmitt's specialty was Renaissance Aristotelianism, but the breadth of his knowledge and interests in Renaissance philosophy—he was one of the editors of the *Cambridge History of Renaissance Philosophy*—encompassed Cardano and much else besides.

After Schmitt's death, it was from his library, thanks to the help of his widow, that I acquired a copy of the facsimile reprint of the seventeenth-century edition of Cardano's *Opera omnia* (as is well known, the works in this set are not actually Cardano's *omnia).* I already knew Cardano as one of the most fascinating and idiosyncratic personalities of the sixteenth century, as well as the author of one of the most revealing of Renaissance autobiographies. But although Cardano was the subject of a considerable historical literature, his medical writings were probably the least known aspect of his work. It was only when I contemplated the ten volumes in folio of his works, and saw that medicine filled almost half the set, that I realized the full extent of those writings and how much of interest they contained.

Cardano the physician proved a richly rewarding, if challenging, subject, not least because of the eclecticism of his own interests. The effort to read his medical writings with understanding required some acquaintance with such subjects as ancient, medieval, and Renaissance ideas about dreams and dreaming; sixteenth-century writers on diet; Renaissance Hippocratism; modern historiography relating to Renaissance "life writing"; and much else besides. Cardano once asserted that medicine required knowledge not only of the human body and its diseases but also of theology, astrology, cookery, natural history, occult sciences, prognostication, and natural philosophy. These claims were exuberant, even for Cardano. Yet the branches of knowledge he mentioned could all in one way or another intersect with Renaissance medical learning. More generally, many physicians of the late fifteenth to early seventeenth centuries were humanistically educated from childhood, and university education in medicine assumed preliminary studies in liberal arts and natural philosophy.

No doubt few university-educated physicians, let alone medical practitioners, engaged equally with all or even some of the branches of knowledge that I have just mentioned. But Cardano himself was a striking example of the range of learning of an erudite physician, combining as he did substantial contributions to mathematics, philosophical writings of considerable originality, and fame as a learned and thoughtful astrologer. By yet another fortunate coincidence, Anthony Grafton was investigating Cardano's astrology in the years during which I was trying to penetrate his medicine. Our common interest in Cardano led to discussions that greatly helped to enlarge my view of Cardano's intellectual universe. Thus, as I worked on Cardano, it became plain to me that many aspects of sixteenth-century medicine needed to be situated not only in social context but also as part of intellectual history, cultural history, and the history of science much more broadly considered.

I had known Grafton and admired his work since meeting him in the mid-1980s at the Institute for Advanced Study. Soon thereafter, he invited me to assist in the selection of materials for and contribute essays to the catalogue volumes of two major exhibitions of which he was the curator. The first, on the intellectual impact of New World voyages and entitled "New Worlds, Ancient Texts," opened at the New York Public Library in 1992 (to coincide with the Columbus centenary). The second, "Rome Reborn: The Vatican Library and Renaissance Culture," opened at the Library of Congress in 1993. The latter, planned with the cooperation of the then Vatican librarian, the late paleographer Fr. Leonard E. Boyle, O.P., brought manuscripts and early printed books on loan from the Vatican Library to the United States. I learned a great deal from the work on both these exhibitions. In particular, preliminary research for the second provided a truly extraordinary opportunity for work in the Vatican Library.

Ten years later, another fortunate collaborative project further expanded my intellectual horizons. Jed Buchwald, then director of the Dibner Institute for the History of Science and Technology, housed for some years at MIT, invited Grafton and me to organize a workshop entitled "Renaissance Natural Philosophy and the Disciplines." This undertaking allowed us to bring together a diverse group of scholars specializing in Aristotelianism, Platonism, medicine, alchemy, natural history, and so-called new philosophies of nature in order to reconsider the roles of reworking tradition, the impact of humanism, and the force of new observations in remaking natural philosophy and the nonmathematical sciences in the Renaissance intellectual world. Credit for the title of the resulting volume, *Natural Particulars*, belongs to one of the participants and contributors, the historian of Renaissance medicine Katharine Park; we recognized her suggestion as a particularly appropriate evocation of an essential feature of Renaissance natural knowledge. That title also spoke to my own growing interest in the development in the late medieval, Renaissance, and early modern centuries of multiple forms of factual, or supposedly factual, medical narrative, among them case histories and autopsy accounts.

To backtrack slightly, work on Cardano had also brought me into contact with the then very flourishing world of Cardano scholarship in Europe. Eckhard Kessler (in Germany) and Guido Canziani and Marialuisa Baldi (in Italy) were active in organizing conferences and publishing new work on Cardano—in the case of Canziani and Baldi, as part of a large project for new editions of Cardano's works. Among the scholars who contributed to these undertakings, the work of Ian Maclean was of especial interest to me, of course for his studies of Cardano, but also more broadly because of the depth of his knowledge of the history of sixteenth-century medicine considered as a branch of academic learning and intellectual history. I was exceptionally pleased, therefore, to be invited to lecture in the Wolfenbiittel international summer course on "Learned Medicine in the Late Renaissance (1530–1630)," organized and led by Maclean in 2000. The contributions of lecturers and participants in this summer program amounted to a valuable overview of medical teaching in universities in many parts of Europe—and of course, the occasion also provided opportunity for work in one of the great European libraries, the Herzog August Bibliothek.

Let me return now to the theme of medical narrative. To work on Cardano was an invitation to think further about the varieties of narrative in sixteenth-century medical writings, for Cardano was a great teller of stories: of himself, his dreams, his patients, and their diseases and of the marvelous cures he brought about. But medical narrative also related to a much larger aspect of the development of medicine from about 1300 to 1600—its increasingly empirical character. The manifestations of this growing empiricism included not only more attention to the particulars of observation but also the multiplication of medical narratives of various kinds. The scholastic methodology and the incorporation of elements of Aristotelian logic and natural philosophy along with Galenic medicine that were salient characteristics—and achievements—of the learned medicine of the high Middle Ages by no means disappeared. But alongside them, newer features emerged or, if already present, became more common: descriptions of individual patients and their symptoms, narratives about epidemic outbreaks, autopsy reports, and so on. Alongside the late medieval genre of the *consilium*, or advice for an individual patient—which often contained only the briefest characterization of patient and disease, and seldom mentioned outcome—began to appear case histories modeled on those in the Hippocratic *Epidemics*. Published collections of medical narratives and observations multiplied, from Antonio Benivieni's accounts of "remarkable" diseases and cures in late fifteenth-century Florence to Theophile Bonet's vast collection of autopsy reports from many times and places published in late seventeenth-century Geneva.

An opportunity to consider these issues further came in the shape of an invitation to participate in a conference at the University of Bologna on the development, from the late Middle Ages to the early modern period, of the concept of the "fact" and the role of empirical evidence. The topic of the conference was inspired by the work on the history of the concept of "fact" by Lorraine Daston, who was one of the participants. (The proceedings appeared as an issue of *Quaderni Storici* in 2001.)

The occasion gave me the chance for extended and, as it turned out, very fruitful discussions with Gianna Pomata, one of the conference organizers. She had recently published an important article calling attention to the significance of the presence of the term *historia* in early modern medical descriptions or records of observation. In thinking about the role of narrative in medicine, I had been equally struck by Cardano's praise of *historia*, in the sense of a record of human experience, as a component of medical literature. Of course, both senses—description of an observation and record of past events—preserved the ancient connotation of *historia* as a narrative presenting the results of an inquiry. Our conversations led us to think that the various uses of the word and idea *historia* by sixteenth-century writers in Latin were a topic that deserved fuller exploration.

To our great delight, Lorraine Daston, in her capacity as director of the Max Planck Institut ftir Wissenschaftsgeschichte in Berlin, invited Gianna Pomata and me to organize a month-long workshop entitled "*Historia:* Explorations in the History of Early Modern Empiricism" in the summer of 2003. The hospitality of the Max Planck Institut enabled us to bring together a group of scholars whose expertise extended across a wide range of early modern learning, from the *artes historicae* and the history of religion to natural history, encyclopedism, and, of course, medicine. The design of the workshop, according to the practice of the Max Planck Institut, began with a discussion of pre-circulated papers, allowing ample time for revision and additional library work, and concluded with further collective discussions. It was an ideal arrangement, both for scholarly interchange and as a method of producing a thematically unified volume; it is a time I remember very fondly, not least for the opportunity it offered to explore Berlin. The results appeared in 2005 as the volume *Historia: Empiricism and Erudition in Early Modern Europe*, jointly edited by Gianna Pomata and me.

But as usual, one thing led to another, and I found I could not leave the subject of Renaissance medical men and their uses of history alone quite yet. While the *Historia* project had taught me a lot about the uses of various types of narrative of past events in sixteenth-century medical writing, I had become increasingly aware of, and interested in, the extent to which physicians participated in writing on history and antiquities in a more general, nonmedical sense. In recent years, Renaissance and early modern historical and antiquarian writing of all kinds has attracted much scholarly interest, and I decided I wanted to bring the contributions of physicians into the scope of the discussion. Of course, as I have already noted, history was certainly not the only discipline other than medicine to which medically trained authors contributed. It would have been perfectly possible—as has been pointed out to me several times—to study such men from the standpoint of their general erudition or, for that matter, to concentrate on the contributions of some of them to, for example, poetry or mathematics.

But in electing to focus my investigation on their writing of history, I had specific considerations in mind. It remains my conviction that for the Renaissance and early modern (or, if you prefer, humanist and Baroque) period, connections between medicine and history were especially far-reaching, if only because both incorporated narrative, the record of human lives, and some essential component of empiricism (as I am by no means the only one to have pointed out). And the more I looked for history-writing physicians, in the sense of nonmedical human history, the more of them I found. Some wrote very well-known histories or chronicles, as in the case of the humanist town physician who compiled the *Nuremberg Chronicle*. Some indeed were given the official title *historicus*, as was the case with several physicians in succession at the imperial court of Vienna. And of course, many more remained relatively obscure.

Examples were indeed so numerous that I soon realized it would be impossible to produce a comprehensive study of history writing by members of the medical profession in Renaissance and early modern Europe (although a bibliography of all such works might be of interest, if anyone has the patience to undertake it). Instead, I began to think of taking selected examples of historical writing by medically trained authors both from within medical literature broadly described (i.e., not only medical texts but also biographies of physicians, histories of the discipline of medicine, and so on) and among general works dealing with history or antiquities. For the general works, I would look for clusters from different centers of regional importance and try to relate them to the local intellectual, cultural, and indeed political contexts. I hoped to get a better sense both of the way historical information played into various genres of medical literature and of any regional patterns in the intellectual interests and the patronage of physician-authors of historical works on subjects other than medicine. In the end, the regionally organized section of my study, which appeared with the title *History, Medicine, and the Traditions of Renaissance Learning* in 2007, looked at physicians with historical or antiquarian interests in two cities in Italy (Milan and Rome) and one in northern Europe (Vienna) and among Venetian physicians active in Egypt and Syria.

The chapters on Venetian physicians in Egypt and Syria represented a considerable extension of the geographical scope of my work, something both desirable and indeed necessary from the standpoint of the book's subject matter, if somewhat rash on my part, given that almost all my earlier work had been on Italy. Even now, the focus remained predominantly Italian, but research for the chapter on Vienna provided a highly enjoyable opportunity to explore some of the manuscript riches of the Osterreichische Nationalbibliothek. But none of the work I have been talking about would have been possible without the rich manuscript and early printed collections and helpful staff of so many libraries in Europe and in this country; perhaps a list of rare book rooms frequented might give a better picture of the feel of the life of learning—and certainly of its pleasures—than any description of the work completed.

Most recently, I have turned my attention to some examples of sixteenth-century medical letters, especially the vogue for publishing large collections of *epistolae medicinales*—those curious compilations into which some authors felt free to insert humanist dialogues, short treatises, forensic reports, diatribes against colleagues, and much else besides that bear little resemblance either to surviving manuscript letters by physicians or the advice of letter-writing manuals. But that is an unfinished project and a topic for another occasion.

As I look back over the many years I have happily spent in historical research and writing, I can see many situations in which chance has offered me favorable opportunities and new ideas. I've already discussed a number of these openings, meetings, and occasions. But I have also been fortunate in other, foundational aspects of life that have nothing to do with scholarship: I have enjoyed excellent health, a happy family life, and constant encouragement in my work from my husband and sons. At the same time, I do have regrets for deficiencies in my formation and missed opportunities in my work both as a historian of premodern Europe in general and, more specifically, as a historian of medicine. In particular, I would have benefited from more and stronger language training. With the decline of access to instruction in Latin, the situation is probably much more difficult for today's students of premodern European history than it ever was for me. Given these circumstances, one begins to wonder about the future of the field outside a few privileged institutions.

Moreover, as a historian of medieval medicine, I should have acquired a fuller knowledge of the Arabic background of so much in Western medieval medicine; as it is, my work on Ibn Sina addresses only its (admittedly extensive) Latin tradition. As a historian of humanist medicine, I would doubtless have benefited from stronger knowledge of the classical tradition. Another methodological limitation may be that while I greatly admire scholars who are able to work with equal facility on archives and texts, most of my own work has been on texts.

And, of course, I have contributed little or nothing to large areas of history or new schools of interpretation that became—some more durably than others, perhaps—the focus of widespread interest and participation in the historical profession during the years in which I have been at work. The long debates on the role of theory in relation to history that went on in the 1970s, 1980s, and 1990s went on without me. On the other hand, I am entirely sympathetic to the expansion of such fields as women's history and gender studies, but I have not contributed to them and, indeed, have directed my attention very differently.

As a historian of medicine, I have added very little to knowledge of the realities, as distinct from the prescriptive literature, of medical practice, and nothing at all to knowledge of patient experience. These are, to be sure, areas particularly difficult to investigate for early periods, although it can be done, as impressive recent studies of letters and archival records relating to the illnesses of individuals, whether from fourteenth-century Aragon or fifteenth-century Milan, testify. Moreover, my training in history (i.e., not in medicine) has governed my selection of topics, chronological period, and research methodology. But I am very far from wishing to denigrate the historical application of professional medical knowledge; rather, I am very conscious of my own limitations. I have focused on the intellectual history of thirteenth- to early seventeenth-century medicine simply because such an approach suited both my interests and my capacities.

But I do not in the least regret concentrating my attention on intellectual history and, especially, on the relation of the history of medicine to aspects of the broader intellectual and cultural history of Renaissance and early modern Europe. Rather, recent historiographical trends have opened new possibilities for the fuller integration of aspects of history of medicine with intellectual and cultural history, as well as with some parts of the history of science. To point to some of the most relevant of these trends: the sixteenth and seventeenth centuries continue to be the focus of much new work in the history of science, but now with much more attention not only to life sciences but also to the persistence of traditional ideas about nature; cultural historians have developed a keen interest in the history of the body; the history of the book and of reading and the history of learning are now prominent fields of specialization in intellectual history. All these developments have helped to arouse further interest among historians in general in the role of medical knowledge and ideas and the means of their transmission. In turn, the history of Renaissance and early modern medicine may benefit if we broaden the scope of inquiry to include the interpenetration of medicine and the surrounding world of learning, education, and intellectual life.

The life of learning, with all its twists and turns and unexpected new paths to follow, has been a constant source of pleasure and interest to me for more than 40 years. I have been exceptionally fortunate, especially in that I have been able to follow my interests wherever they took me without feeling any particular pressure to conform to others' expectations and to have met with so much encouragement and help along the way. As I look back, I am reminded that I've been told that it was a mistake to call my book on Cardano *The Clock and the Mirror* because the meaning of the title is not self-evident (though I did explain it in the preface).

It refers to Cardano's recommendation that "a studious person should always have at hand a clock and a mirror: a clock. . .to keep track of time, especially in the case of someone who is a professor, teaches, or writes," a mirror to be reminded of the aging condition of one's body. Cardano's remark may have been too obscure for a book title, but it still seems to me pretty good advice.

Thank you.

Nancy Siraisi

Part I
Medicine, Commentaries and the Medieval Classroom

Chapter 1
Medicine in Some Thirteenth-Century Biblical Commentaries, with a Flashback on Augustine's *De genesi ad litteram*

Danielle Jacquart

Recent scholarship has drawn attention to the medical information contained in the commentaries on Peter Lombard's *Sentences*, thus renewing our views on the relationship between medicine and theology.[1] Commentaries on the Bible (both on the Old and the New Testament) have been less investigated, perhaps because this kind of information is more scattered therein and may be found almost anywhere in the bulk of these exegetical works. Indeed, apart from the description of the creation of man and the accounts of miraculous healings, which provide the principal places where medical explanations could be inserted, there is a wide variety of occurrences that can only be detected by an exhaustive reading of these commentaries. In the present modest contribution, my intention is not to deal with all kinds of topics related to medicine nor tracking down metaphors, but to focus on traces of theories which may have disturbed religious minds. The notions of corporeal spirits (*pneumata*) and of the virtues or powers they were supposed to convey, as well as the function of the brain and the psychic faculties located in this organ, raised the question of the link between body and soul. More subtle discrepancies between medical theories and theological requirements may be detected: for instance, the physiological explanation of nutrition, which involved the transformation of food into bodily substance, seemed to some theologians incompatible with the dogma of the resurrection of the bodies at the Last Judgment and with the concept of *veritas personae humanae*. Since biblical exegesis did not necessarily imply dealing with these topics, their appearance in this framework, as discreet as it was, reveals a real concern. Each period has from this point of view its proper

[1]For an overview on the relationship between religion and medicine or theology and medicine during the Middle Ages, see Amundsen 1996; Biller and Minnis 1997; Biller and Ziegler 2001; Crisciani 2001; Jacquart 2011; Donato et al. 2013.

D. Jacquart (✉)
l'École pratique des hautes études, Paris, France
e-mail: danielle.jacquart@ephe.sorbonne.fr

G. Manning, C. Klestinec (eds.), *Professors, Physicians and Practices in the History of Medicine*, Archimedes 50, DOI 10.1007/978-3-319-56514-9_1

concerns, according to the level of medical teaching and its diffusion outside the circle of physicians and also according to the contemporary theological concerns. The thirteenth century, with the development of university teaching in both theology and medicine, offers a favourable context for tracking in biblical exegesis the traces of contemporary debates. But before entering this period, a flashback to the very beginnings of Latin Christian exegesis is required.

1.1 Ambrose of Milan and Augustine

It is well known that Ambrose of Milan (ca. 330–397) and Augustine (354–430) both had some medical background, the former to a greater extent and perhaps even including practice.[2] In his biblical commentaries, Ambrose refers frequently to medicine, showing on this field some eclectism depending on his sources. In *De Noe*, he quotes the verse from *Leviticus*: XVII. 11, *Anima totius carnis sanguis est*, which provides the opportunity to distinguish between veins and arteries:

> Cum enim hoc loco dixerit sanguinem animam, utique significavit aliud esse animam, aliud sanguinem, ut sit animae substantia spiritus vitalis, sed ipsum spiritum vitalem non per se tantum et sine sanguine usum adferre vivendi. Sed commisceri sanguini, quia sunt quae appellantur arteriae velut receptacula spiritus, quae non solum aerem purum amplectuntur, sed etiam sanguinem, sed minorem longe sanguinis portionem. Gemina enim cum vasa sint, aliud venae quae "phlebs" graece appelatur, aliud arteriae, vena plus habet sanguinis quam spiritus, hoc est "phlebs," arteria minus sanguinis, multo amplius spiritus.[3]

Considered as the substance of the soul, what is called here "vital spirit" refers probably to the divine breath insufflated on Adam's face, as well as to the air inhaled through respiration. Concerning the latter "spirit," Ambrose had previously stated that if its "nest" was the lung, when joined to the blood its "nest" was the heart.[4] In the *Expositio Evangelii secundum Lucam,* this "spirit," which men inhale through involuntary motion, is said to maintain *collegium animae corporisque contubernio fœderatum.*[5] Although this last assumption might be connected with some kind of medical pneumatism, no obvious support of this theoretical background is to be found in Ambrose's biblical exegesis. In particular, there is no allusion to a psychic spirit, in the manner of Galen, nor to any cerebral ventricle. If

[2]Passarella 2009; Bardy 1953; Rassinier 1991; André 2007. At a more general level, see Young 1997; and Young et al. 2004.

[3]*De Noe* XXV, 92, in *Sancti Ambrosii Opera. Pars prima* 1897, 413–497: quoted in Passarella, 140–141.

[4]*De Noe* VI. 14, quoted in Passarella, 235: "Spiritus quidem nostri, hoc est eius quem carpimus et quo alimur in hac vita, nidus est pulmo, sanguinis autem et spiritus nidus est cor."

[5]*Expositio Evangelii secundum Lucam* VII. 123, in *Sancti Ambrosii Mediolanensis Opera, Pars IV,* 1957 (*Corpus Christianorum Latinorum,* 14), 1–400: quoted in Passarella, 228: "aerius ille spiritus vitale collegium animae corporisque contubernio fœderatum sine nostro labore perpetuat, nec salutaris deficit usus alimenti, nisi cum venerit dies suprema moriendi."

the brain is considered as the origin of sensation, it is because of its softness, which allows it to receive all the senses. As for the nerves that originate in it, some are soft and thus suited to compassion, others are hard in order to be more efficient in action. The distinction between sensitive and motive nerves is discreetly put forward, but there is no allusion to cerebral sensitive faculties other than the reception of external sensations. Nevertheless, the origin of voluntary motion is allotted to the brain.[6]

A few decades after Ambrose's death, Augustine in *De genesi ad litteram* refers to *medici*, when commenting on the verse which deals with the creation of man, particularly its second part: *flavit in faciem eius flatum vitae, et factus est homo in animam viventem.* It is not by chance that Augustine alludes to *medici* when commenting not on the first part of this famous biblical verse (*Et finxit Deus hominem pulverem de terra),* which deals with the sole body, but on its second part, which links the human soul with some insufflated breath. In his long and thorough inquiry concerning the nature of the human soul—*non parva quaestio* in his own words—Augustine asks whether it could have been made by God's breath from the air. At this point he refers to what *medici* not only say, but claim to have proven: even if the flesh seems to have the compactness of the earth, it contains something of the air, which kept in the lungs passes from the heart to the arteries. According to the same *medici*, the flesh contains also fire, which is kept in the liver, and which has besides its warming power a luminous quality. Thanks to this luminous quality, the fire originated in the liver is purified and ascends to the top of the brain, *tamquam in caelum corporis nostri.* From the brain then emanate the rays of the eyes, as well as fine ducts which reach not only the eyes, but the ears, the palate and all the limbs through the marrow inside the backbone.[7] Whatever may

[6]*Hexaëmeron,* VI. 9. 9. 63, in *Sancti Ambrosii Opera. Pars prima,* 3–261; quoted in Passarella, 79, 199, 182: "Itaque propter oculos ferunt medendi periti cerebrum hominis in capite locatum, alios autem nostri corporis sensus propter cerebrum finitimo quodam esse domicilio constitutos. Initium enim nervorum et omnium sensuum voluntariae commotionis cerebrum est atque inde omnis eorum quae diximus causa manat. Initium autem arteriarum et insiti caloris, quo animantur et tepefiunt vitalia, cor esse plerique arbitrantur. Sensuum autem singulorum velut organum nervi sunt, qui cordae et fides quaedam de cerebro oriuntur et per partes corporis in singula quaeque officia derivantur. Ideoque mollius est ceteris cerebrum, quia omnis suscipit sensus. Unde et nervi, qui referunt universa quae vel oculus viderit vel auris audierit vel odor inalaverit vel lingua increpuerit vel os saporis acceperit. Quod enim molle ad compassionem aptius, quod autem durum ex aliquo rigore nervorum ad agendum efficacius."

[7]*De genesi ad litteram* VIII. XIII. 20, in *Œuvres de saint Augustin* 48, 2000 [1972], 534–536: "Deinde—si non est contemnendum, quod medici non tantum dicunt, verum etiam probare se adfirmant—quamvis omnis caro terrenam soliditatem in promptu gerat, habet in se et aeris aliquid, quod et pulmonibus continetur et a corde per venas, quas arterias vocant, diffunditur; et ignis non solum fervidam qualitatem, cuius sedes in iecore est, verum etiam luculentam, quam velut eliquari ac subvolare ostendunt in excelsum cerebri locum, tamquam in caelum corporis nostri. Unde et radii emicant oculorum et de cuius medio velut centro quodam non solum ad oculos, sed etiam ad sensus ceteros tenues fistulae deducuntur, ad aures videlicet, ad nares, ad palatum, propter audiendum, olfaciendum atque gustandum; ipsumque tangendi sensum, qui per totum corpus est, ab eodem cerebro dirigi per medullam cervicis et eam, quae continetur ossibus, quibus dorsi spina conseritur, ut inde se tenuissimi quidam rivuli, qui tangendi sensum faciunt, per cuncta membra diffundant."

have been the medical sources which Augustine alludes to, it is worth noting that his main concern is to stress that this "air" and this "fire" contained in human flesh have nothing to do with the nature of the soul, even if it is through them, because of their subtle quality, near to incorporeality, that the human soul can move the thicker components of the flesh.[8] When dealing with the air and the fire conveyed respectively from the heart and the liver to the brain, Augustine anticipates in some way the medieval theologians' concerns, faced as they happened to be from the twelfth century on with the medical description of three corporeal spirits.

It may also be stressed that, contrary to Ambrose, Augustine in *De genesi ad litteram* avoids using the word *spiritus* even for the aerial substance conveyed in the arteries. In his conclusion on the nature of the soul and on the way of naming it, he points out the ambiguity of this word: *Quid sit autem, non dicitur melius quam anima vel spiritus vitae. Ideo enim additur "vitae," quia et iste aer plerumque dicitur "spiritus."*[9] As we have seen, the expression *spiritus vitalis* used by Ambrose was not without ambiguity. As for Augustine, even nerves contain *aer* and not *spiritus* (i.e. pneuma), although it is highly probable that he is relying here on medical theories, which explained the sensitive and motive function of the nerves by their conveying of psychical pneuma.[10]

In the twelfth book of *De genesi ad litteram,* Augustine deals at length with the meaning of *spiritus*. Even if it refers to the inferior level of the soul, intermediate between its sensitive part and its intellective one, its incorporeal nature is strongly emphasized.[11] When alluding to *medici* in the seventh book and avoiding in this context the word *spiritus*, even in its original meaning, i.e. "breath," Augustine probably intends to remove the ambiguity of medical theories, particularly in their Galenic tradition. Likewise in *De civitate Dei*, he points out that in the verse of Genesis which describes the creation of man, where the Latin text gives *spiritum vitae* in Greek, it is not "pneuma," which is reserved for the Holy Spirit, but "pnoèn." And this is the reason why "some Latin because of this difference prefer using the word *flatus*, instead of *spiritus*."[12] Indeed, the text commented on in *De genesi ad litteram* contains *flatus*.

Without entering the complex question of Augustine's medical sources, his mentioning in *De genesi ad litteram* of cerebral localizations of sensitive and motive faculties as well as "ventricles" of the brain suggests at least some

---

[8]Ibid., VII. XV. 21, 538: "Anima ergo quoniam res est incorporea corpus, quod incorporeo vicinum est, sicuti est ignis vel potius lux et aer, primitus agit et per haec cetera quae crassiora sunt corporis, sicuti humor et terra—unde carnis corpulentia solidatur—quae magis sunt ad patiendum subdita quam praedita ad faciendum."

[9]Ibid., VII. XXI. 30, 550.

[10]Ibid., VII. XIX. 25, 544: "Et aer, qui nervis infusus est, paret voluntati, ut membra moveat, non autem ipse voluntas est."

[11]Ibid., XII. VII. 18—IX. 20, 352–358 and 559–566 ("*Spiritus*" in book XII, *De genesi*). See also Verbeke 1945; and Madec 1996.

[12]*De civitate Dei,* XIII. XXIV. 3, in *Œuvres de Saint Augustin* 35 (*Bibliothèque Augustinienne*), 332.

knowledge, direct or indirect, of post-Galenic theories. When commenting on *in faciem*, Augustine explains the importance of not only the face itself, but also the anterior part of the brain, in connection with sensation. The anterior part rightly precedes the posterior part because sensation precedes movement, *sicut consilium praecedit actionem*. But since the movement of the body does not follow sensation without some interval of time, memory is necessary:

> Et quoniam corporalis motus, qui sensum sequitur, sine intervallis temporum nullus est, agere autem intervalla temporum spontaneo motu nisi per adiutorium memoriae non valemus, ideo tres tamquam ventriculi cerebri demonstrantur: unus anterior ad faciem, a quo sensus omnis; alter posterior ad cervicem, a quo motus omnis; tertius inter utrumque, in quo memoriam vigere demonstrant, ne, cum sensum sequitur motus, non conectat homo quod faciendum est, si fuerit quod fuit oblitus.[13]

As Augustine previously stated, *medici* not only spell out assumptions, they claim to prove them. Concerning the respective attributions of the three parts or ventricles of the brain, they are detected by the fact that a medical treatment applied to the appropriate part repairs the dysfunctioning either of sensation, of movement, or of the memory of bodily movement. Augustine's allusion to the statements of *medici* comes in line with Galen's *De locis affectis*, except that the clear localization of faculties into the ventricles of the brain is post-Galenic.[14] The first author who expressed clearly this theory seems to have been the Greek Church Father, contemporary of Augustine, Nemesius of Emesa in his treatise *On the nature of man*, written between 390 and 400. In order to make the comparison easier between Augustine and Nemesius, the quotation from the latter is given in the literal Latin translation made by Burgundio of Pisa in the twelfth century:

> Quia vero sensuum quidem principia et radices anteriores esse ventres diximus cerebri, discretivi vero eum qui est medius, memorativi vero posteriorem, necesse est demonstrare si haec modo habent, ut non videamus irrationabiliter credere his quae dicuntur. Sufficiens autem demonstratio maxime est quae ex ipso particularum actu assumitur. Nam anterioribus quidem solum ventriculis secundum aliquem unquam modum laesis, sensus quidem impediuntur, discretivum vero adhuc manet servatum; cum autem medium ventriculum solum patitur, discretio quidem fallitur, sensuum vero membra manent custodientia secundum naturam sensum; si autem anteriores et medius ventriculus patiantur, discretio simul cum sensibus absciditur; parenkephalide vero patiente, memoria sola deperit, sensu et discertione nullo leaesis.[15]

---

[13] *De genesi ad litteram*, VII. XVIII. 24, 540–542.

[14] Ibid., VII. XVII. 24, 542: "Haec illi certis indiciis probata esse dicunt, quando et ipsae partes adfectae morbo vel vitio, cum defecissent officia vel sentiendi vel movendi membra vel motus corporis reminiscendi, satis quid valerent singulae declararunt eisque adhibita curatione cui rei reparandae profecerit exploratum est". On the faculties of the brain and its ventricles in Galen, who did not set a clear localization, see Rocca 2003; Donini 2008; Pigeaud 1988.

[15] Némésius d'Émèse 1975, cap. 12, 88. For the Greek text, see Morani 1987 (*Bibliotheca scriptorum Graecorum et Romanorum Teubneriana*); and for an English translation, see Sharples and van der Eijk 2008.

As for voluntary movement, Nemesius, in another part of his work, locates its origin at the bottom of the brain and in the spinal chord.[16] The main discrepancy between Augustine and Nemesius lies in the exclusion by the former of any kind of reasoning from the brain. To the middle ventricle is attributed *memoria* (restricted to the memory of the movement to be impulsed). The question then is raised whether this exclusion was already in Augustine's medical source or he intentionally distorted this source. *De genesi ad litteram* was probably the first Latin writing which set forth the theory of cerebral localizations, and it remained almost the only one which stated it before the translations at the end of the eleventh century of Nemesius' *On the nature of man* by Alphano of Salerno and of Arabic works by Constantine the African. During the twelfth century, religious writers like William of Saint-Thierry had the same concern as Augustine, and endeavoured to restrain medical theories on "spirits" and cerebral faculties within acceptable limits.[17]

1.2 Robert Grosseteste and Bonaventure

Probably written between 1232 and 1235, Robert Grosseteste's *Hexaëmeron* is mainly of a devotional nature. The scholarly background and the scientific interests of its author are nevertheless visible.[18] Robert Grosseteste refers explicitly only twice to *medicus* or *medici*. In quite a common way in the context of biblical exegesis, the art of the physician serves as an example in comparison with the figure of Christ.[19] While commenting on the significance of the biblical tree of the knowledge of good and evil, Grosseteste first states "that there are two kinds of knowledge, a knowledge through wisdom and a knowledge through experience." It is worth noting that the physician's knowledge is considered as only acquired through wisdom, at least if the physician has benefited from a good health since his birth. We have there a testimony of the emphasis put in these first decades of the thirteenth century on learned medicine. As for the knowledge of diseases acquired through experience, it is allotted to the sick man. In the same manner as the

[16]See Debru 2005. Similarities between Nemesius and Augustine have been recently stressed on the topic of God's "mixing" with man. See Boulnois, 2005, 458.

[17]See Jordan 1990; and Jacquart 2013. After Augustine and before the eleventh-century translations, the only author, as far as I know, who has located hegemonic operations into the ventricles of the brain, was Agnellus of Ravenna (sixth century?) in his commentary, written in a very obscure Latin, on Galen's *Ars medica*. Echoing probably Late Alexandrian teaching, he located *fantasia* in the anterior ventricle, *logismos* in the middle ventricle and *mnemi* in the posterior ventricle. See Palmieri 2008, 52.

[18]Grosseteste 1982 (*Auctores Britannici Medii Aevi*, VI); and Grosseteste 1996 (*Auctores Britannici Medii Aevi*, VI. 2). On Grosseteste's biblical exegesis, Smalley 1981; McEvoy 1994; and Grosseteste, *Roberti Grosseteste Exposition in Epistolam Sancti Pauli ad Galatas* 1995. No significant allusion to medicine is made in this *Expositio*, but there is an interesting digression on fascination, with references to Algazel's *Metaphysics* and Avicenna's *De anima*, 72–74.

[19]*Hexaëmeron*, XI. IV. 4, 309–310; trans. by Martin, 316–317.

physician is supposed to have no experience of diseases, Christ had no experience of sin. Grosseteste pursues his comparison by commenting on the disobedience of the first human beings when they ate the fruit of the forbidden tree. Likewise when a sick man eats some food that the physician has forbidden and becomes ill, it would have been better for him to trust the physician's wisdom than to acquire knowledge of disease through experience. Robert Grosseteste uses here the traditional metaphor of "Christus medicus" in an unusual manner, and moreover does not seem to give room to experience in the acquisition of medical knowledge, contrarily to what he suggests to some extent in other writings, at least regarding the testing of drugs.[20] Biblical exegesis has its own requirements, which are not always in accordance with epistemological concerns.

The other explicit allusion to medicine is not included in a comparison, but in a condemnation of astrology. Some *professores* argue vainly that the stars cause many effects on the human body, and that the soul is necessarily affected by that, relying on "what physicians say" (*ut dicunt medici*).[21] Robert Grosseteste puts forward that the rational soul, submitted as it is to God, has more power on its own body than the stars could have:

> Unde quantumcumque moveat Saturnus vel Mars corpus, sive hic sanguinem constringendo sive ille sanguinem accendendo ut proveniat tristicia vel ira in anima, plus potest racio vene ordinata in contrarium operando ut sit in anima gaudium et mansuetudo, et per hoc nulla aut parva et imminuta sit in sanguine et corporeis spiritibus ab accione Saturni vel Martis constrictio vel inflammacio. Plus enim potest vera animi mansuetudo in temperanciam et quietationem sanguinis et spirituum quam possit Saturnus in eorum constrictionem. Quod facile patere potest ex contrario quia videmus quod tristicia vel ira mentis plus et magis subito immutat et constringit vel perturbat et inflammat sanguinem et spiritus quam faciat aliqua siderum vel aeris vel alterius corporis continentis accio [...] Preterea contra disposiciones quas imprimunt sidera, possunt contra operari fortius medicine, consuetudines et studia.[22]

Robert Grosseteste no doubt relies on medical explanations of the physical effects of passions, which physicians attribute to the so-called "vital spirit" conveyed with the arterial blood and to the "vital virtue" located in the heart.[23] He avoids the use of medical terminology, mentioning more vaguely "corporeal spirits." According to him, through its rational part, which is not submitted to bodily deficiencies nor, as a

[20]Despite the fact that Robert Grosseteste is no longer regarded as a forerunner of the experimental method, it is worth referring on this point to one of the landmarks of the history of science. See Crombie 1971.

[21]*Hexaëmeron,* V. X. 3, 167; trans. by Martin, 169. The condemnation of astrology is a major concern of Robert Grosseteste in his *Hexaëmeron.* On his change of mind during his life on this topic, see Dales 1967, 357–363.

[22]*Hexaëmeron,* V. X. 4, 6, 168–169; trans. by Martin, 169–170. I have changed the punctuation of the last quoted sentence, which is in the edition "possunt contra operari fortius medicine consuetudines et studia", translated by Martin as "medical study and practice can prevail." The word *medicine* must certainly be understood here as "drugs" and does not refer to the medical art.

[23]On the medical explanations available in Grosseteste's time, see Gil-Sotres 1994; and Knuuttila 2004.

consequence, to astral impressions, the soul can mitigate some impetuous emotions and their physical effects. As for the supposed individual predispositions caused by the stars, they can be corrected both by physical means with drugs and by moral means with habits and studies.

In Robert Grosseteste's view, the rational human soul commands directly the body, and in *Hexaëmeron* he does not take into account the organic mediations between body and soul. If he alludes vaguely to "corporeal spirits," he does not mention the virtues or faculties they are supposed to convey according to medical physiology. When dealing with the number seven, he states that "the human being is made up by nature according to the number seven, i.e. of four elements as regards the body and of three powers of the soul."[24] No more is said about the connection between the four elements of the body and the three powers of the soul in the nature of man. It is nevertheless obvious that he knew medical theories, as is shown by his description of physical effects of passions. More strikingly, he describes the nutritional action attributed by physicians to a natural virtue, when dealing with plants and *vis vegetativa*. It is likely that he relies here on the chapter of Avicenna's *Canon* in which are enumerated the four *virtutes* which serve the natural faculty, i.e. *attractiva, retentiva, digestiva, expulsiva*. These four virtues act through elementary qualities (cold, dry, hot and moist) and through villosities or fibers (*villi*), which are either longitudinal, latitudinal or transverse according to their intended natural action. Robert Grosseteste transferred thus a medical description of human vegetative life to the plants themselves. In order to be more convincing, he identifies *villi* as described by physicians with the fibers that anyone can see on cut trees, adding nevertheless that these fibers can also be seen in the cooked flesh of animals.[25] Medical theories dealing with the instruments through which the soul animates the body are only taken into account as far as they do not touch the highest levels, sensation and of course the reasoning power.

[24] *Hexaëmeron*, IX. IX. 3, 278; trans. by Martin, 284.

[25] *Hexaëmeron*, IV. XXX. 3, 154–155: "[. . .] eget vita vegetativa necessario virtute attractiva nutrimenti, et virtute retentiva attracti, et virtute digestiva retenti, decoquente retentum et segregante purum eius ab impuro et assimulante substanciam quod prius assimulatum erat secundum complexionis qualitatem. Eget quoque vita vegetativa quarta virtute, videlicet impuri et superflui expulsiva, ne ipsius putredine corpus corrumpatur et ab attrahendo recens nutrimentum prepediatur. Hee autem quatuor virtutes, vite vegetative et nutritive deservientes, operantur suas acciones per quatuor primas naturales qualitates. Attractiva namque attrahit per calidum et siccum, et retentiva retinet per frigidum et siccum. Digestiva quoque decoquit et digerit per calidum et humidum. Expulsiva vero eicit per frigidum et humidum. Frigidum namque constringit, humidum lubricat et mollit, calidum dilatat et decoquit et que unius nature sunt congregat et segregat que extranee sunt nature, siccum vero suggit, sistit et quietat. Item vis attractiva attrahit per villos longitudinales, et retentiva retinet per villos latitudinales, et expulsiva expellit per villos transversales; et hos villos potest quilibet visu discernere in fissionibus arborum et eciam in coctis carnibus animalium". Cf. Avicenna, *Canon*, I. 1. 6. 3, 1507 (reprod. Hildesheim 1964), fol. 23v–24r, which relies here on Galen's *On Natural Faculties*. Robert Grosseteste was among the first readers of Avicenna's *Canon*, which he quotes in his commentary on Aristotle's *Physics*, see Chandelier 2009.

Some other traces of the medical concept of *virtus vitalis* appear along with a metaphorical resort to optics in the explanation of the crucial verse: *et inspiravit in faciem eius spiraculum vite, et factus est homo in animam viventem.*[26] The human soul, at its rational level, is described as "a straight glance reflected on itself." Grosseteste then adds that according to some, life is "a reciprocal spirit," and that "for this reason the vital motion of the whole soul is a kind of spiritual egress from itself and a reciprocal return to itself. This perfect circling exists all at one and the same time in the non-bodily spirit." At this point it was necessary to take the body into account:

> Et quia substantia anime unita est corpori in unitatem persone, corpus humanum necessario sequitur ipsam animam pro modo et possibilitate sua in mocionibus suis naturalibus. Ex reciprocacione igitur spiritalis motus anime in se, sequitur motus cordis qui in se reciprocatur. Movetur enim cor continue per dilatationem et constriccionem, imitans ut potest, per dilatacionem corporalem, extensionem virtutis anime vitalis; et per constriccionem corporalem, reciprocacionem spiritalem. Et sequitur naturaliter ex motu cordis motus inspiracionis et respiracionis, et forte eciam enim motus animalis qui tradit secundum rectum provenit a motu nervorum et musculorum qui fit per constriccionem seu dilatacionem, et ille motus a motibus corporalium spirituum circularibus, et illi motus a motu spiritus incorporei, incorporaliter circulati.

In this passage, which describes the bodily movements as following the movements of the soul, we find again a discreet allusion to corporeal spirits, as well as to "the vital virtue of the soul," in connection with the movement of the heart, the seat of the vital virtue and the origin of the vital spirit according to physicians. Moreover, the movements of nerves[27] and muscles are also related to "corporeal spirits," but without any trace of the animal or psychic virtue as defined by physicians. As a matter of fact, this virtue located into the brain does not appear in Robert Grosseteste's *Hexaëmeron*, and the heart is called, *membrum radicale,* "the principle of sense and motion, and of all the powers, natural, vital and sensitive."[28] As for the brain, it is only mentioned through a quotation from *De genesi ad litteram*, in which Augustine put forward the sensitive role of its anterior part in order to explain why God's breath was insufflated on man's face:

> Et quia anime infuse virtus maxime et evidentissime se exerit in faciem hominis, non dixit: "inspiravit in eum," sed: *in faciem eius*. Hunc sermonis modum notat eciam Augustinus sic dicens: "Quia pars cerebri anterior, unde sensus omnes distribuuntur, ad frontem collocata est atque in facie sunt ipsa velud organa sentiendi—excepto tangendi sensu, qui per totum corpus diffunditur; qui tamen et ipse ab eadem anteriore parte cerebri ostenditur habere

---

[26] *Hexaëmeron,* X. II. 7, 293; trans. by Martin, 299. Robert Grosseteste of course quotes *Genesis*: I. 7 in Jerome's version, which departs, as we have seen, from Augustine's preferred version. On the action of the soul upon its body according to Robert Grosseteste, see Dales 1995, 313–319.

[27] It seems more appropriate to translate here *nervi* by "nerves" than by "sinews" as Martin does (299). Unlike Aristotle, Grosseteste was probably aware of the function of motive nerves, as his allusion to "corporeal spirits" suggests.

[28] *Hexaëmeron,* VII. XIV. 5, 212; trans. by Martin, 216. The heart is defined as the principle of sense and motion in Aristotle's *Parts of animals,* IV. 5, 681b, a text which was available since ca. 1210–1220 in Michael's Scot's translation from the Arabic.

> viam suam, quae retrorsum per verticem atque cervicem ad medullam spine, de qua loquebamur paulo ante, deducitur, unde habet utique sensum in tangendo et facies, sicut totum corpus, exceptis sensibus videndi, audiendi, olfaciendi, gustandi, qui in sola facie prelocati sunt. Ideo scriptum arbitror, quod *in faciem Deus* sufflaverit *homini flatum vite*, cum *factus est in animam vivam*. Anterior quippe pars posteriori merito preponitur, quia ista ducit, illa sequitur; et ab ista sensus, ab illa motus est, sicut consilium precedit actionem." Factus est itaque totus homo in animam viventem, id est in animam vegetabilem viventem per virtutem sentientem, et in animam sentientem per virtutem racionalem.[29]

Contrary to Augustine, Grosseteste does not expand this explanation by introducing the medical theories related to the cerebral faculties and the brain's ventricles. Indeed, he could not quote Augustine's own presentation on this matter, since it was not in accordance with thirteenth-century knowledge: many medical sources as well as Avicenna's *De anima* gave a picture of cerebral localizations, which, even if they sometimes disagreed with each other on the attribution of the faculty located in the middle ventricle, never put memory of movement in it as Augustine did. Was it this discrepancy between Augustine's statement and ordinary thirteenth-century medical knowledge which pushed Robert Grosseteste to avoid this subject? Or, more fundamentally, did he consider medical theories as incompatible with his own views on the link between soul and body in a spiritual context?

In any event, his selected inclusion of the medical knowledge of his time is meaningful, and he probably felt embarrassed about the medical concepts of spirits and virtues, when they were applied to bodily activities which went too far over the limits of vegetative life. In this respect he was in line with the reservations expressed by some twelfth-century religious authors. At least, a closer resort to medical theories, in relation to the animation of human body, would have required thorough explanations, which went far beyond the "devotional" aim of this *Hexaëmeron*. Moreover, at Robert Grosseteste's time, Aristotle's *De animalibus*, stating the primacy of the heart, had only very recently been made available in Latin, and medical discussions had not yet developed in order to reconcile the Aristotelian cardiocentric theory with the Galenic views, or even to compare them in detail. Robert Grosseteste bears witness to this first introduction of Aristotle's philosophy of nature which had not yet completely encountered medical learning. On this point, his reading of Avicenna's *Canon* seems highly selective and partial.

A supplementary allusion to physicians' theories may be found with respect to the origin of semen, even if they are not attributed to *medici*, but to *physici*, which actually in this case could be synonymous. Again, vegetative power is commented on when Grosseteste addresses the creation of plants, in *Genesis* I. 11: *Germinet terra herbam virentem et facientem semen et lignum pomiferum faciens fructum iuxta genus suum cuius semen in semet ipso sit super terram.*[30] He asks: "But why does it say: 'which may have seed in itself'?" He gives then two diverging opinions

[29] Ibid., X. II. 3–4, 291. Cf. Augustine, *De genesi ad litteram*, VII. XVII. 23.

[30] *Hexaëmeron*, IV. XXI. 2, 145; trans. by Martin, 146–147.

about the formation of seed, which undoubtedly are related with human seed.[31] According to *quidam physici*, seed comes from food, already assimilated to the body and prepared for being joined to the body to be nourished. According to others, seed merely separated from food is not *semen seminale*, whether in whole or in part: *semen seminale* must be separated from the substance of the body of Adam "that God made in the beginning from the slime of the earth. This seed thus separated, passes into the substantial body of the child, once food is taken in that supports endurance." Grosseteste adds that those who support the latter opinion make a distinction between "true substantial flesh" or "flesh according to form," and flesh generated from food, or "flesh according to matter."[32] It is hard to guess what Grosseteste's favoured opinion was, since he did not state it explicitly. Related here to the formation of seed, and subtley introduced, the question of nutrition would give rise to intense debates in the universities.

A few decades later, the Franciscan Bonaventure had an even more devotional aim in his *Collationes in Hexaëmeron*, which he read at the University of Paris in 1273 between Easter and Pentecost.[33] The entirely spiritual nature of these speeches limits the place of allusions to scientific knowledge. Nevertheless, there is in the first *collatio* a kind of classification of the sciences, illustrating the idea according to which Christ is the *medium* of all sciences. This *medium* has itself seven forms, each of them corresponding to one science. The second form pertains to natural philosophy, whose range begins with the study of the effect of celestial bodies on the elements and ends with the study of the rational soul. The natural philosopher (*physicus*) deals with a double medium, i.e. the medium of macrocosm and the medium of microcosm:

> Medium maioris mundi est sol, medium minoris est cor. Sol enim est in medio planetarum, secundum cuius delationem in obliquo circulo fiunt generationes, et regulat physicus generationem. Inter omnes autem planetas maioris diffusionis est sol. A corde similiter est diffusio, quidquid dicunt medici. Nam spiritus vitalis ab eo diffunditur per arterias; spiritus autem animalis per nervos, licet complementum recipiat in cerebro; spiritus vero naturalis ab eodem diffunditur per venas, licet compleatur in hepate.[34]

---

[31]This is stated by Aristotle in *De generatione animalium*, I. 19. 726b, but also by medical sources, in particular Avicenna's *Canon*, III. 20. 3: "Sperma est superfluitas digestionis quarte que fit cum dispartitur cibus in membris resudando a venis, tertia digestione iam expleta" (Venice, 1507, fol. 352r).

[32]*Hexaëmeron*, IV. XXI. 2, 145; trans. by Martin, 146–147: "Et nisi esset iste modus decisionis seminalis non solum a nutrimento sed a carne vera substantiali—quam vocant quidam 'carnem secundum speciem', sicut adgeneratam ex nutrimento vocant 'carnem secundum materiam'—non vere, ut aiunt, fuissemus omnes in lumbis Ade, si hec itaque sentencia vera est." The distinction between *caro secundum speciem* et *caro secundum materiam* comes from Aristotle's *De generatione et corruptione*, I. 5. 321a–322a. In James of Venice's translation, *species* translates the Greek word "eidos" and means "form". Martin in his translation of *Hexaëmeron* gives "flesh according to its kind", which, in my view, is not appropriate.

[33]*Collationes in Hexaëmeron*, in Bonaventure 1891. For a recent overview on Bonaventure's life and work, see Cullen 2006. On his reading of the Holy Scripture, see in particular the introduction to Bonaventure 1991.

[34]Bonaventure 1891, 331.

About 40 years after Grosseteste's *Hexaëmeron*, Bonaventure mentions without any reservation the three spirits described by physicians. In the meantime, medical theory has developed as well as natural philosophy, and the nature of bodily spirits has been defined and accepted even by religious minds. Bonaventure shows that he is aware of the main controversy which is supposed to oppose physicians and philosophers about the hegemonic organ of the body (heart or brain?) and even of a more sophisticated question concerning the generation of bodily spirits. According to some medical authors, this aerial substance originates first in the liver from the vapor produced by the making of blood, this "natural spirit" becoming "vital spirit" in the heart through a refinement and the supply of breathed air coming from the lung, and this "vital spirit" after a supplementary refinement at the bottom of the brain and in its ventricles becomes "animal spirit." According to others, the origin of bodily spirits is in the heart, the organ which preserves life, and the "vital spirit" is transformed in "natural spirit" in the liver and in "animal spirit" in the brain. Bonaventure is well informed of the contemporary debates, even if there was no perfect consensus on this topic among *medici*, as he seems to assume. The distinction on this point (as well as others) between philosophers and physicians is more rhetorical than real, and there is a wide range of options even among physicians. This distinction relies actually on the discrepancy between Aristotle and Galen about the seat of sensation, the heart according to the former, the brain according to the latter.[35] Concerning Bonaventure in his *Collationes in Hexaëmeron*, the cardiocentric theory had the advantage of reinforcing his metaphorical development in which the heart is supposed to be at the center of the body as the sun is at the center of the planets.

Discreet and rare as they are, Bonaventure's allusions to medical theories are accurate, without any distortion, and they relay valuable information. Likewise in his commentary on the *Book of Ecclesiastes*, one of the biblical books particularly open to naturalistic explanations, there is only one allusion to medicine, but again very accurate. In his poetic manner, the biblical author describes the decrepitude of old age. The enigmatic verse *et conteratur hydria super fontem* is considered by Bonaventure as a reference to urinary incontinency. He gives then the medical definition of urine, as stated at the beginning of Isaac Israeli's *De urinis*, one of the texts included in the curriculum of the Parisian Faculty of medicine:

> *Hydria* est receptaculum aquae exhaustae a fonte, et vesica est receptaculum urinae procedentis ab hepate tanquam a fonte. Urina enim est colamentum sanguinis et aliorum humorum, qui ab hepate in totum corpus procedunt tanquam a fonte. *Hydria* ergo conteratur super *fontem*, quando vesica rumpitur et colamentum non potest recipere, quod procedit ab hepate.[36]

[35]On this medico-philosophical debate during the thirteenth century, see Siraisi 1981, 166–167, 180–195, 261–262; Pigeaud 2003; Ricklin 2003; Jacquart 2003.

[36]*Commentarius in librum Ecclesiastae,* XII. 5, in Bonaventure 1893, *Doctoris Seraphici*. . .VI, 94. For an English translation, see Bonaventure 2005. Also, see Israeli 1515, fol. 156r. Isaac's work was included into the 1270–1274 Parisian curriculum, see Jacquart 1998, 163.

As in his *Collationes in Hexaëmeron*, Bonaventure echoes in few words accurate medical theories. They are inserted as such in the biblical commentary, without any attempt to integrate them into a spiritual exegesis. There is a clear separation between rational knowledge and moral or spiritual concerns, the former serving as a mere mirror which helps human minds to represent divine realities. The Holy Scripture uses all the things that are in the world as signs which need to be correctly understood, but it is dangerous to deal too much with the study of these signs, at the risk of being removed from the home of Scripture: *melius est enim tenere veritatem quam figuram.*[37] Being one of the instigators of the Parisian condemnations of 13 errors in 1270, Bonaventure was nevertheless intransigent with regard to philosophical or scientific views that contradicted religious orthodoxy. In his *Collationes in Hexaëmeron*, for instance, he condemns firmly the theory of the eternity of the world, one of the thirteen errors listed in 1270. Also pointed out are the psychological determinism of the astrologers and alchemical transmutation.[38] But, while limiting the importance of rational knowledge and condemning it when necessary, he takes it as it is, restituting it faithfully. As far as medical theories are concerned, if some of them are corrected, as we have seen, nothing seems to be firmly condemned.

1.3 Albert and Thomas

Berryl Smalley—the eminent specialist of biblical exegesis in the Middle Ages—stated that Albert the Great considered any commentary on the Bible as an 'inter-disciplinary' work.[39] This statement needs to be qualified. It is indeed true that, compared with other biblical commentators, Albert found more opportunities to give scientific information and endeavoured to find the most accurate ones. But just as in his philosophical and scientific writings, he avoided dealing with spiritual matters and was careful to set strict borders between rational knowledge and articles of faith,[40] his purpose in his biblical commentaries was focused on the spiritual meaning. He in no way gratuitously displayed his large scientific knowledge.

[37]Bonaventure 1891, *Collationes in Hexaëmeron,* 413.

[38]Ibid., 357: "Sed in his omnibus luxuriata est ratio: luxuriata est metaphysica, quia quidam posuerunt mundum aeternum, quia si causa aeterna, et effectus aeternus; et isti male senserunt de causa prima. Similiter mathematici sciverunt numeros et postea ad influentias et secreta cordium venerunt. Naturales sciverunt et de corporibus et mineralibus et dixerunt 'Ars imitatur naturam' et nos scimus secreta naturae, ergo nos faciemus vobis aurum et argentum". On Bonaventure's role in the Parisian condemnation of the idea of an eternal world, see Grant 1996, 67–70. The *Collationes in Hexaëmeron* are not taken into account in Grant's book.

[39]Smalley 1981, 241.

[40]For instance, in his commentary on Aristotle's *De generatione et corruptione*: "Quaeramus igitur, quae causa est, quod generatio sit semper, et illa quae est substantia universaliter, et illa quae est secundum partem ut elementi; numquam enim secundum naturam cessavit nec cessabit generatio. Si autem quis dicat, quod cessabit voluntate dei aliquando generatio, sicut aliquando non fuerit et post hoc cœpit, dico, quod nihil ad me de dei miraculis, cum ergo de naturalibus disseram" (*Alberti Magni Opera Omnia*, V. 2, 1980, I. 1. 22; henceforth *AMOO*).

While dealing, for instance, with miracles, he tried, as much as possible, to detect the natural cause upon which God or Christ had acted. Following the *Liber de causis* and faithful to his own interpretation of it, he stressed the necessity of natural causes for an action of the primary cause in the natural world.[41] As far as medicine is concerned, an example of this search for a natural cause is provided by the comment on the miraculous healing of Peter's mother-in-law, who suffered from a severe fever:

> *Et cum venisset Iesus in domum Petri etc.* Hic incipit agere de ostensione potentiae divinae in restitutione spiritus et caloris vitalis. "Febris enim est calor extraneus accensus in corde et procedens ex eo mediantibus spiritu et sanguine per arterias et venas in totum corpus, inflammatus ex corde inflammatione, quae impedit operationes naturales." Unde patet ex corde habere hunc calorem inflammationem et ita esse ipsum in membro principali et instrumentis eius praecipuis, quae sunt calor et spiritus, et in principali nutrimento, quod est sanguis, et ita potentissimam oportet esse virtutem, quae curat hunc defectum. Est enim hic defectus in eo quod principaliter est sedes animae, quod est cor, a quo fluit actus vitae in totum corpus.[42]

Thanks to his immense power, Christ has thus restored the spirit and vital heat of this sick woman, who was at the point of death, by acting in her heart, the origin of fever as well as the principal seat of the soul. Albert's explanation relies here on Avicenna's definition of fever, which he quotes.[43] Moreover, through his short comment it seems obvious that he considers the heart as the main organ of the body. This is confirmed by his claims in his philosophical works, in which, as Nancy Siraisi put it, he "was unyielding in his defense of the Aristotelian doctrine of the primacy of the heart over the whole body."[44] As I have previously mentioned, it was also Robert Grosseteste's and Bonaventure's opinions, and we may suppose that the Aristotelian thesis seemed more acceptable to theologians: since it was difficult to deny the vital function of the heart, to negate its hegemonic role could lead to the temptation of dividing the soul into three parts. It must be added that beside the authority of Aristotle, according to Jerome on his commentary on Matthew XV. 19, Christ himself seemed to have imposed this view: *ergo animae principale non secundum Platonem in cerebro, sed iuxta Christum in corde est.*[45] Moreover the primacy of the heart over the brain put the emphasis on the vital

[41]De Libera 1990. A clear statement is given in *De causis proprietatum elementorum*, in the chapter (or *digressio*) on the causes of floods: "Sunt autem quidam qui omnia haec divinae dispositioni tantum attribuunt et aiunt non debere nos de huiusmodi rebus querere aliam causam nisi voluntatem dei quibus nos in parte consentimus, quia dicimus haec nutu dei mundum gubernantis fieri ad vindictam maleficii hominum. Sed tamen dicimus haec deum facere propter causam naturalem, cuius primus motor est ipse qui cuncta dat moveri. Causas autem suae voluntatis non quaerimus nos, sed quaerimus causas naturales, quae sunt sicut instrumenta quaedam per quae sua voluntas in talibus producitur ad effectum" (*AMOO*, V. 2, 76).

[42]*Super Matthaeum*, VIII. 14, 1987 (*AMOO*, XXXI. 1), 287.

[43]Avicenna, *Canon*, IV. 1. 1. 1, 1507, fol. 393r.

[44]Siraisi 1980, 401.

[45]Saint Jérôme 1977 (*Sources chrétiennes* 242), 328. On Jerome and medicine, see Vazquez Bujan 2001.

functions of the body. The primacy of the brain, which put the emphasis in sensation and voluntary motion, could lead to bring them nearer to the faculty of reasoning than was permitted by religious orthodoxy—a temptation which Galenic views on the hegemonic role of the brain could easily induce. Augustine himself, when explaining that God breathed into the human face because of the sense organs and the sensitive function of the brain, although he seemed closer to the Galenic views than the thirteenth-century commentators, took care of providing a description of the brain's faculties, which excluded any kind of reasoning or even sensitive judgment.

If the miraculous healing of the sick woman suffering from fever could easily be explained by Christ's action on the main natural cause of this ailment, other miracles did not provide the same opportunity. In this case, Albert resorted to medical explanations in order to praise the extent of divine power, for instance when he commented on the miraculous healing of a paralytic by Christ entering Capernaum.

> Paralysis enim per hoc quod emollitio nervorum, per hoc quod replet nervos, impedit viam sensus et insensibilitatem inducit; et per hoc quod emollit, inducit virtutis motivae detrimentum, quia illa vult habere viam siccam et duram, eo quod non est receptiva formarum sensibilium, sicut vis sensitiva, sed potius activa, et actio potentiae activae melioratur sicco, quia humidum non est instrumentum actionum, sed cedit in seipsum et patitur magis a contrario. Per hoc autem quod humor paralyticus replendo distendit nervos in latum, contrahit eos a longitudine debita, et tunc virtus naturalis nititur eos distendere ad debitum situm; et sic ex lucta morbi et naturae tremor membrorum inducitur et aliquando tortura magna et valde sensibilis afflictionis, praecipue si humor dissolvens et emolliens calidus fuerit; et talem passionem iste paralyticus sustinuisse videtur. Haec autem dicta sint, ut ex magnitudine morbi magnitudo appareat divinae curantis.[46]

Through the extent of the disease, Holy Scripture shows the extent of the divine healing. As Albert stated before this description, paralysis and its healing pertain to *opera sensibilis animae*. In this case, Christ's action did not consist in reinforcing the vital spirit, but in restoring sensation and motion: in both cases, Albert suggests strongly that Christ exerted his healing power on the sick's soul and consequently on its bodily instruments. It is revealing that concerning the actions of sensitive soul, he does not mention at all the brain, but only nerves and primary qualities that operate through softening, contraction or distension. He was certainly guided by his medical sources, in particular Avicenna's *Canon*, in which paralysis is classified among the diseases of the nerves.[47] This situation helped him in not choosing between heart and brain as the seat of sensation, a much debated question at the time, which again opposed Aristotle (plus Averroes) to Galen.[48]

In the same commentary on Matthew's Gospel, another occasion of alluding to a much debated question was provided by the problematic verse which seems to state,

---

[46] *Super Matthaeum*, VIII. 8, 278–279.

[47] Avicenna, *Canon*, III. 2. 2, 1507, fol. 197rv.

[48] It should be pointed out that the recognition of the primacy of the heart did not necessarily lead one to consider it as the origin of sensation, Jacquart 2003.

in Jesus' own words, that all the absorbed food is rejected through excrement: *Nam intellegitis quia omne quod intrat in os in ventrem vadit et in secessum emittitur?* The commentary on this verse was related to the question of the resurrected body and of the *veritas personae humanae,* on the one hand, and to the medical explanations of the transformation of food into bodily substance, on the other.[49] On this difficult issue, initiated by Peter Lombard, Albert the Great in his commentary on the *Sentences*, echoing his lectures in Paris early in his career (in 1240s), agreed with Peter Lombard, stating that nothing coming from outside (like food) entered the truth of human nature inherited from the prime parents of human beings.[50] Albert's views in his commentary on Matthew, written later in his career, after 1262, have changed, and his words are rather harsh against those who ignore natural science. At first, *secessum* is understood as any *poros* (tiny duct, pore), visible or not, through which the body is purged from dissolved or evaporated superfluities. Rabanus Maurus and Bede are then invoked to introduce the difficult point and to act as spokesmen for those "who claim that the Lord here was ignorant of the debates of natural philosophy, when he said that all the food went through excrement and that nothing was converted into the nature of the body."[51] Indeed, any commentator willing to save the literal meaning of this verse, from a naturalistic point of view, had to face severe difficulties and needed a great deal of imagination. After having reported, in Rabanus Maurus' and Bede's names, the opinion of those who thought that the Lord ignored natural science—an opinion no doubt unacceptable for "the universal doctor"—Albert mentions another opinion which claims to save this verse, but which is even more ridiculous, from a naturalistic point of view, than what this verse seems to state. As some say, no eaten thing passes actually into human nature; in nutrition the same process happens as in the fusion of gold, in which lead prevents gold from burning. In the same manner, "the nutrimental humor" preserves the truth of human nature, but is not converted into it. Albert then raises ironically the question: "Why then would it be necessary to spend so much money on food?" If the only usefulness of food was to resist heat, to drink water would be sufficient. In conclusion: "These statements are absurd and come only from a shameless ignorance among those who claim that they know, but know nothing."

At this stage of his long career, Albert, as many commentators on the *Sentences* from the second half of the thirteenth century onwards, considers Peter Lombard's views as indefensible. Finally, he has to give his own explanation of Matthew's verse, an explanation which intends to save both natural science and the Lord's words. A quotation from Aristotle's *De generatione et corruptione* helps him to formulate his answer: "the material flesh continuously flows in and out within what

---

[49]Bynum 1995; Ziegler 1999; Principe 1991 and 1990; and Reynolds 1999.

[50]Ziegler 1999, 215.

[51]*Super Matthaeum*, XV. 17, 442. Bede and Rabanus Maurus actually followed Jerome, from whom the allusion to "poros" is also taken, cf. Saint Jerôme, *Commentaire sur S. Matthieu...*, XV. 17, 442.

it nourishes." Albert endeavors then to relate this quotation, severed as it is from its Aristotelian context, with the Gospel's verse: "The whole nutritive flesh is said flesh according to matter, therefore the whole flows in through food and out through excrement. Nevertheless in the middle of this flow in and out, is the true nature of the body. And the Lord does not deny that."[52] The commentator has also to deal with another question, not directly raised by the literal explanation of the Gospel, but related by theologians to the debate on nutrition. Since the quantity of this material flesh is proportional to the voids of the body it has to fill, what will this measure be at the time of resurrection? Relying on Augustine's words about hair, Albert states that if necessary a complement would be provided *ab opifice*.[53] On this point, he does not agree completely with those who say that what pertains to "the truth of human nature" is only what will appear in resurrection. This statement is somehow true, but not sufficient. So what is "the truth of human nature"? According to Albert, all the matter which shows the form of man may be called the true human nature. But the states induced by corruption or old age do not pertain to it, since without them human nature is not "truer" than with them.[54]

While showing his awareness of both the scientific and the theological debates of his time about the transformation of food, Albert here avoids going into detailed explanations. If he uses the technical expression *nutrimentalis humor*, he does not mention the concept of *humidum radicale*, that any initiated reader would expect in this context. But yet, as Joan Cadden put it, relying on his philosophical writings, "the modern reader may be dismayed at the care Albert took to clarify the concept of nutrition."[55] Indeed, he wrote an entire work *De nutrimento et nutribili*, around the same years as he commented on Matthew. He dealt also at length with the concept of radical moisture, but in a very complex manner.[56] No doubt, this recurrent interest in these matters had to do with his activity as a biblical commentator, which was among his main duties as a teacher. Besides an intellectual

[52]*Super Matthaeum,* XV. 17, 442: "Dicamus igitur, quod si quis bene et interius sciat physicam, domini sermones non habent dubitationem, quoniam Aristoteles probavit in I. de generatione, quod caro materialis influit continue et effluit in id quod nutritur, animal quodcumque; caro autem nutrimentalis tota est caro dicta secundum materiam; ergo tota et omnis influit per cibum et effluit per secessionem. Et tamen in medio influxus et effluxus est vera natura corporis; et hoc non negat dominus." Cf. Aristotle, *De generatione et corruptione,* I. 5. 321b26, 1986 (*Aristoteles Latinus,* IX. 1), 30: "Sic utique augmentatur materia carnis, sed non cuicumque parti omni adgeneratur, sed quidem defluit, hoc autem advenit."

[53]Ibid.: "De hac autem resurget, quod satis est ad staturam resurgentis, sicut etima de capillis, ut dicit Augustinus in XXII de civitate dei. Et si non sit in aliquo sicut in pueris, quod satis est de ista, supplebitur ab opifice, et sicut diximus de carne, ita est de aliis membris similibus. Et haec est de sententia Augustini."

[54]Ibid.: "Hoc enim est de veritate humanae naturae, de quo aliquando possum dicere: hoc est vera et vere natura humana. Et hoc possum dicere de omni materia demonstrata sub forma hominis. Sed hoc non est de veritate humanae naturae, sine quo verior est humana natura quam cum ipso, sicut est corruptio et senectus et huiusmodi quae ex peccato acciderunt. Et per hoc solutione omnium."

[55]Cadden 1980, 334.

[56]On Albertus's views, see Arnald of Villanova 2010, 385–396.

curiosity, for its own sake, which cannot be denied, his concern with biblical exegesis may have played a significant part in his continuous search for reliable philosophical and scientific knowledge. Careful as he was not to mix the two activities, he limits scientific explanations in his biblical commentaries to what seems useful in order to save the literal meaning of the Bible or to praise divine power. On some matters, as on astronomical miracles, he refers his reader to one of his previous writings where the topic is addressed.[57] Concerning the process of nutrition, his obvious embarrassment in commenting on Matthew seems to suggest that, despite the efforts he spent in his non-exegetical works, he did not succeed in finding solutions entirely able to convince him.

Thomas Aquinas' biblical exegesis is, as expected, typical of him. His medical knowledge has been considered limited by modern scholars. This opinion remains, in my view, insufficiently demonstrated,[58] and, anyway, as far as biblical exegesis is concerned, it is not really the point, since in other scientific matters, on which Thomas was obviously well informed, his attitude was almost the same. When he happened to give a scientific explanation in biblical commentaries, he avoided technical terms and remained at a pedagogical level, even when dealing with topics, as cosmology, astronomy or meteorology, in which he showed some expertise in his commentaries on Aristotle's *De caelo* or *Meteora*.[59] As Gilbert Dahan recently stated, Thomas considered biblical exegesis as the main part of his activity. Far from being a quiet meditation, it was "an anxious hand-to-hand encounter with God's word."[60] In other terms, it provided the most serious opportunities for finding a concrete solution to the debate on the relation between faith and reason. Despite this internal tension, Thomas' biblical commentaries were intended for the spiritual edification of a religious audience. There was thus no room for philosophical controversies nor for the expression of doubts concerning the validity of the letter of the Holy Scripture.

When commenting in his *Lectura super Matthaeum* on the same miraculous healings as Albert, Thomas does not feel it necessary to look for the causes of the mentioned diseases. He gives a minimal definition of paralysis: "Paralytics are those who cannot move their members." As for the healing of the woman suffering

[57]For instance, in his *Postilla super Isaiam,* when commenting on the miracle shown in "Achaz horologium," he refers his reader to his commentary on Ps.-Dionysius' *Epistula ad Apollifanium* in which he dealt at length with biblical astronomical miracles (*Alberti Magni Opera Omnia*, XIX, 1952, 397).

[58]Jordan 1988. The main concern of Jordan's paper was to dismiss the legend according to which Thomas received medical training during his youth in Southern Italy. It was also argued that Thomas's medical information came from Albert. But even if Thomas actually had no specific medical training (which is highly probable), it does not mean that he had no medical culture, and there is no reason to assume that he was only indebted to Albert for this knowledge. See also Jordan 1992.

[59]See Litt 1963, 291, 319–321, for reference to Thomas's commentary on Job; and Ducos 1998, 39, for reference to Thomas's commentary on Psalms.

[60]Aquinas 2002, 1.

from fever, it shows that *aliter curat Dominus, aliter natura*. Whereas usually patients of this kind feel weaker when they begin to be cured, in this case full health came back at once. This *Lectura* is preserved in one or two auditors' *reportationes*, reporting a cursive lecture given by Thomas in Paris, probably between 1269 and 1272.[61]

His *Expositio super Iob ad litteram* is of a completely different nature. As its title clearly expresses and as he explains in the prologue, Thomas' aim was not to focus on the moral meaning of this biblical book, which had been done excellently by Gregory the Great in his famous *Moralia in Job*. The point was to provide a very close literal explanation, able to lead faithfully to a spiritual interpretation. In Thomas' mind, a literal explanation is indeed very literal, and any feature of the words must become clear. He applies sometimes this method so scrupulously that a modern reader may be disconcerted by some explanations, which seem beside the point. For instance, when commenting on the poetic expressions *Quis est pluviae pater? quis genuit stillas roris?* and *De cuius utero egressa est glacies?*, Thomas resorts to the distinction, widespread in natural philosophy and medicine, between males' and females' complexions, the former being hotter than the latter. Therefore rain and dew are said to be generated by a father, whereas ice is described as coming out of a womb.[62]

In this *Expositio in Iob*, probably written between 1261 and 1264, during his teaching at the dominican *studium* near the papal court, Thomas gives numerous "scientific" explanations and quotes at many times Aristotle's works, including *De animalibus* in William of Moerbeke's translation, dating from 1260. Compared with astronomical, zoological or meteorological explanations, the medical ones are very meagre. Job's bodily afflictions demanded nevertheless some comment:

> *Induta est caro mea putredine*, quasi dicat: undique corpus meum circumdatum est putredine ulcerum sicut corpus circumdatur vestimento. Et quia vulnera a principio curata ad sanitatem perveniunt, ostendit ulcera sua fuisse neglecta, unde dicit *et sordibus pulveris*: non enim erant debito modo curata quia ad litteram in sterquilinio sedebat, ut supra dictum est. Expectatur autem sanitas aliquando etiam si ulcera sunt neglecta, quando natura est fortis, sed in Iob vigor naturae defecerat, unde dicit *cutis mea aruit et contracta est*, quia scilicet humor naturalis iam consumptus est vel propter senectutem vel propter infirmitatem, unde non videtur locus ut in hac vita ulterius felicitatem expectem.[63]

The medical explanations given about these verses are indeed minimal, but not totally insignificant. Thomas' focusing on literal meaning led him first to explain

[61] *S. Thomae Aquinatis doctoris angelici Super Evangelium S. Matthaei Lectura*, 1951, 109, 111. In this edition, the *Lectura* is erroneously dated between 1256 and 1259, but this dating has been revised by Smalley 1981, 257.

[62] Aquinas 1965, *Expositio super Iob ad litteram* (*Sancti Thomae de Aquino Opera Omnia Iussi Leonis XIII P. M. edita*, XXVI), 205: "Ubi notandum est quod glaciei causa est frigus quae est qualita feminea, pluviarum autem et roris est calor resolvens et non permittens congelari vaporem: calor autem est qualitas masculina, et ideo signanter circa generationem pluviae et roris usus est nomine patris, circa generationem autem glaciei usus est nomine uteri qui pertinet ad matrem."

[63] Aquinas 1965, *Expositio super Iob*, VII. 5, 47.

induta est (clothed), which could seem strange in this context, whereas the meaning of *putredo* (putrefaction) was obvious. It was just felt necessary to add that this putrefaction came from ulcers. More significantly, Thomas adds that these ulcers could have been cured by a treatment *a principio* (at the beginning), and that even if it had not been the case Job could have been cured if his nature was strong. As simple as they are, these additions reflect one of the main trends of medical treatment, which takes into account the strength of the patient's nature. Job's nature was weak, the sign of this being that his skin burnt and was contracted, because of the consumption of "natural humor" due to old age or illness. The expression "natural humor" is vague: does it refer simply to the moisture of the complexion, old people having a dry one, or does it implicitly allude to the consumption of radical moisture? When commenting on *Rugae meae testimonium dicunt contra me*, Thomas mentions the consumption of moisture, which, caused by illness or old age, produces the wrinkles of the body.[64] In this case, *humidum* refers probably to the moist quality of the complexion. More strikingly, Thomas does not mention *humidum radicale* where a thirteenth-century reader would expect it, that is when dealing with the length of life alloted to each individual.

> Sunt autem et termini humanae vitae praestituti ex aliquibus corporalibus causis, puta ex complexione vel ex aliquo huiusmodi, ultra quos vita hominis protendi non potest, quamvis ante possit deficere ex aliqua accidentali causa, sed terminos praestitutos secundum divinam providentiam, sub qua omnia cadunt, nec in plus nec in minus vita hominis potest praeterire.[65]

Thomas' words here are consonant with Avicenna's *Canon,* stating that the length of life is predetermined by the ability of an individual's complexion to resist drying. But in the same chapter Avicenna added among the predetermining factors "natural heat and the quantity of innate moisture (i.e. radical moisture)," which Thomas omits, referring only to complexion and to "something of the same kind." This omission is significant, since in other writings he deals with *humidum radicale* at some length.[66] He considered probably that because of its complexity and its controversial features this notion was out of place in biblical exegesis. Likewise, where the biblical text contains the word *vacuum*, the belief in the existence of void

[64]Ibid., XVI. 9, 102: "ex infirmitatibus enim corrugatur corpus propter humidi consumptionem sicut et ex senectute."

[65]Ibid., XIV. 5, 91.

[66]Avicenna, *Canon*, I. 3. 3. 1, *Capitulum singulare dictionis tertie de causis sanitatis et egritudinis et necessitatis mortis,* 1507, fol. 53r: "Unumquodque preterea corpus terminum habet in quo resistat siccitati necessarie complexione sua et calore suo innato et quantitate sua innati humiditatis a quo non removetur, est tamen cum ante adveniunt cause in siccitatem adiuvantes aut alio modo pernecantes. Et multi quidem hominum debent quod isti sunt termini naturales et termini accidentales sunt alii." On *humidum radicale* in Thomas's works, see Jordan 1988, 243–245; Arnald of Villanova, *Tractatus,* 349–355.

is simply attributed to common people (*vulgares homines*).[67] Thomas did not feel necessary to give in this instance Aristotle's arguments against the existence of void, since the Holy Scripture uses the ordinary langage of common people. In the same way, he tried to use the same language in his own commentary, as far as possible. The main point was to save the letter. Coming back to the topic of nutrition, which does not appear in the *Expositio in Iob*, the same attitude is found in the *Summa Theologiae,* with the question: *Utrum aliquid de alimento convertatur in veritatem humanae naturae*. The verse *Omne quod in os intrat, in ventrem vadit, et per secessum emittitur* is quoted as the first argument *quod non.* Thomas in the solution of this question states first that the Lord has not said *totum*, but *omne*: from any food some impurity needs actually to be ejected. Relying explicitly on Jerome's commentary, he then puts forward that "it might be said that something is generated from food, thanks to the resolution made by natural heat, and is emitted through some hidden pores."[68] In the long development of this question about the transformation of food, in which mention is made of *humidum radicale,* Matthew's verse plays a small part.

The *reportatio* of Thomas' lecture on Matthew does not open the debate on *veritas humanae naturae* nor on the resurrected body. A great deal of Jerome's own commentary is taken up. As for the opinion of those who thought that nothing ingested was converted into human nature, it is clearly rejected as impossible. In the exposition of this opinion, the same comparison with the action of lead on gold appears than in Albert's commentary, but without any mention of "nutrimental humor." It may be noted that in the *Summa Theologiae,* this comparison involves the fusion of silver, and not of gold. Although not exactly in the same terms as in the *Summa*, in the *Lectura* on Matthew the meaning of *omne* is qualified: "sometimes in the Holy Scripture the whole is put for the part."[69]

---

[67] Aquinas 1965, *Expositio super Iob,* XXVI. 7, 145: "*Qui extendit aquilonem super vacuum.* Per aquilonem intelligit superius haemispherium quoad nos [...] *super vacuum,* quia sub superiori haemisphaerio caeli nihil nobis apparet nisi spatium aere plenum, quod vulgares homines reputant vacuum: loquitur enim secundum aestimationem vulgarium hominum, prout est moris in sacra Scriptura."

[68] *Summa Theologiae, Prima pars,* qu. 119, art. 1, 1988, 551: "Ad primum ergo dicendum quod Dominus non dicit quod *totum* quod in os intrat, per secessum emittatur, sed *omne*, quia de quolibet cibo aliquid impurum per secessum emittitur.— Vel potest dici quod quidquid ex alimento generatur, potest etiam per calorem naturalem resolvi, et per poros quosdam occultos emitti, ut Hieronumus exponit." For Jerome's commentary see above, n. 51. On Thomas's conception of *veritas personae humanae,* see Principe 1991.

[69] Aquinas 1951, *Super Evangelium*, XV. 17, 201–202: "Sed, sicut dicit Hyeronymus, contra hoc obiiciunt aliqui, dicentes Dominum esse ignarum naturalis scientiae, quia non totum in secessum transmittitur. Unde quidam volentes sic intelligere, quod totum emittatur, volunt quod nihil convertatur in humanam naturam, sed solum quod tractum ab Adam multiplicetur, et hoc resurget. Unde et artifices plumbum ponunt cum auro, ut plumbum consumatur, aurum conservetur. Sic resistunt cibi, ne calor naturalis consumat illud, quod est de virtute naturae. Sed hoc videtur impossibile [...] Potest etiam sic dici *Omne quod in os, in ventrem vadit,* aliquid: unde aliquando in Scriptura totum pro parte sumitur."

From these few soundings, it seems clear that the exegetical work in Thomas' view is not a proper place for addressing controversial debates on scientific matters. In the *Expositio in Iob*, more abundant in naturalistic explanations than other exegetical works, medicine, as well as other sciences, although at a lesser extent, is convened as far as it clarifies the literal meaning, but to a level which must remain accessible to common people, since it is the level adopted in the Holy Scripture itself. Given these conditions, it is difficult to evaluate the extent of Thomas' medical knowledge from these sources, even though some of his formulations may indicate that it was not as poor as some have said. For instance, when dealing with the causes of pain, the use of the expression *ex solutione continui* may echo Avicenna's *Canon,* directly or indirectly, and, as I already noted about the determination of the length of life, Thomas adds *vel aliquo huiusmodi*, which may suggest that he knows more than he says.[70]

1.4 A Few Conclusive Words

Despite the holy status of the authoritative text, commentaries on the Bible, as every other kind of commentary in the Middle Ages, give room for more freedom than expected. Possibilities of expressing one's own views on all kinds of topics are almost infinite. Compared to the commentaries on the *Sentences*, they are less open to heavy theological debates and the letter of the Scripture can stimulate the commentators' thoughts in different directions. As far as scientific explanations are concerned, cosmology, astronomy, meteorology, speculative arithmetic, geometry or optics are certainly more represented than medical theories. These are not always introduced by *ut dicunt medici* and they are more often than not scattered among more general statements. When they can be detected they allow us to measure the commentators' knowledge (or ignorance) in this field. Moreover, through obviously deliberate omissions or distortions, they reveal the major conflicting points between medical theories and religious requirements. When these conflicting points coincide with controversies between "philosophers" and "physicians," i. e. between Aristotelian and Galenic traditions, for instance about the primacy of the heart, does the more frequent adhesion to the former reflect the predominance of these commentators' philosophical training, while their medical information remains marginal? Or does it suggest more fundamentally that this choice was considered more fitting to the Christian message? Rarely taken into

[70]Aquinas 1965, *Expositio in Iob,* XVI. 7, 101: "Est autem duplex dolor: unus quidem interior qui tristicia nominatur, proveniens ex apprehensione alicuius mali inhaerentis; alius autem est dolor exterior qui est dolor secundum sensum, puta ex solutione continui proveniens vel ex aliquo huiusmodi." See also Avicenna, *Canon,* I. 2. 2. 1. 19, 1507, fol. 38r: "Dicemus igitur quod dolor est sensibilitas rei contrarie. Omnes vero doloris cause in duobus comprehenduntur generibus, scilicet genere mutationis complexionis cite facte et est malitia complexionis diverse et genere solutionis continuitatis [...] dolere non est nisi contrarie rei conrarietatem sentire."

account by historians of medicine or natural philosophy, biblical commentaries sometimes deliver concisely the personal opinions or doubts of a medieval author, out of their usual context. But as I have tried to suggest through a few examples, it is only at the expense of a literal reading and even sometimes between the lines, that these biblical commentaries meet the investigations of an historian of medieval medicine.

Bibliography

Albertus Magnus. 1952. *Alberti Magni Opera Omnia*, XIX, ed. F. Siepmann. Münster: Aschendorff.

———. 1980. *Alberti Magni Opera Omnia*, ed. P. Hoßfeld, vol. 2, 2nd edn. Münster: Aschendorff.

———. 1987. *Alberti Magni Opera Omnia*, XXXI, ed. B. Schmidt. Münster: Aschendorff.

Ambrose. 1897. *Sancti Ambrosii Opera. Pars prima*, ed. K. Schenkl. Vienna: F. Tempsky.

———. 1957. *Sancti Ambrosii Mediolanensis Opera, Pars IV*, ed. M. Adriaen, 1–400. Turnhout: Brepols.

Amundsen, Darrel W. 1996. *Medicine, Society and Faith in the Ancient and Medieval Worlds*. Baltimore: Johns Hopkins University Press.

André, Jean-Marie. 2007. Saint Augustin et la culture médicale gréco-romaine. *La cultura scientifico-naturalistica nei Padri della Chiesa (I-V sec.)*, 597–604. Rome: Institutum Patristicum Augustinianum.

Aquinas, Thomas, 1951. *S. Thomae Aquinatis doctoris angelici Super Evangelium S. Matthaei Lectura*, ed. P. Raphaele Cai. Turin-Rome: Marietti.

———. 1965. *Sancti Thomae de Aquino Opera Omnia Iussi Leonis XIII P. M. edita, Tomus XXVI*. Rome: Sancta Sabina.

———. 1988. *Summa Theologiae*, ed. P. Gianolio. Turin: Editiones Paulinae.

———. 2002. *Commentaire de la première épître aux Corinthiens*, intro. G. Dahan and trans. J. E. Stroobant de Saint-Eloy. Paris: Cerf.

Aristotle. 1986. *De generatione et corruptione*, trans. Burgundio of Pisa and ed. Joanna Judycka. Leiden: Brill.

Arnald of Villanova. *Tractatus de humido radicali*, 2010. *(Arnaldi de Villanova Opera Medica Omnia, V. 2)*, ed. M. McVaugh, C. Crisciani and G. Ferrari, 385–396. Barcelona: Publications of the University of Barcelona.

Avicenna. 1507. *Canon*. Venice: per Paganinum de Paganinis.

———. 1964. *Liber canonis*. Hildesheim: G. Olms.

Augustine. 2000 [1972]. *Œuvres de saint Augustin* 48, trans., intro., and notes by P. Agaësse and A. Solignac. Paris: Institut d'études augustiniennes.

———. 1959. *Œuvres de Saint Augustin* 35, introduction générale et notes par G. Bardy, traduction française de G. Combès. Paris: Desclee De Brouwer.

Bardy, G. 1953. Saint Augustin et les médecins. *Année théologique augustinienne* 13: 327–346.

Biller, Peter, and Alistair J. Minnis, eds. 1997. *Medieval Theology and the Natural Body*. New York: York Medieval Press.

Biller, Peter, and Joseph Ziegler, eds. 2001. *Religion and Medicine in the Middle Ages*. New York: York Medieval Press.

Bonaventure. 1891. *Doctoris Seraphici S. Bonaventurae Opuscula varia theologica* V. Florence: Ad Claras Aquas (Quaracchi), Collegii S. Bonaventurae.

———. 1893. *Doctoris Seraphici S. Bonaventurae Opuscula varia theologica* VI. Florence: Ad Claras Aquas (Quaracchi), Collegii S. Bonaventurae.

———. 1991. *Les six jours de la creation*. Trans. M. Ozilou. Paris: Desclée.

———. 2005. *Works of St. Bonaventure 7. Commentary on Ecclesiastes*. Trans. C. Murray and R. J. Kanis. New York: Franciscan Institute Publications.

Boulnois, M.O. 2005. L'union de l'âme et du corps comme modèle christologique. In *Les Pères de l'Église face à la science médicale de leur temps*, ed. V. Boudon-Millot and B. Pouderon, 451–475. Paris: Beauchesne editeur.

Bynum, Caroline W. 1995. *The Resurrection of the Body in Western Christianity, 1200–1336*. New York: Columbia University Press.

Cadden, Joan. 1980. Albertus Magnus' Universal Physiology: The Example of Nutrition. In *Albertus Magnus and the Sciences: Commemorative Essays*, ed. J. Weisheipl, 321–339. Toronto: Pontifical Institute of Medieval Studies.

Chandelier, Joel. 2009. *La réception du* Canon *d'Avicenne. Médecine arabe et milieu universitaire en Italie avant la peste noire,* thèse de doctorat en histoire, Paris, École pratique des Hautes Études, 2007. Lille: Atelier national de Reproduction des Thèses.

Crisciani, Chiara, 2001. Religione e medicina. *Storia della scienza,* IV, Istituto della Enciclopedia Italiana Treccani. Rome: Institute of the Italian Encyclopedia.

Crombie, Alistair C. 1971. *Robert Grosseteste and the Origins of Experimental Science, 1100–1700*. 3rd ed. Oxford: Clarendon Press.

Cullen, Christopher M. 2006. *Bonaventure*. Oxford: Oxford University Press.

Dales, Richard C. 1967. Robert Grosseteste's Views on Astrology. *Mediaeval Studies* 29: 357–363.

———. 1995. Robert Grosseteste on the Soul's Care for the Body. In *Robert Grosseteste. New Perspectives on His Thought and Scholarship*, ed. James McEvoy, 313–319. Turnhout: Brepols.

Debru, Armelle. 2005. Christianisme et galénisme: le mouvement volontaire chez Némésius d'Émèse. In *Les Pères de l'Église face à la science médicale de leur temps*, ed. V. Boudon-Millot and B. Pouderon, 89–103. Paris: Beauchesne editeur.

Donato, M.P., L. Berlivet, S. Cabibbo, R. Michetti, and M. Nicoud, eds. 2013. *Médecine et religion, Collaborations, compétitions, conflits (XIIe-XXe siècles)*. Rome: École française de Rome.

Donini, Pierluigi. 2008. Psychology. In *The Cambridge Companion to Galen*, ed. R.J. Hankinson, 184–209. Cambridge: Cambridge University Press.

Ducos, Joëlle. 1998. *La météorologie en français au Moyen Âge: XIIIe-XIVe siècles*. Paris: Champion.

Grant, Edward. 1996. *Planets, Stars and Orbs, The Medieval Cosmos, 1200–1687*. Cambridge: Cambridge University Press.

Gil-Sotres, P. 1994. Modelo teorico y observacion clinica: Las pasiones del alma en la psicologia medica medieval. *Comprendre et maîtriser la nature au Moyen Âge, Mélanges d'histoire des sciences offerts à Guy Beaujouan*, 181–204. Geneva: Droz.

Grosseteste, Robert. 1982. *Hexaëmeron* (*Auctores Britannici Medii Aevi,* VI), R. C. Dales and S. Gieben. Oxford: Oxford University Press.

———. 1995. *Roberti Grosseteste Exposition in Epistolam Sancti Pauli ad Galatas* (*Corpus Chritianorum, Continuatio Mediaevalis,* CXXX). Turnhout: Brepols.

———. 1996. *On the Six Days of Creation*, a translation of *Hexaëmeron*, trans. C. F. J. Martin. Oxford: Oxford University Press.

Israeli, Isaac. 1515. *De urinis*. Lyon: Johannis de Platea.

Jacquart, Danielle. 1998. *La médecine médiévale dans le cadre parisien, XIVe-XVe siècle*. Paris: Fayard.

———. 2011. Medicine and Theology. In *Crossing Boundaries at Medieval Universities*, ed. S.E. Young, 213–226. Leiden/Boston: Brill.

———. 2003. Cœur ou cerveau? Les hésitations médiévales sur l'origine de la sensation et le choix de Turisanus. *Il cuore, Micrologus* 11: 73–95.

———. 2013. La reconstruction médicale de la nature de l'homme aux XIe et XIIe siècles. *Revue de Synthèse* 134: 445–472.

Jérôme, Saint, 1977. *Commentaire sur S. Matthieu, (Sources chrétiennes* 242), ed. E. Bonnard. Paris: Editions du Cerf.

Jordan, Mark D. 1988. Medicine and Natural Philosophy in Aquinas. In *Thomas von Aquin: Werk und Wirkung im Licht neuerer Forschungen*, ed. A. Zimmermann, 233–246. Berlin: W. de Gruyter.

———. 1990. The Construction of a Philosophical Medicine, Exegesis and Argument in Salernitan Teaching on the Soul. *Renaissance Medical Learning, Evolution of a Tradition*, ed. M. R. McVaugh and N. G. Siraisi, *Osiris*, sec. ser., 6: 42–61.

———. 1992. The Disappearance of Galen in Thirteenth-Century Philosophy and Theology. In *Mensch und Natur im Mittelalter*, ed. A. Zimmermann and A. Speer, 703–713. Berlin: W. de Gruyter.

Knuuttila, Simo. 2004. *Emotions in Ancient and Medieval Philosophy*. Oxford: Oxford University Press.

De Libera, Alain. 1990. Albert le Grand et Thomas d'Aquin, interprètes du *Liber de causis*. *Revue des sciences philosophiques et théologiques* 74: 347–378.

Litt, Thomas. 1963. *Les corps célestes dans l'univers de saint Thomas d'Aquin*. Louvain-Paris: Louvain Publications Universitaires.

Madec, Goulven. 1996. *Augustin et la philosophie*. Paris: Institut d'études augustiniennes.

McEvoy, James. 1994. *Robert Grosseteste, Exegete and Philosopher*. Aldershot: Variorum.

Némésius d'Émèse. 1975. *De natura hominis*, trans. Burgundio of Pisa, ed. G. Verbeke and J. Moncho. Leiden: Brill.

———. 1987. *De natura hominis*, ed. M. Morani. Leipzig: Teubner.

———. 2008. *On the Nature of Man*, trans. Robert W. Sharples and ed. Philip van der Eijk. Liverpool: Liverpool University Press.

Palmieri, Nicoletta. 2008. Tempéraments du corps et caractères de l'âme dans le commentaire à l'*Ars medica* d'Agnellus de Ravenne. *L'*Ars Medica (Tegni) *de Galien: lectures antiques et mé diévales*, ed. N. Palmieri, 31–66. Saint-Étienne: Publications de l'Université.

———. 2009. *Ambrogio e la medicina: le parole e i concetti*. Milan: LED, Edizioni Universitarie di lettere economia diretto.

Pigeaud, Jackie. 1988. La psychopathologie de Galien. In *Le Opere psicologiche di Galeno*, ed. Paola Manuli and Mario Vegetti, 153–184. Naples: Bibliopolis.

———. 2003. Cœur organique, cœur métaphorique. *Il cuore, Micrologus* XI: 9–36.

Principe, Walter H. 1990. *De veritate humanae personae:* Theology in Conversation with Biology, Medicine, and Philosophy of Nature. In *Knowledge and the Sciences in Medieval Philosophy: Proceedings of the Eighth International Congress of Medieval Philosophy*, ed. D. Follesdal, R. Tyorinoya, A.I. Lehtinen, S. Knuuttila, and S. Ebbesen, 486–494. Helsinki: Yliopistopaino.

———. 1991. Truth of Human Nature' According to Thomas Aquinas: Theology and Science in Interaction. In *Philosophy and the God of Abraham: Essays in Memory of James A. Weisheipl*, ed. R.J. Long, 162–177. Toronto: Pontifical Institute of Medieval Studies.

Rassinier, Jean-Paul. 1991. Le vocabulaire médical de saint Augustin. Approche quantitative et qualitative. In *Le latin medical: La constitution d'un langage scientifique. Réalités et langage de la médecine dans le monde romain*, ed. G. Sabbah, 379–395. Saint-Étienne: Publications de l'Université de Saint-Étienne.

Reynolds, Philip Lyndon. 1999. *Food and the Body. Some Peculiar Questions in High Medieval Theology*. Leiden: Brill.

Ricklin, Thomas. 2003. Le cœur, soleil du corps: une redécouverte symbolique du XIIe siècle. *Il cuore, Micrologus* 11: 123–143.

Rocca, Julius. 2003. *Galen on the Brain: Anatomical Knowledge and Physiological Speculation in the Second Century A.D.* Leiden: Brill.

Siraisi, Nancy G. 1980. The Medical Learning of Albertus Magnus. In *Albertus Magnus and the Sciences, Commemorative Essays 1980*, ed. J.A. Weisheipl, 379–404. Toronto: Pontifical Institute of Medieval Studies.

———. 1981. *Taddeo Alderotti and His Pupils: Two Generations of Italian Medical Learning*. Princeton: Princeton University Press.

Smalley, Beryl. 1981. *The Gospels in the Schools, c. 1110–c.1280*. London/Ronceverte: Hambledon Press.

Vazquez Bujan, Manuel E. 2001. Connaissance directe ou tradition commune? Pour une relecture des références de Jérôme à Hippocrate et à Galien. In *'Docente natura': Mélanges de médecine ancienne et médiévale offerts à Guy Sabbah*, ed. Armelle Debru and Nicoletta Palmieri, 293–312. Saint-Étienne: Publications de l'Université de Saint-Étienne.

Verbeke, Gerard. 1945. *L'évolution de la doctrine du pneuma du stoïcisme à saint Augustin*. Paris/Louvain: PERSEE.

Young, Francis. 1997. *Biblical Exegesis and the Formation of Christian Culture*. Cambridge: Cambridge University Press.

Young, Francis, Lewis Ayres, and Andrew Louth, eds. 2004. *The Cambridge History of Early Christian Literature*. New York/Cambridge: Cambridge University Press.

Ziegler, Joseph. 1999. *Ut dicunt medici:* Medical Knowledge and Theological Debates in the Second Half of the Thirteenth Century. *Bulletin of the History of Medicine* 73: 208–237.

Chapter 2
In a Montpellier Classroom

Michael McVaugh

2.1 Teaching at Montpellier

Not surprisingly, historians' perceptions of the medical faculty of Montpellier c. 1300 have been shaped by the accomplishments of the two great figures active there at that time, Arnau de Vilanova and Bernard de Gordon. Both wrote a considerable number of long works on the theory and practice of medicine, works whose character and scope are virtually unparalleled in Western Europe at that moment, and in the ensuing half-century several other Montpellier masters—Stephen Arlandi, Jordan de Turre, Gérard de Solo—produced similar writings. Arnau and Bernard, it would seem, launched a self-conscious intellectual and literary tradition at Montpellier at the beginning of the fourteenth century, one that set the school well apart from its northern counterpart, Paris.[1]

The achievements of Arnau and Bernard are so remarkable that we tend to identify the school with them, and indeed they may actually have been vying with one another for their fellow masters' attention and endorsement right down until the moment when the medical activity of both ended, about 1310.[2] Those others, the contemporary Montpellier masters who have left us almost no writings of their own, remain largely characterless, at best mere reflections of the famous pair. Master Jacques Egidii is mentioned in scattered documentation as a master there in 1319 and as chancellor of the faculty between 1324 and 1332, but we would have no knowledge of what might be called his "medical personality" if in 1310 he had not proudly signed and dated his own copy of Arnau's great *Speculum*

[1] Jacquart 1998, 160–61, offers explanations for the lack of medical writing from medieval Paris.

[2] Bernard's last known work was finished in 1309; Arnau died in 1311, but his last medical work, the *Speculum medicine*, had been completed in 1308 or conceivably early 1309.

M. McVaugh (✉)
University of North Carolina, Chapel Hill, NC, USA
e-mail: mcvaugh@live.unc.edu

G. Manning, C. Klestinec (eds.), *Professors, Physicians and Practices in the History of Medicine*, Archimedes 50, DOI 10.1007/978-3-319-56514-9_2

medicine, which had been finished only a year or two before.[3] There were of course not many masters teaching at Montpellier anyway. Only ten were present in a 1313 meeting of the faculty in the church of Saint-Firmin (though to be sure there may have been others who did not attend).[4] Of them, Aimon de Mazeriis, Barthélemy de Lunello, Guillaume Brocha, Guillaume Martini, Hugues de Montebusserio, and Ponce de Trellia remain no more than names in the archives; Bernard de Bonahora and Jean Massati were at least remembered by later members of the school for their recipes.[5] Pierre de Capitestagno (Capestang) is a little more of a personality, remembered like Bernard and Jean for his recipes but also the instigator in 1299 of a Latin translation of a work by Avenzoar, and the author of a surviving set of *questiones* on Hippocrates' *Regimen acutorum.*[6] The last of the ten, Jordanus de Turre, was still in 1313 a young man, though in later decades he contributed significantly to the intellectual production of the school.[7] Perhaps historians can be forgiven for discussing Montpellier medicine in this period as if it were wholly embodied in Arnau and Bernard.

Yet it must be remembered that the business of the masters was to teach, not to write long treatises for the delight of historians. In the 1250s or thereabouts we know that they were teaching the curriculum that had been a standard for the past century, the *ars medicine* or *articella*, for a master named Cardinalis has left us his commentaries on most of its ingredients: the *Isagoge*, the *Aphorisms* and *Prognostics* and *Regimen acutorum* of Hippocrates, Galen's *Tegni*, and Philaretus on the pulse. But again historians can be forgiven for neglecting the school's teaching practices in Arnau's and Bernard's day, forty years later, for neither of the two has left much evidence of how they taught. There are no Bernardian commentaries surviving, and Bernard did not refer to his teaching in his writings, while Arnau told a friend that he had chosen deliberately *not* to publish his commentaries.[8] None of the three surviving commentaries ascribed to the latter is on an introductory school text like those of the *articella*—the Hippocratic *Aphorisms*, for example.[9] To be sure, Arnau did prepare elaborate commentaries on certain individual Hippocratic aphorisms, I.1 and II.34, but rather than reflecting the routine teaching of the faculty, aimed at helping students assimilate entire texts, these are really extended

[3]Schuba 1981, 204.

[4]*Cartulaire de l'Université de Montpellier* 1890, vol. 1, 232–33 (no. 31).

[5]McVaugh 1976.

[6]Wickersheimer 1936, 620.

[7]McVaugh 1987.

[8]Vilanova 1988, 30.

[9]One possible exception to this might be the Hippocratic *Regimen acutorum.* Three different texts purport to be Arnau's notes on this work. (a) "Nota quod quinque sunt consideraciones" (five manuscripts), which is a summary of material drawn from the book, not a commentary thereupon; (b) "Hic liber Hippocratis idcirco regimen," a fragmentary text in a style not unlike Arnau's (found in MS Paris, Arsenal 709); and (c) "Intentio Ypocratis in libro regimenti" (in MS Erfurt, Q. 368). The question of the authenticity of any or all remains to be explored. See Paniagua 1994, 86 n. 13, 111.

discourses that use the texts primarily as a starting point for developing Arnau's own ideas.[10] Only his commentary on the short (but non-*articellan*) Galenic work *De malitia complexionis diverse* might let us expect a sight of Arnau in the classroom.[11]

However, if we can tear our gaze away for a moment from Arnau and Bernard, and look instead at scattered manuscript sources, we may in fact get a few glimpses of Montpellier's teaching around the beginning of the century. We can get a first sense of this by observing how the thirteenth-century school presented Hippocrates' *Regimen acutorum* to its students. The *Regimen* had been a late addition to the *ars medicine*: emphasizing the importance of diet in therapeutics, and recommending *ptisana* or gruel and syrups like oxymel (a honey-vinegar mixture), it begins to turn up in early thirteenth-century manuscripts of the collection.[12] When Cardinalis taught this work at mid-century, he divided its text into a few big blocks—just nine in the whole of its first book—and explained their meaning straightforwardly, a phrase or two at a time; at the end of each explanation he usually posed one or two simple questions that could be answered summarily.[13] Cardinalis's approach can be contrasted with that of master Pierre de Capestang, at the end of the century, whom we have already met. At the moment when we found Pierre meeting with his Montpellier colleagues in 1313, he had already been teaching there for over a dozen years; he would soon leave the faculty for the service of the royal family and a canonry at Narbonne that he enjoyed until at least 1330, suggesting that his *questiones* on the *Regimen* were not the work of a particularly old man.[14]

Pierre's treatment of the Hippocratic treatise survives in an Oxford manuscript, MS Bodl. Laud. Misc. 558. Unlike Cardinalis's, it is not a careful outline and analysis of its object, but a very selective series of questions linked sometimes only loosely to portions of the original text.[15] The whole of the first book of the *Regimen* gave rise to just two *questiones* in Pierre's collection: whether *ptisana* was truly *viscosa*, as Hippocrates had said; and whether it was really *artificialis*, medicinally useful, to mix syrups with water. Neither of these questions had been raised by

[10]Vilanova 2014.

[11]Vilanova 1985.

[12]O'Boyle 1998, 103–4.

[13]I have consulted the text of Cardinalis's commentary on the *Regimen acutorum* as contained in MS Kues 222, fols. 167r–181r.

[14]Wickersheimer 1936, 620. Some of the evidence presented here is given in the *Dictionnaire* a second time, in my view mistakenly, in a subsequent article concerning a second medical "Pierre de Capestang" (ibid., 640). My reasons for supposing that Pierre's *questiones* on the *Regimen* were composed at Montpellier rather than later, at Paris, are set out in "Averroes Comes to Montpellier," a contribution to the forthcoming *Mélanges* in honor of Danielle Jacquart (in press).

[15]Pierre was not the only Montpellier master c. 1300 concerned to explain the *Regimen* to a wider audience, but he is the only one we know to have taught the work via a commentary. Bernard de Gordon drew up two works built around its content, his *De regimine acutorum morborum* of c. 1294 and a *Compendium de regimine acutorum* that may be somewhat earlier (see Demaitre 1980, 37–40), and Arnau may have done something similar (see above, n. 9), but there is no proof that either actually expounded the text and its problems to students.

Cardinalis, and both were answered by Pierre at much greater length and far more elaborately than had been the earlier master's practice. An exceptionally intricate *questio* drawn from the second book of the *Regimen*—"Utrum flegma convertatur in sanguinem ad hoc ut nutriat sanguinem?"—takes up most of four leaves (fols. 221r–224v) and concludes with the statement that "hec determinatio istius questionis est data a magistro Petro de capitestangno"; it is perhaps the earliest scholastic determination (a master's authoritative resolution of a *questio disputata*, a problem formally posed for debate) that has survived from medieval Montpellier.[16] It may of course be that these general differences in treatment are simply a function of what has survived of each master's instruction, but the pattern is consistent with what has been noted in other areas of medieval instruction: that as school texts became more familiar, close-grained commentaries were produced less often by teaching masters and were indeed supplanted by the increasingly sophisticated *questiones* that had originally been a subsidiary feature.[17] With Pierre's commentary, therefore, we begin to see that the chance-surviving works of even minor masters have the potential to help us understand how a Montpellier classroom operated.

2.2 At the Threshold of the Classroom

In MS Munich, Staatsbibliothek CLM 534, at fol. 43r, there is a text whose title the catalogue reports as "Supra libellum de mala conditione diversa dubitata, secundum Mag. Bernardum de Hangarra, cancellarium Montispessolanum."[18] This must surely be the same man as the *phisicus* Bernard Engarran who in 1294 witnessed the will of the seigneur of Lunel, some 25 km. east of Montpellier; I will refer to him henceforth as "Bernard de Angarra" or simply as "Angarra."[19] His name does not figure in the faculty's archival record—he was not present at the 1313 assembly, when in any case Hugues de Montebusserio is identified as chancellor—but it is

[16]On the *questio disputata* (in arts commentaries), see Weijers 1996, 62–70.

[17]Weijers 1994; and also Weijers 1996, 42–46.

[18]*Catalogus Codicum Latinorum Bibliothecae Regiae Monacensis* 1892, tom. 1 pars 1, rev. ed., 151, where the manuscript is dated to the fourteenth century. I have given a fuller account of this manuscript in "A Miscellany? Or the Evolution of a Mind? MS Munich CLM 534," forthcoming in *Micrologus*.

[19]The will was published in Roüet 1878, 409–17. (Wickersheimer 1936, 74, reports his name as "Engarra" from another publication of the same document.) In a third source, a late fourteenth-century copy of his commentary on Hippocrates' *Aphorisms* (MS Erfurt, Ampl. F. 290, fols. 40r–115r (old numeration); see below, n. 25), he is named "magistrum B. de angrarra cancellarium." Engarran was a possession of the bishops of Montpellier, and as a place name could easily have been Latinized as "de (H)angarra" or perhaps "de Angrarra." The Munich manuscript is of the early fourteenth century, copied quite possibly at Montpellier, and my references to him follow its spelling (without the aspirate), despite the fact that previous scholarship has consistently referred to him as "de Angrarra."

probably safe to date his teaching to the period 1290–1320, thus making him a contemporary of Pierre de Capestang and, of course, of Montpellier's two most famous masters. The title "chancellor" may not be a gratuitous scribal assumption, since another manuscript independently calls him that as well. We might immediately suspect from the title given in the catalogue that the book Angarra is examining is Galen's *De malitia complexionis diverse*, and so it proves to be when we look at the text in question; but when we study it more closely, we discover that these *dubitata* or questions are limited to fol. 43ra–b and that they are followed immediately by quite different sets of *dubitata* (unrecognized by the cataloguer) that are apparently also Bernard's, first on Galen's *De complexionibus* (43rb–44va), then briefly on Johannitius's *Isagoge* (44va), and then (44vb–50v) on the *Aphorisms* of Hippocrates.[20]

This discovery is of great interest, for there is every reason to think that it shows us a new stage in the evolution of the Montpellier curriculum, involving formal instruction in new texts going beyond the *ars medicine*: instruction, specifically, in a "new Galen," as Luis García Ballester called it.[21] When Pierre de Capestang formulated his *questiones* on Hippocrates' *Regimen acutorum* ca. 1300–1310, he incorporated into them opinions drawn from Avicenna's *Canon*; from Aristotle's *Physics, De anima*, and *De generatione*; and from a great number of works by Galen: above all, *De simplici medicina, De complexionibus, De crisi, De creticis diebus, De malitia complexionis diverse*, and *De morbo et accidenti*.[22] Although Cardinalis had cited no authorities in his mid-century commentary on that same Hippocratic text, this may have been due to the comparative unfamiliarity of the *Regimen acutorum* in his day, for when we examine his careful expositions of the more traditional texts in the *ars medicine*—Johannitius's *Isagoge*, for example, which had been the object of student study in medical schools for more than a hundred years—we find him much more likely to refer to the *Canon* and to some of

[20]The four sets of *dubitata* are written in the same hand, one after another, with no spatial divisions. "Bernardus de Hangarra cancellarius" is named directly as the originator only of the *dubitata* on *De malicia complexionis diverse*, within which work he is frequently referred to as "C [ancellarius]." While the *dubitata* on *De complexionibus* and the *Isagoge* do not mention him by name, they too refer to the author of the questions as "C." The questions on Hippocrates' *Aphorisms* neither refer to Bernard by name nor identify the author as "C.", but we will see below that there are solid reasons for attributing these questions to him as well.

[21]García Ballester 1982. Vivian Nutton has questioned the appropriateness of the term, pointing out that a large Latin corpus of "new" Galenic writings was already circulating in the middle of the previous century; see Galen 2011, 91–100. However, García Ballester's conception seems useful to me as referring to a small subset of these works (in his words, a "decantación académica") that became the focus of instruction in European medical faculties about 1300. See McVaugh 2013, 132–33.

[22]He also reveals his acquaintance with the *Regimen sanitatis, De elementis, De virtutibus naturalibus, De differentiis febrium, De interioribus*, and *De tremore*. The last-named of these (cited at fol. 203r of MS Oxford, Bodl. Laud. Misc. 558) is of additional interest because it testifies to the accessibility at Montpellier of Arnau's translation of that work within a decade or two of its production at Barcelona in 1282.

the same works of Galen that would be mentioned by Pierre: *De crisi*, *De simplici medicina*, *De complexionibus*, *De malitia complexionis diverse*. Evidently these specialized Galenic works were already beginning to be studied by mid-century teachers even if they were not yet being required of their students.

Pierre de Capestang's references in his *Questiones* on the *Regimen* bear out the continuing importance of these works to the Montpellier masters fifty years later, but in themselves they give us no sign that the students were as yet being formally exposed to them. On the face of it, this is surprising, for at Paris in the last quarter of the century Jean de Saint-Amand was preparing summaries of exactly the same new Galenic works, summaries that were addressed explicitly to students.[23] Yet hitherto the first proof that Montpellier also recognized direct study of these works as necessary to medical education has been their appearance in 1309 as requirements in Clement V's new curriculum for the students there[24]—a curriculum, incidentally, that Arnau de Vilanova had a hand in drawing up—though because changes in teaching practice often precede the codification of change, it might be reasonable to suspect that the "new Galen" was already being taught at Montpellier as well as at Paris. And now in fact Bernard de Angarra's *dubitata* on *De complexionibus* and *De malitia complexionis diverse* in the Munich MS reveal that those works were indeed being explained to Montpellier students, probably well before a formal requirement came into place. To the extent that we can begin to identify the content and manner of medical teaching there, we might be listening at the doorway to the classroom.

But our Munich manuscript has even more to contribute to a broadened understanding of teaching at Montpellier c. 1300. As we have noted, it contains not only *dubitata* by Bernard de Angarra on two Galenic works but others by him on the *Isagoge* and on the Hippocratic *Aphorisms* as well. Now it happens that another set of *questiones* on the *Aphorisms* ascribed to Bernard has long been identified in an Erfurt manuscript,[25] and the two collections seem at first markedly different. The Erfurt *questiones* are long and detailed, with arguments and rebuttals carefully set out and with repeated appeals to other authorities, Galen and Avicenna, while the Munich *dubitata* are terse and far less carefully structured. The analysis of aphorism I.2 (which deals with purgation) runs to something over 1000 words in the elaborately organized three-question Erfurt version; Munich's, in contrast, is a mere 116 words long. (The texts are printed below in Appendix 1.) After raising each question, Munich immediately gives Angarra's response in the third person (*dicit*); whereas, in good scholastic form, Erfurt follows the question instead with potential objections (*Videtur quod non*), a counterargument (*Oppositum*), a resolution (*Dicendum quod*), and not infrequently a response to the initial objections.

Yet when the texts are compared closely, they appear to go back to the same source. Erfurt's commentary on aphorism I.2 explores exactly the same three issues as Munich's, though they are titled slightly differently, and the conclusions are

[23] Jacquart 1998, 257–75.

[24] *Cartulaire de l'Université de Montpellier* 1890, vol. 1, 219–21 (no. 25).

[25] Cited above, n. 19. The manuscript is described by Schum 1887, 199.

essentially the same as well: the connection between the two is unmistakable. Munich abandons the scholastic framework entirely and starts outright with the conclusion it attributes to Angarra, that from purgation "realiter sequitur alleviatio" (Erfurt in fact said "alleviatur sensibiliter et bene"); it concedes that purging can do harm (it illustrates this with one of the examples that Erfurt had used to test its conclusion), but only *accidentaliter* (Erfurt used the phrase *per accidens*), and declares that the harmful consequences can be anticipated and prevented (Erfurt went on to explain what specific medicines could be mixed with the purge for this purpose). Munich has not only abandoned the formal structure so conspicuous in Erfurt and reorganized the material, it has cut out much of Erfurt's examples and detail, and as a result it has been able to reduce 235 words to fewer than 40, but it has not lost the core meaning of the original. Obviously the two texts deserve someday to be compared more thoroughly and systematically, in their entirety. But it seems likely that Erfurt represents Bernard de Angarra's own full master's version of a professional commentary on the *Aphorisms*, and that Munich shows us a student's digest of what Angarra had said or written,[26] not infrequently using his master's own language in boiling down a complicated argument to a simple question-and-answer format that could be assimilated easily for purposes of study and perhaps preparation for an examination.[27] In that case, the very similar Munich *dubitata* on the texts of *De malitia* and *De complexionibus* are presumably records of how a student reacted to these constituents of the "new Galen." Together, therefore, the Erfurt and Munich manuscripts appear to bring us across the threshold into the heart of a Montpellier classroom, a classroom of which Arnau's and Bernard de Gordon's more polished and more ambitious texts can give us no hint.

2.3 Classroom Debates

Or is that last statement really true? We have already seen that Arnau de Vilanova himself wrote a commentary on *De malitia complexionis diverse*, and Bernard de Angarra evidently did something of the same sort. Can a comparison of these two works tell us more about Montpellier's teaching? Why, to begin with, did the school take an interest in *De malitia* anyway? It is certainly not one of Galen's more famous books, and it might seem odd that it should have become part of the "new Galen" along with such better-known books as *De crisi* or *De morbo et accidenti*.

[26]Weijers 1996, 146–48, comments on student note-taking from oral presentations.

[27]On the basis of a broader comparison of the two texts, I have recently argued that the Munich version probably contains a student's notes on an early version of Bernard's lectures on the *Aphorisms*, and the Erfurt manuscript a later and somewhat modified version of those lectures: McVaugh, "Hippocrates at Montpellier," paper delivered to "*Sicut dicit*... A Methodological Workshop on the Editing of Commentaries on Authoritative Texts," LECTIO, Leuven (Belgium), 11 March 2016, and now being prepared for publication.

Although it is short, it is not easy to summarize, but it comprises elements of both nosology and pathology.[28]

Many diseases, perhaps most, Galen says here, are caused by a complexional imbalance or *discrasia*—that is, an excess of one or more of the four primary qualities (the hot, cold, dry, or moist). Such a *discrasia* can arise from an external cause; exercise or climate, for example, can overheat the body as a whole. It can also come from an internal cause, as when a humoral residue dominated by one or more of the four qualities flows into one of the fundamental constituents of the body, the homoiomerous members (that is, those composed of one and the same type of material: bones and muscles and veins and flesh and skin and fat, rather than the hand or finger built up from those simplest members). *Discrasias* vary in kind, depending not only on the nature of the inflowing humoral qualities but also on the member into which they flow. Along with a qualitative *discrasia*, such humoral influxes also typically produce a swelling in the member, phlegmon or cancer or erysipelas, and different influxes produce different kinds of swellings: for example, phlegmon (a hot abscess or aposteme) arises from hot blood.

Local hot *discrasias* (which can arise not only from humoral influxes but also from localized humoral putrefaction, which does not entail swelling) are the causes of fevers. In such cases the *discrasia* is not universally distributed throughout the member, it is "unequal," heterogeneous: most intense at the site of the humoral influx, weaker at the periphery of the member, weaker still in the simple members with which the originally affected member is connected (as muscles communicate with veins); and in fact it is this heterogeneity or unequal intensity that we perceive as pain: thus the pain of the fever called "epiala" arises out of the juxtaposition of cold and hot in the body. If a hot *discrasia* should succeed in extending itself throughout the whole body, so that the body is fully and perfectly heated, a hectic fever is the result, in which the patient feels no pain since all parts of his body are equally hot. In this way, though not as systematically expressed as here, Galen has in *De malitia complexionis diverse* unified different pathological conditions, both fevers and swellings (including tumors and abscesses), within a single explanatory framework.

Galen ended the work by explaining that it could serve as a background for reading others of his books, including *De ingenio sanitatis*. While this promise could have led medieval readers to explore its content with particular seriousness, they would in any case have been attracted to it from the outset by the promise of its title and the word *complexio*, for *complexio* seemed to the later thirteenth century to be a concept of enormous explanatory power, a concept at the very center of health and disease. It was the physical expression of a being's nature, whether that being was an individual human, a body part, or a plant or drug—a precise combination of hot or cold, dry or moist powers in measurable proportion. The normal or natural qualitative proportion entailed health, and its disturbance caused illness, which could be treated by using medicines with an opposite *complexio* to restore health. A full understanding of *complexio* seemed likely to unlock the secrets of medical

[28]What follows has been taken from the introductory material to Bos et al. 2015, viii.

practice, and it is surely no coincidence that among the other works of the "new Galen" taken up by thirteenth-century schools were his *De complexionibus* and *De simplici medicina*. In the former, Galen developed the concept of *complexio* and explained its different forms (e.g., hot and wet), including the state of complete neutrality or temperancy; how these different complexions corresponded to the natures of individuals differing, say, in age or body type; and how the physician could infer a normal or abnormal complexion by examining his patient. In the latter work, Galen explained how to determine the *complexio* or qualitative nature of individual medicines, how they acted upon the body (and were acted upon by it), how their primary qualities could vary through four degrees of measurable intensity, and how they could be understood as entailing further secondary properties (e.g., hot medicines are aperitive). Both books were established as required reading by the Montpellier curriculum of 1309, but even earlier, as we have seen, Bernard de Angarra had exposed his students to *questiones* on *De complexionibus* as well as on *De malitia*.

Arnau's commentary on *De malitia complexionis diverse* was more or less contemporary with Bernard de Angarra's *dubitata*, for it seems to have been composed before 1295; it is about ten times as long as the Galenic object of its attention. Its modern editors have shown that he constructed his work carefully in the scholastic form, with first a *lectio* or word-by-word analysis of a given passage in the text and then a discussion of its meaning, introducing occasional *questiones* to explore its implications; they have concluded that "the present text is notes taken down by students from Arnau's oral delivery, a text that he did not have time to revise personally, although he did at least have it in his hands and may have made a few changes."[29] Considered as mere student notes, however, it seems unusually smoothly written, and there are other difficulties with interpreting the work as an actual classroom presentation. For one thing, it is at odds with Arnau's statement a few years later that he had chosen not to publish his commentaries on the works of Galen and Hippocrates.[30] For another, it is clear that the work is not merely a simple exegesis of Galen's discussion of qualitative *discrasia*: it is planned as an exposition and defense of a particular theory of the nature of fever, one that sees it as resulting from an abnormal intensification of the natural heat, and an attack and refutation of the contention of contemporary masters—including, at Montpellier, his rival Bernard de Gordon (left unnamed in the text)—that fever results from a combination of two heats, that it is a mixture of the natural and an unnatural heat.[31] Galen's work, then, served Arnau as a kind of vehicle for developing an elaborate scientific structure only loosely connected with the original text, in just the way, for

[29]"Creemos que el presente texto son apuntes tomados por estudiantes durante la exposición oral de Arnau, el cual ... no tuvo tiempo de revisarlos personalmente, aunque sí de tenerlos en sus manos y quizás hacer algunos cambios"; Vilanova 1985, 38.

[30]"Admirari non debes si ea que ad exposicionem librorum Galieni et Ypocratis scripsimus communicare publice denegemus"; Vilanova 1988, 30.

[31]Vilanova 1985, 104–10.

example, that aphorism II.34 served Arnau in the commentary he prepared on that Hippocratic passage.[32] Might not this Galenic commentary have been, not a transcription of a series of lectures aimed at medical students who were preparing for a degree in the near future, but a carefully designed polemic with a wider audience in mind, written under the guise of a commentary? The words with which Arnau closes the work tend to confirm this:

> So we urge our readers, *bachelors and masters alike*, that they read this treatise through frequently… We ourselves have not been able to observe that care in its exposition that its contents deserve, on account of our preoccupation with current troubles that have gravely disturbed our studies; yet we know that in that exposition we have opened a path for those who are adequately prepared.[33]

This would seem to indicate that in the form in which we have it the *Commentum* is not a transcript of Arnau's lectures delivered to medical students, but a text whose preparation had required even more time and care than he had been able to give it, and which he intended to be read and reread, not just by his students but by his colleagues.

It would naturally not be surprising if Arnau's *Commentum* had been based to some extent on his prior classroom exposition of the Galenic work, especially since we now know from the Angarra *dubitata* that *De malitia* was actually a school text in Arnau's day. One way to explore this possibility is by comparing the *questiones* that it contains with Bernard de Angarra's *dubitata* on the same treatise. They are not directly comparable, to be sure. Bernard's twelve *dubitata* on *De malitia*[34] (edited below in Appendix 2) presumably have the same origin as the *dubitata* on the *Aphorismi* that we have already discussed, and are probably a second-hand student summary of *questiones* that Bernard himself had originally formulated and determined more elaborately, perhaps in the course of a full commentary on the work. The *questiones* that the editors of Arnau's *Commentum* have identified in that work are not really the same thing: rather than posing problems for formal determination, they respond to possible objections to or misunderstandings of his argument; rather than beginning "Utrum," they begin with phrases like "Si quis autem obiciat" or "Circa predictam queri consuevit."[35] Thus our task is to compare these informal questions with the more formal *questiones* by Bernard, which we cannot study directly but must reconstruct from the *dubitata* in the Munich manuscript that they gave rise to.

---

[32] Vilanova 2014, 93–102.

[33] "Hortamur autem lectores, tam bacalarios quam magistros, ut frequenter perlegant hunc tractatum…. Nos autem circa exposicionem ipsius non potuimus illam diligenciam observare quam exigebat materia, propter varios labores instancium tribulacionum qui studium nostre mentis valde perturbaverunt; sed scimus quod in predicta exposicione aperuimus viam sufficienter intelligentibus"; Vilanova 1985, 296.

[34] A thirteenth title has been included by mistake in the list given in Vilanova 1985, 67.

[35] Vilanova 1985, 66.

Arnau's editors identified and extracted eighteen such questions in the *Commentum*, which they viewed as generally meant "to explain his personal concept of fever and the mechanism of its production ... to a student audience."[36] In contrast, Bernard de Angarra's *questiones* on *De malitia* have no such thematic coherence (to judge from their reflection in the *dubitata*), and only one or two explore issues related to fever; but it is significant that six of them (nos. 2, 4, 5, 6, 9, 10) coincide with those few queries raised informally by Arnau that do not concern general issues of fever and instead explore a more restricted point in *De malitia*.[37] These typically arise when the Galenic text is in disagreement with other established medical authorities. For example, both pointed out that Galen had said in *De malitia* that *lepra* was a disease of an individual member, whereas Avicenna had famously declared in the *Canon* that it was a disease of the whole body (IV.3.3.1); here Bernard agreed with Galen, whereas Arnau tried to reconcile the two viewpoints.[38] Again, Galen had said that choler was the hottest thing in the body, whereas Avicenna had said that that was true of the heart; once more Arnau tried for a middle ground, while Bernard agreed with Galen, though the student seems to have had some question about his master's resolution of the difficulty. It seems reasonable to interpret these six debates as representing a common store of questions evolved by the Montpellier masters as they dealt with the new Galenic text in the light of standard authorities, and to conclude that Arnau had carried them over in his commentary, which he later reworked in such a manner as to promote his own theory of fever.

2.4 Through a Student's Eyes

One of Angarra's *dubitata*, however, apparently innocuous in itself, becomes controversial when read in the light of Arnau's commentary. Bernard's immediate doubt had arisen from Galen's statement that in cases of *ethica* (that is, hectic fever) "omnia prima corporis membra integre et perfecte sunt calefacta"[39]: the intense overheating should corrupt and kill, yet some patients are able to survive the disease. In his seventh *questio* on *De malitia*, Bernard had asked "Whether in hectic fever all the members are *supercalefacta*," and he is reported by his student to have said that the statement is not true if by that you understand them to be filled completely by a *calor extraneus*, since then they would indeed be corrupted. However, if you suppose them instead merely to have "some" (*aliquod*) *calor*

[36]Ibid., 67: "Para exponer su personal concepción de la fiebre y de su mecanismo de producción ... ante un auditorio de estudiantes."

[37]Bernard's *questiones* 2, 4, 5, 6, 9, 10 seem to me to correspond to Arnau's Ib, IVa, Vb–c, VIIb, Xa, as enumerated at ibid., 65–66.

[38]Their disagreement is put into a wider context in Demaitre 2007, 117–20.

[39]Vilanova 1985, 219.

extraneus, the statement can be accepted, because then they can still possess some of their own natural heat as well—that is, the heat which makes it possible for a member to live and function.

Behind this response lie assumptions that are sharply at odds with the model of fever that Arnau developed, a model largely consistent with Avicenna's definition that "Febris est calor extraneus accensus in corde et procedens ab eo . . . in totum corpus," except that Arnau understood the *calor extraneus* specifically as an intensification of the *calor naturalis*.[40] In his commentary on *De malitia* Arnau attacked certain contemporaries who espoused a very different position, that in fever two separate heats were carried together to all parts of the body, one natural and one unnatural. Though he does not say so, those contemporaries were evidently basing their arguments on the very different dualistic definition of fever given in Averroes' *Colliget*, translated only a decade or so earlier (1285): "Febris . . . est calor compositus ex calore naturali et extraneo putredinali misso a corde ad totum corpus."[41] Arnau rejected their interpretation as an "evidens et stulta falsitas." Galen, he declared, had shown that

> there are not two heats in separate subjects carried from the heart to all the body, instead just one in a single subject, in which a single form of heat is present. That subject of the natural heat, which previously was sent out to the members informed by a temperate heat, is now sent out superheated [*incensum*]. Thus he plainly means that, whether in subject or number, there is in fever *qua* fever just one heat.[42]

Arnau's editors have argued plausibly that Bernard de Gordon was one of the people whom Arnau was attacking here, on the basis of statements scattered through Gordon's writings, since he never wrote a commentary on *De malitia* (as far as we know) where such ideas could have been developed at length.[43] Now we can see that Bernard de Angarra is quite likely to have been another of Arnau's targets, for his answer to the seventh *questio* seems to set aside the Avicennan understanding of fever—he virtually quotes the *Canon*'s definition in dismissing it—in favor of the Averroistic position that Arnau ridiculed, by saying that in hectic fever the members receive not only their own natural heat but a

[40]Some of the fourteenth- and fifteenth-century debates arising out of the Avicennan definition are surveyed in Jacquart 1998, 379–91.

[41]Averroes 1562 (III.3), fol. 34vb. In his *Aphorismi de gradibus* Arnau wrote that "in omnibus erravit in quibus [Averoys] invectus est contra Galienum; et ideo contra eum specialiter scripsimus tractatum de intencione medicorum, et tractatum de consideracionibus operis medicine, ac epistolam de dosi tyriacalium medicinarum, ne laberentur in errorem debiles occasione dictorum"; Vilanova 1975, 201. To Arnau's list of his consciously anti-Averroistic medical writings can thus now be added his commentary on *De malitia*.

[42]"A corde ad totum corpus non procedunt duo calores secundum numerum, scilicet distincti per diversa subiecta, sed unus est tantum numero per unum subiectum, in quo una forma caloris fundatur. Illud enim subiectum caloris naturalis, quod antea temperato calore informatum mittebatur ad membram, nunc incensum mittitur. Unde plane vult quod, secundum rem et numerum, non sit nisi unus calor in febre, sed in quantum febris"; Vilanova 1985, 269.

[43]See, above all, García Ballester 1987, 317–32; and García Ballester and Gil Sotres 1986.

separate *calor extraneus* (which would be carried by a different vehicle, hence a *vapor extraneus*). The issue is likely to have been a matter of general discussion by the masters, a discussion that generated a shared terminology (both Arnau and Angarra deploy the term *supercalefactum*, which does not appear in *De malitia complexionis diverse*), though we will surely never be able to understand the dynamics and alliances within the faculty, and whether Gordon and Angarra had made common cause.[44]

The remainder of Bernard de Angarra's *questiones* on *De malitia* pursue statements in Galen's text that apparently reveal his own interests at that moment—at any rate, Arnau had not seen them as problems and simply passed each over without special comment—and they therefore convey a certain sense of his personality as a teacher. Bernard's tenth question—"How can those suffering pain caused by the cold suffer even more when they approach the fire?"—is an Arnaldian one too, resolved not very differently. But Bernard goes on to an eleventh, related, question that Arnau had not seen fit to pose: "Whether heat existing in one part and cold in another are each felt in the same part?" Galen says explicitly in *De malitia* that this is the case, that the two are felt simultaneously and together, but Bernard evidently wanted to convince his class that this apparently self-contradictory statement could actually be true. So he supported Galen's statement by offering a homely thought-experiment that stuck in the student-reporter's mind: "The chancellor says this is because the senses cannot separate them because of their closeness, in the way that a whirling firebrand seems to show a circle [*motus circularis*]."

Another *questio* (no. 3) apparently arose out of Galen's explanation that a phlegmon was caused by the presence of a superfluity of blood that heated the tissues to excess. The unspoken difficulty was, since blood is temperate, how could it be a cause of overheating? Arnau recognized the problem but resolved it in passing without a formal *questio*: the blood was trapped in a confined space and hence became abnormally hot itself, causing the phlegmon.[45] Bernard, however, seems almost petulantly to have dismissed the matter as a non-problem, denying that the phenomenon required an explanation: "The chancellor declared," wrote the student, "that this question wasn't a rational one, because [the blood] couldn't increase in quantity without its quality being changed for the worse." One wonders why, if he felt that the question was not a rational one, he chose to bring it up. Perhaps it was another traditional staple of discussion that Bernard felt obliged

[44]Another subject of contemporary debate within the Montpellier faculty seems to have been whether the natural heat (*calor naturalis*) or its material substrate, the radical moisture (*humidum radicale*), was the true foundation of physiological life. Bernard de Angarra took the former position in his *Aphorisms*-commentary ("dicendum quod humidum radicale non est primum principium vite sed calidum naturale"), while Bernard de Gordon seems to have favored the latter; see Demaitre 1992, esp. 272–73, who suggests that Angarra's position may eventually have prevailed at the school. Arnau de Vilanova's *Tractatus de humido radicali* is another witness to the importance of this theme to the faculty in the early 1290s; see the editorial introduction to Vilanova 2010.

[45]Vilanova 1985, 170.

at least to raise before the class, failing to understand what fundamental question was involved. Or perhaps, to be sure, the student failed fully to understand his teacher's response—for students are as much individuals as teachers.

Indeed, it is not always easy to decide how much the anonymous student may have been adding to the substance of his teacher's report. In no. 5, raising the question whether choler is the hottest thing in the body, it is clear that Bernard de Angarra agreed (with Galen) that it was. His flat statement is followed by the student's assertion that the problem was a difficult one, but in the end the initial statement is confirmed by another thought-experiment—perhaps Bernard's, conceivably this time the student's own: "choler brings on fever by igniting the heat of the heart, which it could not do if it were of less heat than it or the same, which is clear because if a hand equal in warmth to my own is placed on my hand, it will not change its hotness"—and therefore implicitly choler must be hotter than the heart. We have already seen that complexion theory was of great interest at Montpellier, and this passage reflects wider discussions in *De complexionibus* and *De simplici medicina* of the difference between qualitative and quantitative measure, and of what happens when qualitative intensities are added together: a medicine hot in the second degree, added to another of the same degree, will not increase the heat of the mixture.[46] That understanding underlies the argument here about body heat, whether it is Bernard de Angarra's argument or (as is quite possible) that of his student. It may not be coincidence that it is the same understanding which we found underpinning part of Arnau's criticism of the two-heat theory of fever, when he pointed out that the hot putrid vapor arriving at the heart could not be heated still further there, since the heart was of a temperate and appropriate warmth, not *supercalefactus* as it would need to be.

Perhaps most intriguing of all these questions, this time certainly telling us as much about the student as the master, is the last (no. 12). The *dubitata* follow almost exactly the sequence of Galenic passages as the reader proceeds through *De malitia*, and both this one and the one that precedes it were suggested by Galen's explanation late in the work of why we sometimes seem to feel heat and cold at the same time. It is because the infinitesimal parts of the body, which are either hot or cold, are packed so closely together that we cannot sense one without sensing the other. Angarra went on to ask whether hot and cold could coexist in the *same* subject (rather than in adjoining subjects), and seems to have given two reasons why they could, which the student reported perhaps a little confusedly. Then, showing a certain intellectual independence, he went on to say "but I think this question is the same as or subsidiary to a larger question, whether elements persist in a mixture in their original forms."

[46]Arnau felt it necessary to spell out that point for any reader of his *Aphorismi de gradibus* (composed in the mid-1290s) who might still be confused about it: "Si distemperata equaliter in eisdem qualitatibus commisceantur, compositum resultabit distemperatum in terminis componencium"; Aphorism 35, in Vilanova 1975, 196.

Here the student was evidently reacting, not to his teacher's resolution(s) of the question, but to something else, something that in the later thirteenth century had become a topic of lively discussion in natural philosophy and subsequently in medicine: "utrum maneant in mixto?"[47] When two substances are mixed together and produce a new substance, can the substantial forms of the original ingredients persist in any way in the product? The weight of philosophical opinion was against this, holding that the prior forms had to be destroyed if a new form were to come into existence, though Avicenna had held out the hope that they might somehow "virtually" persist. Philosophically alert physicians soon recognized that this issue was of great importance to them, for their compound medicines were just such mixtures. When the famous antidote theriac was compounded, for example, did the particular powers of each of its dozens of individual ingredients continue to have the ability to act upon the patient too, over and beyond the power specific to the new theriac to combat poison? As one might have expected, the two great Montpellier masters of the 1290s both considered the issue seriously, though they were not perhaps in entire agreement. In his *Pronostica* of 1295 Bernard de Gordon took a rather Avicennan tack in dealing with the matter:

> When the elements are combined [*confracta sunt*] in a mixture so that the least parts of one touch the least parts of another, then the elements do not exist in their substantial forms yet do not entirely lose them either; then from the mutual action of hot with cold arises a certain quality, and from the action of dry with wet results another proportion. Hence the quality that arises from a proportion of active and passive qualities is called a complexion, and because this combination can have many different forms, therefore complexions can have many different forms.[48]

For his part, Arnau de Vilanova was engrossed by the problem during all his career, and in his last great work, the *Speculum medicine* (1308), he treated it exhaustively, arguing that experience showed how in many cases the ingredients in a compound were so loosely fused that they could easily separate and act individually, "et propter hanc considerationem asserit medicus quod forme substantiales miscibilium salvantur in mixto."[49] We have known for some time that the issue attracted the attention of these most visible and productive members of the Montpellier faculty; but in the student notes in the Munich manuscript we see that the interest of their seniors had filtered down to at least some of the students as well, and that they had assimilated it well enough to apply it to situations where their teachers had not appreciated its relevance.

Few though they are, therefore, the Montpellier commentaries that survive in manuscript from the school c. 1300 have a considerable amount to tell us about

---

[47]Maier 1952, esp. 3–35. Pages 3–22 have been given English translation in Maier 1982, 125–42.

[48]"Cum elementa confracta sunt in mixto et minimum in virtute tangit minimum alterius, tunc elementa non sunt formis substancialibus nec omnino amiserunt eas, et tunc ex mutua accione calidi cum frigido resultat quedam qualitas, et ex accione sicci cum humido, resultat alia proportio. Qualitas igitur que resultat ex proporcione activarum et passivarum, vocatur complexio, et quia hec confraccio est multiplex, ideo complexio est multiplex"; Alonso Guardo 2003, 246.

[49]Vilanova 1520, fol. 11rb.

medical instruction there. In the Oxford and Erfurt manuscripts we can get a sense of how the texts of the old *ars medicine* were being taught, as exemplified in commentaries on Hippocrates' *Regimen acutorum* and *Aphorisms* by two of the "ordinary" Montpellier masters who have always been overshadowed by Arnau and Bernard de Gordon. The latter manuscript not only shows us Bernard de Angarra dissecting one of the foundational texts of medieval medicine, aphorism by aphorism, explaining to his class the specific issues that each raises, it also establishes that the school's masters were now expounding the "new Galen" to them as well, works like *De malitia complexionis diverse* and *De complexionibus.* Arnau de Vilanova's exegesis of *De malitia* is a commentary in the same line, though it seems to have more to tell us about the particular scientific interests of an unusually intellectually ambitious figure than it does about what he taught to his students. Even so, in the background to Arnau's treatise we can discern the process of lively discussion and debate of Galenic themes being carried on within what was clearly an interactive community of teachers: whether famous, like Arnau, or today comparatively obscure, like Bernard de Angarra. And we are especially fortunate that in the Munich manuscript we can apparently identify one of Bernard's students responding to his master's exploration of these new Galenic problems, interpreting it, rephrasing it, and assimilating it to what he had already learned. Here, in the end, it is as if we were actually looking over the student's shoulder in the medieval classroom.

Appendix 1

Bernard de Angarra, *questiones* on Hippocrates, *Aphorisms* I.2:

In perturbationibus ventris et vomitibus que sponte fiunt, siquidem qualia oportet purgari purgentur, et confert et bene ferunt; si vero non, contrarium. Sic et inanitio, siquidem qualem oportet fieri fiat, et confert et bene ferunt; si vero non, contrarium. Contemplari igitur oportet et tempus et regionem et etates et egritudines in quibus oportet aut non. (*Galeni in aforismos Hippocratis*, in *Articella* [Venice, 1523], fol. 3v: "translatio antiqua").

A. In MS Erfurt, Ampl. F. 290 (fols. 40vb–41rb)

[1]. **Queritur** utrum natura semper evacuet quale est evacuandum ...

[2]. **Queritur** utrum si quale oportet quod purgari purgetur conferat. Videtur quod non. Ex tali evacuatione contingit intestinorum excoriatio, ut cum colera superflua a corpore purgatur; hec autem non confert, quare et cetera. Item ex tali purgatione contingit effymera ex commotione frigida in corpore in evacuatione humoris nocitivi; hoc autem non confert, quare et cetera. Item videtur illorum inculcatio cum dicitur "confert et bene ferunt"; unum enim videtur alterum ponere, quare et cetera. Oppositum dicit littera. Dicitur quod quale ex purgari est nocivum, tale dicitur semper quod huiusmodi peccans in corpore facit nocumentum, quare eductio eius a

corpore faciet iuvamentum nature, pro quanto confert nature, unde ipsa alleviatur, eo quod a gravitate ipsam exoneratur et sic virtus invalidatur, propter quod etiam eductio illius nocitivi a natura faciliter toleratur; et hoc videmus ad sensum, quod cum aliquis purgatur ab humore nocitivo, ex hoc alleviatur sensibiliter et bene ac faciliter coleram absque molestis fert. Intelligendum enim quod per accidens hanc laudabilem purgationem aliquando sequitur nocumentum per accidens, ut excoriatio intestinorum ex mordicatione humorum purgari et debilitate intestinorum, pro quanto artificialiter in medicinis laxativis ponuntur quedam confortativa huiusmodi membra, ut dragagantum, bdellium, et similia. Similiter ex commotione humorum in evacuatione inflamantur spiritus unde contingit effymera, sed hoc accidit evacuationi nocitivi secundum quod huiusmodi, ista tamen accidentaliter. Nocumenta minus nocent quam purgatio nocitivi, et per hoc ad rationes; bene enim ostenditur quod aliquando per accidens aliqualiter non conferunt, per se tamen confert.

[3]. **Queritur** utrum medicus debeat imitari actionem nature...

B. In MS Munich, CLM 534 (fol. 44vb)

[1]. **Utrum** natura evacuat quod non oportet. Dicit sic: aliquando enim expulsiva in stomaco expellit iuvativum cum nocitivo, et hoc convenit propter cibi malam qualitatem vel nimis quantitatem, unde expellitur ante digestionem et divisionem.

[2]. **Utrum** purgatio nocitivi conferat. Dicit sic, quia si quod lesit evacuatur realiter sequitur alleviatio; tamen accidentaliter nocet, ut pote quia aliquis habet intestina debilia, ideo leditur per transitum colere. Sunt ergo prius confortanda vel alibi purgandus est.

[3]. **Utrum** medicus debet sequi actionem nature. Dicit sic, cum sit minister suus. Si arguitur de statu, dicit quod non prohibetur nisi per accidens, propter accidentium fortitudinem, et fit aliquando in crisi incompleta, fortitudo autem talis accidentium in purgatione nature non ita timetur.

Appendix 2

Student notes on Bernard de Angarra's questions on Galen, *De malitia complexionis diverse* (transcribed from MS Munich, CLM 534, fol. 43ra–b)[50]

[50]References given below in square brackets at the end of each *dubitatum* are to the pages in Vilanova 1985 bearing the Galenic passage from which it appears to have been derived. I have had difficulties reading a word or two in the Munich manuscript; they have been left as they appear to me with a question mark attached.

Supra libellum de mala complexione diversa hec sunt dubitata. Quorum positiones secundum magistrum Bernardum de Hangarra montis pessolani quondam cancellarium recita<n>tur.

(1) Utrum in omni specie ydropisis sit mala complexio diversa. Quidam dicunt quod sic. G[alienus] autem expresse secundum unam literam vult oppositum, secundum aliam non expresse. Sciendum ergo quod in carnosa possibile est quod mala complexio diversa primo sit in carne, postea in epate. Quia vult G. vi.^{o} de morbo cum membra nimis avide attrahunt cibum, replentur humiditatibus crudis indigestis, unde hec species ydropisis causatur; in aliis licet epar patiatur cui totum corpus communicat, hoc tamen non fit statim sed successive. [147]

(2) Utrum lepra sit mala complexio totius corporis. Dixit Cancellarius quod non, quia ordo nature est ut membris nobilibus detur nutrimentum nobilius, unde extremitatibus cedit malum ex quo primo inficiuntur. [147]

(3) Utrum sanguinis augmentatio in quantitate tantum faciat malam complexionem in illo membro. Dixit C[ancellarius] quod illa questio non erat rationalis, quia non augmentatur in quantitate nisi mutetur ad malam qualitatem. [163]

(4) Utrum resolutio in apostemate insensibilis sit melior evacuatione sensibili. Dicit in quantum arguit fortitudinem virtutis quod meliora est, tam febribus quam apostematibus, tamen quia in febribus sepe fit cum permutatione occulta et inducit dubietatem, ideo ibi magis diligitur sensibilis, etiam apostematibus aut adhuc est periculum de sanie ne corrodat et ne vulnus male curatum inducat fistulam, ideo etc. [171–72]

(5) Utrum res calidior in corpore sit colera. Dicit C. quod credit quod sic. Sed declaratio huius est difficilis. Sed sicut G. declarat frigiditatem flegmatis extat hu. (?), ita dicit Avicenna quod egerenti coleram patet quod ignis transeat intestina. Et quia colera inducit febrem incendens calorem cor(**43rb**)dis, quod non est si est [*MS*: etiam si etiam] minoris caloris quam eorum vel equalis, quod patet quia si ponatur manus equalis calore cum mea super meam, non mutat eius caliditatem. [196]

(6) Utrum flegma sit frigidius quod est in corpore. Patet quod sic, quia diversitas huius partium chyli sequitur dominium elementorum; ergo quod operatur frigidissimo ut aqua, hoc erit frigidius. Hoc etiam patet ex parte efficientis, quia efficiens melancolie est calore temperatus, flegmatis autem calore abbreviatus. Hoc etiam patet ex tactu. [196]

(7) Utrum in etica omnia membra sint supercalefacta. Dicit C. quod si intelligatur quod totaliter sint plena calore extraneo in omni parte, non est verum, quia tunc erunt corrupta. Sed si intelligatur quod nullus est, quoniam habeat aliquod de calore extraneo, sic potest stare, quia adhuc potest habere proprium calorem quo vivit et operatur. [219]

(8) Utrum possit esse mala complexio equalis et diversa simul. Dicit C. quod sic virtualiter, ut in etica est equalis, cum qua si ponatur putrida, ea recedente, materiam adhuc virtualiter. [230–31]

(9) Utrum membra sub naturali dispositione existentia alterent se. Dicitur quod sic, quia sunt contraria. Sed tamen unde non corrumpit aliud propter resistentiam. Et quia diversis mutationibus subiacet corpus, ideo non semper vincit unum ut alia ad se convertat. [230–31]

(10) Utrum dolentes a frigore magis debent dolere cum approximantur calori. Dicit quod sic, quia redit sensus a frigore debilitatus, vel quia unum contrarium iuxta aliud fortificatur. [275]

(11) Utrum caliditate existente in una parte et frigiditate in alia sentiantur in eadem. Dicit G. quod sic. Et ait C., hoc est propter vicinitatem quam sensus non distinguit, sicut propter velocitatem motus ligni ardentis estimatur ille motus circularis. [279–80]

(12) Utrum in eadem parte possunt esse caliditas et frigiditas simul. Dicit C. quod sic, quia omnis dilectio preter intellectualem mixta est tristicia. Cum ergo calor inducat dilectionem, frigus inducet tristitiam. Et in alia re quam ponit, dicit quod non sint in ultimo, nam omnis calor et omne frigus. Sed ego credo istam questionem equalem et subalternatim illi, utrum elementa sint in mixto sub propriis formis. [279–80]

Bibliography

Alonso Guardo, Alberto. 2003. *Los pronósticos médicos en la medicina medieval: El* Tractatus de crisi et de diebus creticis *de Bernardo de Gordonio*. Valladolid: University of Valladolid.

Averroes. 1562. *Colliget*. Venice: Juntas.

Bos, Gerrit, Michael McVaugh, and Joseph Shatzmiller. 2015. *Transmitting a Text Through Three Languages: The Future History of Galen's Peri anomalou dyskrasias; Transactions of the American Philosophical Society*, vol. 104 part 5. Philadelphia: American Philosophical Society Press.

Cardinalis. Commentary on Hippocrates, *Regimen acutorum*. MS Kues 222, fols. 167r–181r.

Cartulaire de l'Université de Montpellier. 1890. 2 vols. Montpellier: Ricard Frères.

Catalogus Codicum Latinorum Bibliothecae Regiae Monacensis. 1892. Munich.

Demaitre, Luke E. 1980. *Doctor Bernard de Gordon: Professor and Practitioner*. Toronto: PIMS.

———. 1992. The Medical Notion of 'Withering' from Galen to the Fourteenth Century: The Treatise on Marasmus by Bernard of Gordon. *Traditio* 47: 259–307.

———. 2007. *Leprosy in Premodern Medicine: A Malady of the Whole Body*. Baltimore: Johns Hopkins University Press.

Galen. 2011. On Problematical Movements. In *Cambridge Classical Texts and Commentaries*, ed. Vivian Nutton and Gerrit Bos, vol. 47. Cambridge: Cambridge University Press.

García Ballester, Luis. 1982. Arnau de Vilanova (c. 1240–1311) y la reforma de los estudios médicos en Montpellier (1309): El Hipócrates latino y la introducción del nuevo Galeno. *Dynamis* 2: 97–158.

———. 1987. La recepción del *Colliget* de Averroes en Montpellier (c. 1285) y su influencia en las polémicas sobre la naturaleza de la fiebre. In *Homenaje al professor Darío Cabanelas Rodríguez, O.F.M., con motivo de su LXX aniversario*, vol. 2, 317–332. Granada: University of Granada.

García Ballester, Luis, and Pedro Gil Sotres. 1986. *Teorías sobre la fiebre y Averroismo médico en Montpellier: Bernardo de Gordon y Arnau de Vilanova*. Cuadernos de Historia de la Medicina 2. Santander/Barcelona.

Jacquart, Danielle. 1994. L'oeuvre de Jean de Saint-Amand et les méthodes d'enseignement à la faculté de médecine à Paris à la fin du XIIIe siècle. In *Manuels, programmes de cours et techniques d'enseignement dans les universités médiévales*, ed. J. Hamesse, 257–275. Louvain-la-Neuve: IEMUCL.

———. 1998. *La médecine médiévale dans le cadre Parisien*. Paris: Fayard.

Maier, Anneliese. 1952. *An der Grenze von Scholastik und Naturwissenschaft*. Rome: Edizioni di Storia e Letteratura.

———. 1982. *On the Threshold of Exact Science: Selected Writings of Anneliese Maier on Late Medieval Natural Philosophy*, ed. and trans. Steven D. Sargent. Philadelphia: University of Pennsylvania.

McVaugh, Michael. 1976. Two Montpellier Recipe Collections. *Manuscripta* 20: 175–180.

———. 1987. The Two Faces of a Medical Career: Jordanus de Turre of Montpellier. In *Mathematics and Its Applications to Science and Natural Philosophy in the Middle Ages*, ed. Edward Grant and John Murdoch, 301–324. Cambridge: Cambridge University Press.

———. 2013. The Future of a Disease: The Impact of Galen's *De Crisi* on Medieval Medical Thought. In *Die mantischen Künste und die Epistemologie prognostischer Wissenschaften im Mittelalter*, ed. A. Fidora, 131–150. Cologne-Weimar-Vienna: Böhlau.

———. in press. A Miscellany? Or the Evolution of a Mind? MS Munich CLM 534. Forthcoming in *Micrologus*.

———. in press. Averroes Comes to Montpellier. Forthcoming in *Mélanges* in honor of Danielle Jacquart.

O'Boyle, Cornelius. 1998. *The Art of Medicine: Medical Teaching at the University of Paris, 1250–1400*. Leiden/Boston/Köln: Brill.

Paniagua, Juan A. 1994. *Studia Arnaldiana*. Barcelona: Fundación Uriach 1838.

Roüet, A[dolphe August]. 1878. *Notice sur la ville de Lunel au moyen-âge*. Montpellier/Paris: Séguin.

Schuba, Ludwig. 1981. *Die medizinischen Handschriften der Codices Palatini Latini in der Vatikanischen Bibliothek*. Wiesbaden: Reichert.

Schum, Wilhelm. 1887. *Beschreibendes Verzeichniss der Amplonianischen Handschriften-Sammlung zu Erfurt*. Berlin: Weidmannsche Buchhandlung.

Vilanova, Arnau de. 1520. Speculum medicine. In *Opera Arnaldi*. Lyons: Guilhelmi Huyon.

———. 1975. *Aphorismi de gradibus. Arnaldi de Villanova Opera Medica Omnia*, ed. M.R. McVaugh, vol. 2. Granada/Barcelona: University of Granada.

———. 1985. *Commentum supra tractatum Galieni de malicia complexionis diverse*, *Arnaldi de Villanova Opera Medica Omnia*, ed. L. García Ballester and E. Sánchez Salor, vol. XV. Barcelona: University of Barcelona.

———. 1988. *Tractatus de consideracionibus operis medicine*, *Arnaldi de Villanova Opera Medica Omnia*, ed. P. Gil-Sotres and L. Demaitre, vol. IV. Barcelona: University of Barcelona.

———. 2010. *Tractatus de humido radicali*, *Arnaldi de Villanova Opera Medica Omnia*, ed. C. Crisciani, G. Ferrari, and M. McVaugh, vol. V.2. Barcelona: University of Barcelona.

———. 2014. *Expositio super In morbis minus*; *Repetitio super Vita brevis*, *Arnaldi de Villanova Opera Medica Omnia*, ed. M.R. McVaugh and F. Salmón, vol. XIV. Barcelona: University of Barcelona.

Weijers, Olga. 1994. L'enseignement du *trivium* à la faculté des arts de Paris: La 'questio'. In *Manuels, programmes de cours et techniques d'enseignement dans les universités médiévales*, ed. J. Hamesse, 57–74. Louvain-la-Neuve: IEMUCL.

———. 1996. *Le maniement du savoir: Pratiques intellectuelles à l'époque des premières universités (xiiie –xive siècles)*. Turnhout: Brepols.

Wickersheimer, Ernest. 1936. *Dictionnaire biographique des médecins en France au Moyen Age*. Paris: Droz.

Part II
Medicine, Writing History and the Republic of Letters

Chapter 3
Medicine as Queen: The *Consilia* of Bartolomeo da Montagnana

Chiara Crisciani

3.1 A Famous Physician. Bartolomeo da Montagnana

Like many other writers in the Quattrocento, the physician-historian Michele Savonarola wrote a *Libellus* to praise the history, merits, monuments, and beauties of his town, which in his case was Padua. Among the accolades for its most distinguished citizens, who contributed to different fields, he lists famous physicians who adorned Padua with their contributions to medicine. The last physician in the list is Antonio Cermisone, but his disciple and colleague, Bartolomeo da Montagnana, does not appear.[1] Perhaps Savonarola thought that Montagnana's fame had not yet been fully confirmed. Or perhaps Savonarola's very high praise of Cermisone indicates he considered Cermisone the pinnacle of the Paduan *studium* and its medical learning. Regardless, Montagnana's studies and his whole academic career, along with his life as a practitioner and his fame, developed entirely within the town of Padua until his death in 1452. Montagnana is a prime example of how, in the fifteenth century, there is a growing tendency for teachers in main Italian Universities to remain in the same place, and for descendants of physicians to occupy the same professional and teaching positions. Indeed, Montagnana's father[2] (Giacomo) was a surgeon, and at least one of his children became a physician trained and subsequently practicing in Padua.[3]

For Nancy the "queen", with gratitude and all my best wishes.

[1]Savonarola 1902; on the *Studium* of Padua the basic reference still is Siraisi 1973.

[2]See Pesenti 1994, 141–57; Bacchelli 2010, 738–741; and we find a very useful and recent treatment in Peloffy 2010/2011.

[3]See Micca 1941, 26–34.

C. Crisciani (✉)
Università di Pavia, Pavia, Italy
e-mail: chiara.crisciani@gmail.com

G. Manning, C. Klestinec (eds.), *Professors, Physicians and Practices in the History of Medicine*, Archimedes 50, DOI 10.1007/978-3-319-56514-9_3

Bartolomeo da Montagnana joined Padua's distinguished list of educated physicians when he obtained his degree from the university in 1403. He had been a disciple of Marsilio Santasofia,[4] who also belonged to another renowned family of physicians. Starting from 1403, his name regularly appears in the academic records until 1450. He began teaching *medicina pratica* in Padua in 1409. Beginning in 1422 he was professor of *medicina pratica straordinaria* and then of *medicina pratica ordinaria* (perhaps the most requested, and well remunerated chair). Although Montagnana's intense professional activity is well documented by his *Consilia*, we have no evidence of any political assignments or activity. He appears as a witness in several notarial and civil documents, and he was given several professional-civil assignments, but always in his capacity as a physician. For example, in 1406 he was already syndic of the *Fraglia* (which means corporation) of the Paduan physicians, and as such he brought legal proceedings against those who had practised as physicians in Padua without the necessary authorisation. Much later (1445) we find him as syndic of the College of Philosophers and Physicians.

Consistent with Montagnana's teaching and active career are his written works, which are all centered on *medicina practica*. These works are not particularly long, apart from the *Consilia*, but this is not unusual. Nor does it indicate a lack of deep and ample theoretical research on Montagnana's part. Indeed, it would be a mistake to believe that theoretical research is not an integral part of Montagnana's practical medicine (or *medicina operativa*), which is a form of *scientia* as well.[5] In addition to the *Consilia*, and like his contemporaries, such as Michele Savonarola,[6] he wrote a text, *De balneis Patavinis*,[7] on the baths, i.e. the famous thermal springs (which he also mentions in the *Consilia*, 26va) present in the Paduan area. Written for Malatesta Malatesti of Pesaro, this work may be interpreted as a very large *consilium*. In it, Montagnana describes the morphological and mineral features of the various locations and prescribes the rules for the thermal cure – e.g. the ailments for which the cure is suitable, the proper dietary laws and the expected healing effects.

Montagnana's other works include *De urinarum iudiciis*, which was probably written in collaboration with his disciple Zaccaria dal Pozzo the senior of Feltre. It is a more traditional diagnostic treatise; and even in the *Consilia*, Bartolomeo duly considers the importance of such signs-symptoms. His *De compositione et dosi medicinarum*, which mostly refers to the analogous work by Mondino de' Liuzzi, is also within the specific competence of a professional physician. And in the *Consilia*

[4]See Pesenti 2003.

[5]Avicenna, *Liber canonis*, I, fen I, *doctrina prima*. A book entitled *Practica* is ascribed to him, though it was not written by him but more likely by his great-grandson Bartolomeo da Montagnana junior. See Ms Venezia, Biblioteca Nazionale Marciana, Marc.lat. VII, 66 (=9684).

[6]About this subject, which has been widely examinated recently, I confine myself to citing only a few important essays: Park 1999; Nicoud 2011; Boisseuil and Nicoud 2010; Guérin-Beauvois and Matrin 2007.

[7]*De aspectu, situ, minera, virtutibus et operationibus balneorum in comitatu Patavino repertorum.* This work has been published many times -either alone or together with the *Consilia*- and is inserted in the collection, *De balneis omnia quae extant.*

too, Montagnana is very keen to discuss the several problems and features of pharmacological therapy. Also worth mentioning is the *Diffiniciones terminorum medicinalium*, a lexicon of medical terms written for (or collected by) his German disciple Brocardo de Horneck (one of many who came to Padua to study medicine at the time). In this work, the treatment is quite elementary, but the strong interest for the analysis of terms and synonyms, one of the main features of the medical literature in Italy at the time,[8] is especially apparent in Montagnana's *Consilia*. Moreover, related to this are the *Recepte*. These are often taken from the *Consilia* and were presumably collected by the students themselves during the lessons or during professional visits.

At any rate, the *Consilia* are the most conspicuous product of Montagnana's studies and activity.[9] This is acknowledged with real bursts of admiration by Geraldo Bolderio,[10] his faithful disciple, who specifically singles out this work (*consiliorum divinum opus, consiliorum opus inauditum*) among the others of the fifteenth century. In comparison with works written by his contemporaries, says Bolderio, Montagnana's *Consilia* are more structured and richer in authoritative support. It may be said – Bolderio concludes – that ancient and classical authors, and even not yet well known works of Galen, are now fully alive and vividly speaking on Montagnana's side. As we shall see, Bolderio's assessment is substantially correct. We dedicate ourselves below to a presentation of Montagnana's *Consilia* and their significance for understanding his career and the state of scholastic medicine in Italy during the fifteenth century.

3.2 The *Consilia* Litterature

The *consilia*[11] is a genre typical of practical medicine in the Late Middle Ages.[12] It expresses the links between teaching and professional activity, that is, between the *magisterium* of the physician as a teacher[13] and the *ministerium* of the physician as

[8]See, for example, various works of Weijers, esp. 2009; and Crisciani, Lambertini, Tabarroni, eds., *Due manoscritti con questioni mediche. Note e schede (prima metà del secolo XIV)*, 2015.

[9]I am using here Bartolomeo da Montagnana, *Consilia*, 1525.

[10]See his 'prefatory epistle', which introduces the edition of 1525 of the collection of Montagnana's *Consilia*.

[11]I rework in this paragraph some observations which are more fully developed in Agrimi and Crisciani 1994. See also Pesenti 1983; and Baader 1987.

[12]On the practical medicine in the Middle Ages, see Demaitre 1975; eadem 2013; Garcia-Ballester et al. 1994; and Siraisi 1990. On the now huge literature about the scholastic medicine, I confine myself to citing some important essays collected by Grmek 1998.

[13]About the epistemological and institutional aspects of the physician's formation, see Agrimi and Crisciani 1988; Ottosson 1982; and more recent contributions on this theme, see O'Boyle et al. 2000; and Crisciani 2010.

a professional figure. Indeed, the *consilium* is primarily a professional act, even though its emergence coincided with the institutionalization of medical learning in the universities and, even though, it remained linked in several ways to the various phases of evolution in university teaching.[14]

The *consilium* of physicians (which differed somewhat from the juridical *consilium*[15]) was a prescriptive evaluation of an ongoing pathological condition of a specific individual patient. When this opinion is in written form and involves more than one physician,[16] we have a *consilium* as a genre or literary form. Thus, a *consilium*, properly understood, is a text written by a physician (and usually addressed to another less famous physician) following the rules of a prescriptive code of writing,[17] at the request of another physician (less famous), relating to an ongoing disease. It is the role of the more famous physician to diagnose the disease and prescribe the proper treatment.

The written *consilium* usually has three sections. The first is an examination of the reported pathological situation (*casus*); the second and the third prescribe what should be done. The first section gives a detailed description of the patient's symptoms and the proper definition of his disease, along with its causes. The second part provides suggestions as to the *regimen* appropriate for that patient and focuses not only on the diet but also on the patient's lifestyle as a whole,[18] based on the proper use of the six non-natural (*res non naturales*).[19] In the third section, we can find pharmacological remedies (together with recipes and doses), plus any other more invasive treatment (plebotomy, enema, inhalations etc.), with details about the proper order and frequency for applying the latter.

[14]Further indications of this relationship, and, more generally, about the role of *consilium* in various medieval contexts can be found in Casagrande et al. 2004.

[15]Among his many studies on this theme, Ascheri 2004; Chiantini 1996; and Belloni 1995.

[16]We must point out that this characterization better describes the *consilia* written between the end of the fourteenth century and the fifteenth century, and does not suit equally well the previous and early *consilia*: for example, Alderotti's *Consilia* do not have hints of letters between physicians, while on the contrary, they present a number of clues which suggest actual visits and consultations at the patient's bedside.

[17]See Agrimi and Crisciani 1994, esp. 27–38; and Lokwood 1951.

[18]On the *regimen*, Sotres 1993; eadem 1996, "introduction" to Arnaldo da Villanova, *Regimen sanitatis ad regem Aragonum*; and Nicoud 2007.

[19]See Rather 1968; Garcia Ballester 1993. The six *res non naturales* (climate, food and drink, rest, exercise, evacuation, passion) are conditions which are not properly parts of the organism, even though the organism cannot exist without them; they differ from *res naturales* (the natural components of the organism: elements, members, etc.) and from *res contra naturam* (diseases and their causes). *Res non naturales* constitute the only domain where patients can really choose how to behave and physicians can give good advice, which oftens overlaps with ethical suggestions. The *Consilium* is intended exactly to orient the choice about what has to be done. On this aspect, see Aquinas 1964, 476, 134: "Omne autem consilium est questio, idest inquisitio quaedam, etsi non ommnis quaestio, idest inquisitio, sit consilium. . .Sola enim inquisitio de operbilibus est consilium."

Certainly, not all *consilia* follow precisely this model. However it does allow us to see similarities, differences and points of contact between the *consilium* and other forms of writing found in practical medicine. For example, it allows us to consider the *consilium* in connection with the *tractatus*, or specialized monograph on an illness, the personalized *regimen sanitatis*, the *exemplum*-report, and even the *experimentum*-recipe. From an epistemological point of view, it may be noted that the *consilium, exemplum*, and *experimentum* are similar because they are connected to the concrete operation and concern for the individual. Indeed, the frequent links found between *consilium* and *tractatus* (especially in the fifteenth century) show the author's need to relate the individual's symptoms to more general theoretical approaches capable of explaining them.

The *consilium* thus shows a set of relations, which can also be seen in other anthropological and sociological fields. Indeed, the *consilium* shows us, as if it were a summary, the range of figures, roles, and relationships involved in a therapeutic process. These would obviously include the patient and his personal physician, but also the more famous physician consulted by the latter. We can also find the disciples of the *magister* and, sometimes, mention is made of the surgeon required for a certain operation. Other figures include the *nuntii*, who transmitted information, data, *cedule* and letters from one physician to another. Finally, in the patient's house, we can find assistants and nurses who were *adstantes* at the bedside; and perhaps the patient's relatives and friends. When dealing with all these figures the physician had to weigh his words and suggestions carefully, paying due attention to social and family relationships through a consolidated etiquette and a persuasive rhetoric. These were required not only in the consultations *in presentia*,[20] but also when writing *consilia in absentia*.

3.3 The Evolution of the Italian *Consilia*

The *Consilia* genre was particularly well developed in Italy. From the time of Taddeo Alderotti[21] onward, this genre obviously changed, both in terms of the development of the medical doctrines it espoused, and in terms of the structure and the style of the writing itself. We have no space here to consider the nature of these changes in depth. Suffice it to say that, by the fifteenth century, the structure of the *consilium* was perfectly defined according to its various parts: first, *casus in terminis*, i.e., the diagnosis formulated according to the famous logical rules proposed by Antonio Cermisone, Bartolomeo's collegue; second, the dietetic

[20] As is for instance remarked by Henri de Mondeville, the typically Italian practice of the written *consilium in absentia* is strongly refuted in Paris and in France. On the contrary, Mondeville himself, in his *Cyrurgia*, 1982, gives plenty of useful instructions and rules of etiquette about the correct behaviour of physicians and surgeons during direct consultations. In addition, see Crisciani 2004a, "Ethique de la consultation."

[21] Alderotti 1937.

section; and, third, the pharmacological one (the treatment). Indeed, this was the structure of the most famous *Consilia* of this period written by Matteo Ferrari da Grado (Agrate),[22] Baverio Baviera,[23] and Montagnana himself. We will next outline some specific and general aspects of these *consilia*, before providing a closer analysis of Montagnana's *Consilia*.

Several stabilized features of these *consilia* are worth noting. First of all, references to the *auctores*, especially in the section of the *casus*, became essential. This is hardly surprising, in view of the importance that reference to specific points in the texts of famous authors had been given in the official model for formulation of the *casus* proposed by Antonio Cermisone. By then, the texts and the authors had actually become a requirement for diagnosis and a guarantee of its accuracy and validity.[24] Moreover, often the *consilium* was greatly expanded, with a well organised structure, divided into *capituli, sectiones* and *particulae*. This order was required precisely because of the growing length of the *consilium*, since, by this point, the *consilium* was as extensive as a *tractatus* and often had the same aims. The number of internal references to past chapters or sections also grew exponentially, and would have made the *consilium* hard to understand had it not followed the pre-established order. Additionally, the interest in nomenclature and in the use of synonyms was particularly marked: here this attitude in practice introduced a test to show range and breath of the writer's erudite knowledge, a proof of his sophisticated magisterial doctrine.

Unlike earlier *consilia*,[25] these were not always aimed at obtaining a more effective and more cartain practical result. The argumentation-cum-scholastic structure in the discussion of the diagnosis was often quite complex, and, in any case, always strongly evident. The number of internal *questiones* grew apace. A brief review of the lexis used in this section is sufficient to illustrate that these later *consilia* kept strictly to a scholastic intellectual style. We can observe a recurrent use of terms and expressions such as discussions *solemnes, theorice contemplans, intelligens speculans, speculatores, rationabiliter speculantes, contemplatio theorica, declarare, concludere, concedere, confirmare, deducere, inducere, inferire, demonstrare, convenire, probare per rationes, conclusiones, dubia, problemata, premisse, corellaria, puncta, verificatio, consequentia* and so on.[26]

In these *consilia*, indications that the patient had actually been physically examined by the writer are very rarely to be found. Instead, we see a far greater

[22]Ferrari de Gradi 1501.

[23]Baviera 1489.

[24]See, for instance, Baviera 1489, f.1rab; and Ferrari 1501, ff. i1rb, 15vb, n6va.

[25]Gentile da Foligno 1498, f.66rb–va, for example, expressed his concern about possibly misinterpreted strange medical words (perhaps wrongly translated) in the following way: "Licet hic error sit solum in vocabulo non tamen est parvum, quia ex ipso forsitan resultavit maximus error in opere."

[26]It is worth mentioning some titles of Montagnana's *Consilia*: *Disceptatio theoricalis de disposicione capitis et nasi*; *consilium theoricum de fetore in ore*. Moreover, he often divided his *consilia* in *pars theorica* and *pars practica*.

number of signs that the patient is 'distant': in fact, most of these *consilia* had been formulated *in absentia*. This 'distance' implied a specific need for theoretical and doctrinal thoroughness, including the need to list, compare and discuss all the possible causes of a disease. This approach is certainly indicative of an intellectual style, but it was also rendered necessary by the fact that, generally speaking, the more famous physician who was consulted did not have a flesh and blood patient to examine, but only a *cedula*, a letter describing the patient. Therefore, even if the *cedula* was detailed and complete, the consulted physician had to give all the possible hypotheses which might account for, or be connected with, the symptoms described by his colleague. This may be one reason why so many 'probabilistic' expressions and arguments are found[27]: the *iudicium* provided is often called a *coniectura*; some conclusions, which are declared *verisimiles*, were obtained only *probabiliter*.

Summing up, the *consilium*, although it preserved the aspects of a professional act, began increasingly to take on the features of a doctrinal and teaching text, as indicated by the characteristics listed above. Although the *consilium* is always rooted in the physician's profession and in his fame as a professional, it sometimes seems to approach the goals of a teacher rather than those of a practicing medical doctor: indeed, the practical treatment, the *actum medicationis*, was delegated to others, to the physicians present at the patient's bedside (*physici adstantes*).

3.4 The *Consilia* by Montagnana

Many of the characteristics described in the previous section are evident in Montagnana's collection of *Consilia*. His masterly knowledge of authorial texts was unequalled in excellence, breath of vision and variety, just as Bolderio had claimed. Hippocrates, Galen and Avicenna are the most widely quoted authorities, along with the main authors of recipes and practical texts. Montagnana acknowledges the advantages of working in a well equipped *Studium* where he could make use of "variis auctorum voluminibus in studio Padue" (130vb). In fact, reading his *Consilia* often gives the impression he was writing with many texts in front of him, moving from one volume to another, comparing his sources (204ra). To one of his correspondents, Montagnana recommends this research *in varia volumina* as

[27]It may be remembered that, in any case and from an epistemological Aristotelian point of view, medicine is part of naturalistic science, which, as is well known, deals with phenomena and rules related to what happens most of the time. However, it is also legitimate to connect this 'probabilistic' trend in later *Consilia* with the probabilistic currents of natural philosophy in the fourteenth and fifteenth centuries, and with the contemporary debates concerning the 'method' in medicine and the values of certitude and truth here obtainable. On this, see McVaugh 1990; and Siraisi, "Views on the Certitude of Medical Science Among Late Medieval Writers", a paper presented at the Conference "Science and technology in the Middle Ages" held at Barnard College, New York, 1983.

necessary because he felt that many 'moderns' neglected to make the effort. Indeed, it is to their shame that they fail to consult the texts not because they already know the right solution but because they find it boring to do so or are totally unable to perform this indispensable task (203ra).

Montagnana recognized the various difficulties inherent in this exercise worth specificing. The text (in this case one by Avicenna) could be corrupted due to "defectus exemplaris vel scriptoris" (93b). Moreover, one may also find difficulty in literal comprehension. Montagnana himself admits that (93va) as regards one passage in Avicenna "usque ad hos dies non potui plenam habere expositionem licet multum vigilaverim ad id" and that "etiam intelligo litteram illam sed non nisi cum magna difficultate et temporis prolixitate" (11v). Sometimes these problems could not be resolved by comparing the texts with other books, but only by means of daily experience (203rb). Indeed, in one *consilium* (11vb), Montagnana mentions the case of a deaf friar of Padua, whom he had cured, a case which evidently and practically contradicts a statement in Avicenna's *Canon*.

As with many others before him, Montagnana was not unaware of the contradictions between medical authors and between physicians and philosophers. For example (31rab), as regards the function of sperm, "est discordia non parva inter philosophos et medicos illustres": and in two lengthy columns of the *consilium* he lists the various positions *tantorum virorum*, finally admitting that "conatus tamen fuit Avicenna talem discoliam conciliare caute in his sermonibus." However he acknowledges shortly after that, in the end, considering the choice of practical treatment, these different opinions have little effect on the type of remedy to be used. The list of varying and divergent opinions on a topical theme of disagreement between physicians and philosophers – e.g., what is the governing organ in the body? (21vab) – is equally long and is closed, yet again, by recourse to Avicenna's position. The point is that Montagnana was not afraid to stress disagreement among physicians in many passages in his *Consilia*.[28]

In numerous cases, this appears to be little more than an ornamental device designed to show Montagnana's extensive knowledge of the authors as well as his competence and erudition. In particular, he exhibits erudition when comparing 'ancients' and 'moderns', where the latter nearly always fare worse.[29] Sometimes, however, Montagnana attempts to provide a plausible solution, as Pietro d'Abano and Arnau of Villanova had done.[30] Yet it could happen that three different opinions that Montagnana himself expresses regarding problematic aspects of diagnosis (this is, therefore, an 'internal' and 'personal' discrepancy) were all supported by Avicenna's text (21b). In his case even Avicenna's text partially loses its function as a reliable and discriminating source. In other cases of contrast and disagreement, it is indispensable to have recourse to logical argumentations:

[28]For example, Montagnana 1525, ff. 11vab, 15va, 17rb, 21rb, 21vab, 24vb, 25rab, 28va.

[29]Ibid., ff. 8va, 27va, 28rab, 28vb, 81vb, 84va 196rab, 319ra.

[30]See d'Abano 1565b (ripr. 1985); Arnau of Villanova 2000.

some *questiones*[31] contained within various *consilia* can be used for this purpose. In fact, we must remember that, in general, following Cermisone's rules, the whole structure of the fifteenth-century *consilium*, especially as regards the structuring of *casus*, was essentially based on logical arguments, as the above-mentioned terminology clearly indicated.[32] Finally, even if rarely, it can be seen that the authors' considerations agree harmoniously.[33]

Despite Montagnana's great ability in dealing with authoritative texts and authors, his writing is not dominated by his erudition. Indeed, he distinguishes himself amongst the various texts with great skill, demonstrating masterfully his knowledge of the contents, including their ambiguities and limits, which he manipulates with textual precision but also with a critical eye. On the one hand, he sings the praises of classical writers, to whom we must be grateful for their labors, which are of great benefit to us: indeed, "est enim honestum laudes conferre illustribus qui pro nobis insudaverunt assidue contemplando, in quorum inventis oportet adhibere grates" (12rb). On the other hand, Bartolomeo acknowledges that it is very stimulating and gratifying for the true scholar, such as the physician, who knows how to speculate, to encounter new, difficult and unexplored cases.[34] Indeed, "esset speculatio pulchra in inveniendo", i.e. when the case appears not to have been dealt with by any author (130ra), or, at least, when "non valde frequenter hec dispositio videtur ab auctoribus capitulata."

3.5 Epistemological Features

Montagnana's intimate familiarity with the texts of others is repeatedly demonstrated, as is his thoroughness and the variety of approaches available to him. These features distinguish his *consilia* from those of his contemporaries. The same may be said as regards his attention to the etymological and cultural analysis of terms and synonyms and the names of ingredients expressed in vulgar languages, especially in the pharmacological section of the *consilium*.[35] Here too his display of erudition (he draws on mythology, geography and the poets for his information) does not have the straightfoward aim of greater understanding, but seeks to exhibit his vast knowledge of culture. This would, in fact, contribute to enhancing his reputation whilst exciting the interest and gaining the respect of students and the rich, whose high rank and sophisticated tastes encouraged them to request his *consilia*.

---

[31]For example, Montagnana 1525, ff. 20va, 21vab–ra, 195vb.

[32]See here, 10–11.

[33]For example, Montagnana 1525, ff. 12ra, 15vb, 25ra, 30rb and passim.

[34]Ibid., f.130a: "Oblata est mihi dispositio que multum diligenti consideratione indiget quia rarus satis et pergrinus est eventus eius."

[35]Ibid., for example, ff. 4ra, 6va, 8va, 9ra, 17rb, 27rb–27va, 29va, 33rb,87va, 91rb, 93vb, 94vb, 97ra, 195rb.

This is not to suggest that efforts to enhance a physician's reputation were entirely unique to Montagnana or absent from the main corpus of fifteenth century *consilia* even if in Montagnana's *consilia* these efforts are particularly evident and frequent. On the contrary, it seems to me that in Montagnana's approach there are two epistemologically important and very typical features of the *consilia* genre. These are: first, the tendency to turn doctrinal digressions within various *consilia* into little *tractatus* on specific topics and, second, the so called 'reduction of causes,'[36] either because they may derive from a single root cause or, more frequently, because Montagnana concludes that they are relatively unimportant in choosing the proper practical remedy. Let us examine these aspects more closely.

The first feature is found both in the more theoretical part of the *consilium* (the *casus*) and in the practical part (the *regimen*). Indeed, when constructing his diagnosis, Bartolomeo often extends the amount of doctrinal information provided about the illness in question much further than necessary.[37] Thus, at least in two *consilia* (18r and 25rab), he presents minutely detailed considerations concerning the theory of vision and its physiology. His reflections, together with the information provided on coitus, sexual orientation and habits, and the psychological and emotive causes of inconveniences and appropriate hygienic measures are equally expansive (14vb). It is not surprising then that because of the wealth of information, *consilium* XV (30vab) reads as a treatise on love sickness and its possible remedies.

Montagnana's exhaustive approach is even more evident in the prescriptive parts of the *consilium*. Almost all six *res non naturales* in the *regimen* are extensively developed but the examination of some of them standout even by comparison. For the *res* 'food', for example, Montagnana proposes minutely detailed diets, giving classifications of the meat of various animals divided up according to the seasons and their ages. He also provides lists of fishes classified on the basis of the different types of water in which they live and the way in which they are caught and preserved. He also describes wide selections of vegetables and pulses permitted for the consumption of each patient according to the time of day and the seasons, with lengthy digressions relating to their nomenclature and their proper place of cultivation: these sometimes almost turn into little botanical-herbalistic treatises. Moreover, his treatment of the item 'food' generally leads not so much and to healthy suggestions on the appropriate diet but rather translates into a set of detailed and complex culinary instructions approximating treatises on haute cuisine. The item 'air' (which includes, by metonymy, instructions on clothing) also reaches an extreme degree of subtlety in Montagnana's *consili*a when he gives advice about the type of headwear appropriate for each of the four seasons and precisely how the headwear should be made (33vb).

There are at least three observations worth making about Montagnana's propensity to take an exhaustive approach, which may give rise to results that appear both

[36]A deeper analysis of this epistemological process can be found in Crisciani 1996, esp. 18–19, 27–28; and also Savino 1995, esp. 228–236.

[37]I do not take into account here the *Consilia* on the plague, and those at the opening of the collection, devoted to impersonal *regimina*: they are obviously more extended.

useless or bizarre. First and foremost, we should consider the kind of patients who benefited from his *consilia*. They were certainly very rich and mostly high ranking individuals, even princes, magnates, and religious prelates. These patients would definitely appreciate such a fully explained diagnosis and attention to detail when it came to their personalized treatment. Often they were not only rich but also cultivated, thus well versed in culinary sophistication and curious to learn about the marvelous facts of nature.

Second, Montagnana's propensity confirms that his *Consilia*, which were written in a professional sphere and designated for rich clients, were also composed in this way – and perhaps mainly – for didactic purposes. They were written for students, future physicians, who could use them to learn more about doctrinal topics, e.g., coitus, love sickness, vision and so on. In addition, these future physicians would have at their disposal a good example of how to adroitly mix erudition and technical knowledge.

Finally, and from a more general point of view, in these *consilia* we are witnessing a progressive liberation of some of the *res non naturales* as part an overall trend of practical medicine between the end of the Middle Ages and the early modern time. Proof of this process can be found in the large number of specific monographic treatises – on food, diet, baths and so on – composed during these years. In addition, there was a progressive increase not only in the number of treatises relating the *res* 'food-drink' (dietetics), but, within this (which becomes a sort of general autonomous container) culinary treatises and even systematic texts devoted to classes of food (pasta, fish, dairy products and fruit) analyzed according to their provenance, their therapeutic uses, ways of cooking and serving them, and, not infrequently, their 'presence' in classical texts. The same goes for other *res*. 'Air', for example, which includes clothing, or love sickness, which is part of the *res* 'passions' were gaining ever more specialized attention. This trend towards liberation and specialization also occurs in the case of baths (it is no coincidence that Bartolomeo wrote a treatise on balneotherapy), which goes along with the autonomous development of another area, the *res* 'exercise', i.e. suitable physical movement and gymnastics.

We are thus dealing with an important and generalized epistemological phenomena, which involved the development of specializations.[38] Specializations became more and more well defined and restricted by the fifteenth century, but also far more complex with the development of theoretical and doctrinal justifications. At the same time, this process of liberation and specialization reveals the increasing attention and interest for a type of description, 'historialis' for the particulars, the details, careful identification and exhaustive description.[39] In other words, it was a process that involved a reassessment of naturalistic disciplines at the beginning of the modern period. Applying this epistemological context to Montagnana, and to his propensity to take an exhaustive approach in the detailed

[38]For an institutional point of view, Lines 2001.

[39]Crisciani 2001; and, more in general, Pomata and Siraisi 2005.

little treatises that he sometimes includes in his *Consilia*, his efforts and his little treatises appear entirely well placed.

As regards the 'reduction of causes', Montagnana's *Consilia* show the following approach. He declares that a certain number of causes are responsible (five or seven or more, defining them as concurrent, connected or parallel), and that these contribute to determining certain effects and help explain certain symptoms. He lists them punctiliously, generally with references to the authorial texts in which they are described. He examines them carefully, corroborating them with suitable logical arguments. However, he often remarks that, in the end, it would be useless to choose carefully between them, as no precise identification is required in order to select the right pharmacological treatment he is going to prescribe.

At first glance, this may look like some sort of empirical operative 'reduction', perhaps corresponding to an interpretative rule inspired by some 'principle of economy', although we should not be too quick to accept this conclusion. Indeed, Montagnana dwells on a theoretical examination of the causes he lists, even though he says doing so is not really necessary when considering the choice of remedies (a point he makes only after the analytical doctrinal discussion). So, this attitude actually constitutes further proof of Montagnana's doctrinal commitment in his *Consilia*. Moreover, his way of proceeding highlights the divergence between the section on diagnosis and the part in which the remedies are prescribed, even if, in both of these parts, Montagnana allows his writing to expand considerably: in the first case with analytical and argumentative procedures and, in the second, abandoning himself to an obsessive description down to the most insignificant detail.

3.6 A Sympathetic Phisician

I have shown elsewhere that 'giving good advice' was not only an expected and typical function of a physician's activities extending well beyond the technical writing of *consilia* as a genre.[40] In fact, it consisted of a wide ranging service, which – especially in the fifteenth century – could include suggestions as to lifestyle, studies to be undertaken, existential choices and ethical behaviour, extending even to advice about political stances, especially in the frequent cases of physicians who were also court advisers.[41]

We can find traces of attention which go beyond specifically medical considerations in Montagnana's *Consilia* too. This attention can be seen in the lengthy

[40]Crisciani 2004b, "'Consilia', responsi, consulti."

[41]See, among others, Crisciani and Zuccolin 2011; Park 1985; Nutton 1990; Pesenti 2003; Crisciani 2003; Naso 2000; Rotzoll 2000; Cuomo 1987; De'Reguardati 1977; Deffenu 1955; and Pedralli 1996. See also various essays in *I saperi nelle corti* 2008 and the very recent and comprehensive book by Nicoud 2014.

prologues to certain *consilia* or else in some parts of the *consilium* itself. Thus, in *consilium* XV (love sickness), Bartolomeo devotes space to the possible trauma which may have conditioned and upset the patient in the past. The *consilium* XVI[42] (defined *solemnis*) for the Count Palantine Ludovico opens with a prologue in which the physician promises to pay his utmost attention to all the disorders the Prince is complaining of, including the terrible sadness (causing serious insomnia) from which he is suffering: "Ad hoc autem non parum conferre possunt impressiones melancolice aut alia animi accidentia tristia a causis primitivis originem habentia"-the original cause, for example, might be the loss of friends and relations or the miscarrying of financial operations. For this, he continues: "que omnia per serenissimum principem moderari debere cum humilitate et patientia consulo." In the case of a sick priest, Montagnana strongly recommends (18rb) that gloomy readings should be avoided: so the priest should not dwell, for example, on the ghastly details of terrible instances of martyrdom. Prescribing for a man of the Church, Montagnana is careful to suggest a diet which takes into account every aspect of his religious duties (11va).

In another case of inutterable sadness (9vb), Montagnana determines the main cause to be the death of the patient's beloved daughter. The result has been an excessive agitation of the spirits: too many tears have been shed, and so the humors of the brain have become irritated and unbalanced. In addition to suggesting some useful drugs, Montagnana takes pains to show how to mitigate the severity of the patient's grief by following wise advice:

> memoriam suam, voluntatem et animam conformet voluntati Altissimi. Iudicia enim Dei abyssus multa. Fortasse enim vir iste tristitiam maiorem valde recepisset ex longa vita filie sue, que ab humanis periculis fere infinitis non erat tuta, et maxime in hoc tempore nostre infelicis etatis subiecte tribulationibus insidiarum, odii, belli et aliorum discriminum. Equo itaque animo tolerandum est quod previsum et factum evenit. Aliter enim periculum sue vite grande valde preparat sibi vir iste ex forti passione anime. (10rb)

Montagnana nevertheless believed that the best comfort and, more importantly, practical relief were undoubtedly the help and treatments which medicine could provide, especially if the practitioner – like Montagnana himself – was gifted in his standard of care. He has qualities of diligence, theoretical attention, and above all, a capacity for speculative depth and a tireless dedication for the study of the subject. We have already remarked on his intellectual appreciation of particularly rare and complex cases, which offered exciting topics for research.[43] These cases were worthy of *ardua contemplatio*. They did not allow the physician's mind (*mens*) to remain idle and become blunted by routine; they obliged him to contemplate the structure of the human organism "mentis acuitate profunda, studii contemplatione sublimi."[44]

[42]Montagnana 1525, ff. 33rb–33va.

[43]Op. cit. n. 35 and Montagnana 1525, ff. 20ra, 26rb.

[44]Ibid., f. 12ra.

Montagnana places medical science at the top of the hierarchy of sciences, surpassed only by natural philosophy (upon which medicine is dependent and with which it sometime overlaps), and by theology (which is a different kind of knowledge upon which medicine does not appear to be dependent).[45] The *medicinalis contemplatio* is rightly "omnium aliarum contemplationum preter divinam et naturalem", 'regina' and 'domina sacratissima et altissima.' Therefore the physician has the greatest "necessitas... contemplari corpus et animam domitam facere in continua speculatione circa ea que necessaria sunt in curatione languorum humani corporis."

In these considerations of the relationship between *speculatio* and *curatio*, Montagnana highlights a conception of medicine which, although shared by his contemporaries, is not usually expressed in their shorter and less discursive *consilia*. These considerations of Montagnana describe medicine as separate from natural philosophy, which was intended as a more theoretical science, even if the two were connected. Moreover, they also presented medicine as dependent on the theoretical principles of natural philosophy and, as it was based on these naturalistic theories, it could legitimately claim to be a *scientia*. But medicine was also, at the same time, a specialized science, devoted to a specific, more determinate intervention on a specific 'subject', namely the human body. In other words, it could enjoy its own legitimate and relative autonomy. For medicine was a *scientia operativa* or an *ars scientifica*, in which the 'general' *speculatio* was designed to produce the 'specific' *opus* and *curatio*. The *Consilia* of masters such as Montagnana, but also Ferrari and Baviero Baviera, in which they define 'general' causes starting out from 'particular' *signa* and 'specific' accidents, in order to suggest 'general' rules and *canones* by means of which a 'particular' course of treatment for each individual patient could be determined, was (perhaps) the literary genre in which the concept of medicine as a *scientia operativa* is best expressed in the fifteenth century.

The pioneering criticism of Francesco Petrarca and of other Humanists in the Quattrocento against this kind of knowledge, this style and these goals of research are well known.[46] Just as well known is the picture of the physician-*magister* lecturing from his chair and commenting on passages from a text, while the assistent-disciple at his feet points out the relevant parts of the corpse.[47] This depiction has often been used as an icon to symbolically describe the abstract and textual nature of late scholastic medical science and to evaluate it in critical or negative terms. Whether or not we agree with this critical judgement, it is however certain that the format of the *Consilia* in the fifteenth century was just as representative, if not more so, regarding the nature of scholastic medicine, especially in Italy. The later *Consilia*, those of Montagnana in particular, present the *medicus*

[45]Ibid., f. 95rb; also Antonio Cermisone, 1498, f. 34vb, praised the theorical highness of medicine in these terms: "Medicinalis scientia et ars summe speculativa omnibus aliis scientiis et artibus totius universi difficilior."

[46]See Berté et al. 2006; Mammola 2012, esp. ch. 1.

[47]About this image, see Carlino 1988, esp. 38–39.

doctoratus who worked with his books and coordinated the eyes, the hands and the operative ability of others at a distance. His tools were medical practitioners (the *medici adstantes* at the bedside) acting as an extension of his senses and as intermediaries between the authors' texts and the actual body of each individual patient.

Bibliography

Agrimi, Jole, and Chiara Crisciani. 1988. *Edocere medicos. La medicina scolastica nei secoli XIII–XV*. Milano/Napoli: Guerini e Associati.

———. 1994. *Les 'Consilia' medicaux*. Trans. Caroline Viola. Turnhout: Brepols.

Alderotti, Taddeo. 1937. *I 'Consilia'*, ed. Giuseppe Michele Nardi. Turin: Minerva medica.

Aquinas, Thomas. 1964. *In decem libros Ethicorum Aristotelis ad Nicomachum expositio*. ed. Raimondo M. Spiazzi, 3rd ed., III, l.VII. Turin: Marietti.

Arnau of Villanova. 1996. *Regimen sanitatis ad regem Aragonum*, ed. Pedro Gil Sotres. Barcelona: Publicacions de l'Universitat de Barcelona.

———. 2000. *De intentione medicorum*, ed. Michael McVaugh. Barcelona: Publicacions de l'Universitat de Barcelona.

Ascheri, Mario. 2004. Il 'consilium' dei giuristi medievali. In *Consilium. Teorie e pratiche del consigliare nella cultura medievale*, ed. Carla Casagrande et al., 243–258. Florence: SISMEL-Edizioni del Galluzzo.

Baader, Gerhard. 1987. Konsilienliteratur und der neue Naturalismus in Italien im Spaetmittelater und Fruehumanismus. In *Die kunst und das Studium der natur vom 14.zum 16 Jahrhundert*, ed. W. Prinz, A. Beyer, and G. Baader, 127–139. Weinhein: Acta Humaniora/VCH.

Bacchelli, Franco. 2010. Montagnana, Bartolomeo. In *Dizionario Biografico degli Italiani*, vol. 75. Rome: Istituto dell'Enciclopedia Italiana.

Baviera, Baverio (De Baveriis, Bonetti). 1489. *Consilia*. Bologna: Platone de' Benedetti pour Marcus Antonius et Ludovicus de Baveriis.

Belloni, Annalisa. 1995. 'Quaestiones' e 'consilia'. Agli inizi della prassi consigliare. In *'Consilia' im spaeten Mittelalter: Zum historischen Aussagewert einer Quellengattung*, ed. Ingrid Baumgaertner, 19–32. Sigmaringen: Thorbecke.

Berté, Monica, et al., eds. 2006. *Petrarca e la medicina*. Messina: Centro interdipartimentale di studi umanistici.

Boisseuil, Didier, and Marilyn Nicoud, eds. 2010. *Séjourner au bain*. Lyon: Presses Univ. de Lyon.

Carlino, Andrea. 1988. The Book, the Body, the Scalpel. Six Engravede Title Pages for Anatomical Treatises of the First Half of the Sixteenth Century. *RES* 16: 32–50.

Casagrande, Carla, Chiara Crisciani, and Silvana Vecchio, eds. 2004. *Consilium. Teorie e pratiche del consigliare nella cultura medievale*. Firenze: SISMEL-Edizioni del Galluzzo.

Chiantini, Monica. 1996. *Il 'consilium sapientis' nel processo del secolo XIII. San Gimignano 1246–1312*. Siena: Il leccio.

Crisciani, Chiara. 1996. L' 'individuale' nella medicina tra Medioevo e Umanesimo: i 'Consilia'. In *Umanesimo e medicina. Il problema dell' 'individuale'*, ed. Roberto Cardini and Mariangela Regoliosi, 1–32. Rome: Bulzoni.

———. 2001. Fatti, teorie, 'narratio' e i malati a corte. Note su empirismo e medicina nel tardomedioevo. In *Fatti: storie dell'evidenza empirica*, ed. Simona Cerutti and Gianna Pomata, special issue of *Quaderni storici* 108(3): 695–717.

———. 2003. Tra università, corti e città: note su alcuni medici 'pavesi' del sec. XV. *Annali di storia delle università italiane* 7: 55–69.

———. 2004a. Ethique de la consultation: à propos de la cohésion morale de la profession médicale. *Médiévales* 46: 23–44.

———. 2004b. 'Consilia,' responsi, consulti: i pareri del medico tra insegnamento e professione. In *Consilium*, ed. Carla Casagrande et al., 259–279. Florence: SISMEL-Edizioni del Galluzzo.

———. 2010. La formazione del medico nel Medioevo: dottrina ed etica. In *Formare alle professioni. Figure della sanità*, ed. Monica Ferrari and Paolo Mazzarello, 36–57. Milano: Franco Angeli.

Crisciani, Chiara, and Gabriella Zuccolin, eds. 2011. *Michele Savonarola. Medicina e cultura di corte*. Florence: SISMEL-Edizioni del Galluzzo.

Crisciani, Chiara, Roberto Lambertini, and Andrea Tabarroni. 2015. *Due manoscritti con questioni mediche. Note e schede (prima metà del secolo XIV)*.

Cuomo, A.M. 1987. *Ambrogio Varese. Un rosatese alla corte di Ludovico il Moro*. Rosate: Amministrazione comunale.

d'Abano, Pietro. 1565. *Conciliator controversiarum quae inter inter philosophos et medicos versantur*. Venice: Giunta.

———. 1565 (ripr. 1985). *Conciliator controversiarum quae inter inter philosophos et medicos versantur*, ed. E. Riondato and L. Olivieri. Padua: Antenore.

De'Reguardati, Fausto M. 1977. *Benedetto de' Reguardati da Norcia: "medicus tota Italia celeberrimus", pagine inedite di storia sforzesca*. Trieste: Lint.

Deffenu, Gonario. 1955. *Benedetto Reguardati, medico e diplomatico di Francesco Sforza*. Milan: Hoepli.

Demaitre, Luke. 1975. Theory and Practice in Medical Education at the University of Montpellier in the Thirteenth and Fourteenth Centuries. *Journal of the History of Medicine* 30: 103–123.

———. 1976. Scholasticism in Compendia of Practical Medicine, 1250–1450. *Manuscripta* 120: 81–95.

———. 2013. *Medieval Medicine. The Art of Healing from Head to Toe*. Santa Barbara: Praeger.

Ferrari de'Gradi, Giovanni Matteo. 1501. *Consilia*. Papia: per Andream de Boscho, impensis d. Bernardin de Scotis Modoetiensis.

Foligno, Gentile da. 1498. *Consilia*. Venetiis.

Garcia Ballester, Luis. 1993. On the Origin of Six Non-Natural Things in Galen. In *Galen und das hellenistische Erbe*, ed. Jutta Kollesch and Diethart Nikel, 105–115. Stuttgart: Franz Steiner Verlag.

———. 1996. Introduction. In Arnaldo da Villanova, *Regimen sanitatis ad regem Aragonum*, (AVOMO, X.1.). Barcelona: Publicacions de l'Universitat de Barcelona.

Garcia Ballester, Luis, Roger French, and Iain Lonie, eds. 1994. *Practical Medicine from Salerno to the Black Death*. Cambridge: Cambridge University Press.

Grmek, Mirko. 1998. *Storia del pensiero medico occidentale*, vol. 1. Rome/Bari : Laterza.

Guérin-Beauvois, Marie, and Jean-Marie Matrin, eds. 2007. *Bains curatifs et bains hygiéniques en Italie de l'antiquité au Moyen Age*. Rome: Ecole française de Rome.

I saperi nelle corti/Knowledge at the Courts. 2008, *Micrologus* 16.

Jacquart, Danielle. 1998. *La médecine médievale dans le cadre parisien*. Paris: Fayard.

Lines, David. 2001. Natural Philosophy in Renaissance Italy: The University of Bologna and the Beginnings of Specialization. *Early Science and Medicine* 6: 267–323.

Lokwood, Dean P. 1951. *Ugo Benzi. Medieval Philosopher and Physician. 1376–1439*. Chicago: University of Chicago Press.

Mammola, Simone. 2012. *La ragione e l'incertezza: Filosofia e medcina nella prima età moderna*. Milan: FrancoAngeli.

McVaugh, Michael. 1990. The Nature and Limits of Medical Certitude at Early-Fourteenth-Century Montpellier. *Osiris* 6: 62–84.

Micca, A. Botto. 1941. *Una famiglia di medici: i Montagnana, in Scritti in onore del prof. P. Capparoni*. Turin: Minerva medica.

Mondeville, Henri de. 1982. *Die Chirurgie des Heinrich von Mondeville*, ed. J.-L. Pagel Hirschwald. Berlin: Verlag.

Montagnana, Bartolomeo da. 1525. *Consilia*. Venice: summo studio et arte impensis nobilis viri domini Lucae Antonii Iunta Florentini excussa.

Ms 2030Venezia, Biblioteca Nazionale Marciana, Marc.lat. VII, 66 (=9684).

Naso, Irma. 2000. *Università e sapere medico nel Quattrocento. Pantaleone da Confienza e le sue opere*. Cuneo/Vercelli: Società per gli studi storici, archeologici ed artistici della provincia di Cuneo.

Nicoud, Marilyn, 2007. *Les régimes de santé au Moyen Age*, 2 vols. Rome: Ecole Française de Rome.

———. 2011. 'Inventio', 'experimentum' e perizia medica nel 'De Balneis' di Michele Savonarola. In *Michele Savonarola. Medicina e cultura di corte*, ed. Chiara Crisciani and Gabriella Zuccolin, 83–112. Florence: SISMEL-Edizioni del Galluzzo.

———. 2014. *Le prince et les médecins. Pensée et pratiques médicales à Milan (1402–1476)*. Roma: Ecole française de Rome.

Nutton, Vivian, ed. 1990. *Medicine at the Court of Europe, 1500–1873*. London/New York: Routledge.

O'Boyle, Cornelius, Roger French, and Fernando Salmón, eds. 2000. El aprendizaje de la medicina en el mundo medieval: las fronteras de la eseñanza universitaria. *Dynamis* 20: 17–393.

Ottosson, Per-Gunnar. 1982. *Scholastic Medicine and Philosophy. A study of Commentaries on Galen's 'Tegni' (ca. 1300–1450)*. Naples: Bibliopolis.

Park, Katherine. 1985. *Doctors and Medicine in Early Renaissance Florence*. Princeton: Princeton University Press.

———. 1999. Natural Particulars: Medical Epistemology, Practice, and the Literature of Healing Springs. In *Natural Particulars: Nature and the Disciplines in Renaissance Europe*, ed. Anthony Grafton and Nancy Siraisi, 347–367. Cambridge: MIT Press.

Pedralli, Monica. 1996. Il medico ducale milanese Antonio Bernareggi e i suoi libri. *Aevum* 70 (2): 307–349.

Peloffy, Audrey. 2010–2011. *Les 'Consilia' de Bartolomeo da Montagnana*, Mémoir de Master 1 (sous la direction de Marilyn Nicoud). Université d'Avignon.

Pesenti, Tiziana. 1983. Generi e pubblico della letteratura medica padovana nel Tre e Quattrocento. In *Università e società nei secoli, Atti del Nono Convegno Internazionale del Centro Italiano di Studi di Storia e d'Arte, Pistoia*, 523–545. Bologna: Centro Italiano di Studi di Storia e d'Arte.

———. 1994. *Professori e promotori di medicina nello studio di Padova dal 1405 al 1509. Repertorio bio-bibliografico*. Trieste: Lint.

———. 2003. *Marsilio Santasofia tra corti e università. La carriera di un 'monarcha medicinae' del Trecento*. Treviso: Antilla.

Pomata, Gianna, and Nancy Siraisi, eds. 2005. *'Historia': Empiricism and Erudition in Early Modern Europe*. Cambridge: MIT Press.

Rather, L.J. 1968. The 'Six Things Non-Natural': A Note on the Origins and Fate of a Doctrine and a Phrase. *Clio Medica* 3: 337–347.

Rotzoll, M. 2000. *Pierleone da Spoleto. Vita e opere di un medico del Rinascimento*. Firenze: Olschki.

Savino, E. 1995. Umanesimo scientifico o Medioevo umanistico? *Schede umanistiche: Rivista semestrale dell'Archivio umanistico rinascimentale bolognese*, 228–236.

Savonarola, Michele. 1902. *Libellus de magnificis ornamentis regie civitatis Paduae*. In *Rerum Italicarum Scriptores*, ed. Arnaldo Segarizzi, vol. 24, part 15. Città di Castello: S. Lapi.

Siraisi, Nancy. 1973. *Arts and Sciences at Padua. The 'Studium' before 1350*. Toronto: Pontifical Institute of Mediaeval Studies.

———. 1983. *Views on the Certitude of Medical Science Among Late medieval Writers*. Paper presented at the Conference 'Science and technology in the Middle Ages' held in the Bernard College, New York, unpublished.

———. 1990. *Medieval and Early Renaissance Medicine. An Introduction to Knowledge and Practice*. Chicago: University of Chicago Press.

Sotres, Pedro Gil. 1993. Le regole della salute. In *Storia del pensiero medico occidentale*, ed. Mirko Grmek, vol. 1, 399–438. Rome/Bari: Laterza.

Weijers, Olga. 2009. *'Queritur utrum': Recherches sur la 'disputatio' dans les universités médié vales*. Turnhout: Brepols.

Chapter 4
John Caius, Historian

Vivian Nutton

In her recent books, Nancy Siraisi has investigated the wider cultural world of Renaissance physicians.[1] She has introduced us to a great range of historians and letter-writers in Italy and Germany, and has shown the extent to which they flourished in a society where their professional medical expertise at times took second place to their other activities and interests. They gained leading positions at court or in the life of a city state because of their ability as doctors, but this was supported, and sometimes surpassed, by their social and intellectual skills. Participation in a circle of antiquarians, exchanging letters across Europe, or composing a history of a family or a state brought access to potentially wealthy patients, for the sixteenth century was a time when culture mattered, both politically and socially. A display of literary skills marked off the gentleman physician from humbler healers; an ability to conjure up a neat Latin phrase or to allude to a past historical event showed that the physician was a truly learned man and worthy to be considered a member of polite society. That both enhanced his status and gained him access to fame and fortune.

One celebrated doctor of this period whose historical interests have never been fully investigated is the Englishman, John Caius (1510–1573). He is known to the general public today, if at all, only as the (second) founder of the Cambridge College that bears his name, Gonville and Caius College, although members of the Royal College of Physicians of London will be familiar with the display of gifts that he made to it during the many years of his vigorous presidency.[2] More recently,

---

[1]Siraisi 2007, 2013.

[2]The most accessible accounts of his life are Venn 1912; Clark 1964, 107–24; Brooke 1985, 55–78; Nutton 2004: [http://www.oxforddnb.com/view/article/4351, accessed 14 April 2014].

V. Nutton (✉)
Department of the History of Medicine, I. M. Sechenov First Moscow State Medical University, Moscow, Russia
e-mail: v.nutton@ucl.ac.uk

G. Manning, C. Klestinec (eds.), *Professors, Physicians and Practices in the History of Medicine*, Archimedes 50, DOI 10.1007/978-3-319-56514-9_4

scholars have drawn attention to his role in propagating the works and ideas of the ancient Greek physician, Galen of Pergamum (129-ca. 216) as an editor, translator and collector. In addition to forming his own important collection of Greek medical manuscripts, mostly of Galen, he filled the margins of his editions with annotations and corrections, some the result of his own talent for emendation, but most readings copied in Italy and England from books and manuscripts he had seen. Several of his sources have been identified, but many appear now to be lost, leaving his notes as the only record of them. In particular, his preservation of emendations associated with Thomas Linacre, John Clement and other members of the London College of Physicians allows us to trace the introduction of Italian medical ideas into English medicine.[3]

By contrast his historical writings are almost entirely forgotten today, although they occupy some 410 pages in E. S. Roberts' edition of his collected works.[4] Two of them are devoted to the University of Cambridge, the first and longer being a defence of its superior antiquity against Oxford aspersions, *De antiquitate Cantaebrigiensis Academiae*, and the second the more sober *Historia Cantaebrigiensis Academiae ab urbe condita.* Roberts also printed Caius' *Annalium Collegii medicorum Londini Liber*, which he had begun in 1555 and which runs until 1572. For all its gaps, it still remains a valuable resource for the history of the College from its foundation in 1518 and, in particular, during the tumultuous years of his presidency in the reigns of Mary and Elizabeth. Caius also began, and insisted that his successors in office maintain, the *Annales* of his Cambridge College, a long and detailed account from its foundation.[5] Like his juvenile, and never completed, history of his native Norwich, all are concerned with institutions with which Caius was long and intimately connected.[6] They are both records of the past and, whether explicitly or implicitly, a defence of Caius' involvement in the affairs of these institutions. By comparison with his medical writings, which made little use of medieval authors, and which privileged classical Greek sources, his histories, when they do not describe events in his own lifetime, depended on his investigation of medieval sources, many still in manuscript and most only recently introduced to the world of learning. Putting Caius' work into context is thus difficult, not least because, as this essay will show, there are gaps in the evidence and, perhaps more importantly, because the great development of both British antiquarian studies and a critical approach to sources occurred in the decades immediately following his death. As a result his naïve interpretations are exposed to easy ridicule, while his innovations are subsumed into the larger and more accurate productions of later authors.

---

[3]Nutton 1987. One of Caius' major sources, newly rediscovered, is discussed by Gundert 2006.

[4]Roberts, ed., *The Works of John Caius, M.D.* References are to the pagination of this edition. Not forgotten, though, by Siraisi 2007, 131, 197. Caius also includes historical sections in his other writings.

[5]Venn 1904.

[6]Caius, *De libris suis*, *Works*, 72; *A boke againste the Swete*, *Works*, 8. No trace remains of his draft of this early work.

The earliest of Caius' surviving historical writings is also the longest. The *De antiquitate* was hastily written in 1566 or 1567 in reply to an *Assertio* by his Oxford namesake, Thomas Caius, claiming that Oxford was by far the older university. Caius had been shown the book, and been encouraged in his response, by Archbishop Parker, to whom Caius sent a near-complete draft in April 1567. The volume first appeared in London in 1568 under a pseudonym, *Londinensis*, and was revised and reprinted by Parker as Caius' executor in 1574.[7] Parker's kindness was misplaced. Few treatises have attracted so much scorn. James VI, presented with *De antiquitate* on a visit to Cambridge, immediately declared that he preferred Caius' *De canibus, a* much shorter and livelier work. The historian F. W. Maitland, who had at least opened the volume, thought that both antagonists were simply indulging in a lying match. Even the generous Christopher Brooke, in his history of Caius college, was moved to declare it and the subsequent *Historia* "the largest and most calamitous of its author's works: a farrago of invention and credulity."[8]

From the perspective of a modern historian of the university such a condemnation of the *De antiquitate* is entirely justified. The work is long, rambling and incoherent. Caius oscillates between a point-by-point demolition of his opponent's case and an ordered exposition of his own. He combines stories of the remote past with snippets of good information taken from his own experience or from the university archives, all placed hugger-mugger together.[9] Following the train of his thought is far from easy as he moves from a quotation of the Paduan professor Nicoletto Vernia on the superior nobility of the medical faculty compared with that of lawyers to a discussion of Erasmus' stay in Cambridge.[10] Quotations from poetry are mingled with passages from the Anglo-Saxon Chronicle, Alfred's laws, and Aelfric's Anglo-Saxon grammar, which would have appealed to Parker, a celebrated collector of Anglo-Saxon manuscripts.[11] Stories of the age of the giants are retailed as well as reminiscences of Caius' own time in Padua and Pisa.[12]

The faults of organisation are only magnified by Caius' erudition and the tradition of humanist historiography in which he worked. Straightforward exposition is interrupted by rhetorical questions or learned digressions. A discussion of a letter of King Alfred on the poor state of learning in Wessex is interrupted by a long discussion of the monetary value of the *mancus*a which quotes a variety of authors from Alfred's own preface and Aelfric's grammar (both in Anglo-Saxon) down to Georg Agricola and Leland in his own day.[13] Its relevance is doubtful, as even

---

[7][John Caius] 1568; repr., 1574. Roberts reprints the second edition; Venn 1904, 52, gives the evidence for the date.

[8]Brooke 1985, 75, gives the references to earlier verdicts.

[9]Caius, *De antiquitate*, 24–25 on the University Black Book; 57, archives of Corpus Christi College; 58, archives in the Tower of London; 41, a MS in the possession of Lord Derby.

[10]Ibid., 19–21.

[11]Ibid., 70, 135, 167–8, 195.

[12]Ibid., 14, 79, 147.

[13]Ibid., 167–8.

Caius appears to realise, but it emphasises his authority as a scholar dealing with ancient matters.

That authority is strengthened by Caius' ability to call on many writers, from Antiquity onwards, to support his case. The list of writers he claims to have consulted in manuscript covers three pages of double columns, divided into British ("nostri") and non-British ("externi"), and his list of printed books a further two.[14] They include poets as well as historians, classical authorities as well as moderns, some consulted at length, others mentioned only briefly and possibly only at second hand. Together they proclaims a breadth of learning that few among his contemporaries could match. It invites comparison with the similar lists in the *De arte gymnastica* of the Paduan professor Girolamo Mercuriale, whose first edition appeared in Venice in 1569.[15] Both authors display a coruscating range of erudition, but Mercuriale's work has a clearer focus, a better organisation and a minimal reliance on medieval sources. Mercuriale draws on his classical sources to correct a long-standing error, the medieval physician's disdain for physical exercise as part of medicine, but such a strategy was rarely available to Caius, for whom the medieval chroniclers provided the most striking, and often the only, evidence for what he believed had taken place in Antiquity.

Caius' list of printed works cited raises a variety of problems. His modern continental authorities are a very mixed bunch. Historians familiar throughout Europe like Johannes Nauclerus, Sebastian Munster, Paolo Giovio, and Johannes Carion sit alongside others much less famous today such as Albertus Crantzius, Christianus Massaeus, Lodovico Guicciardini. and Achilles Pirmin Gasser.[16] His most recent citation comes from Jacob Middendorp's *De celebrioribus academ*iis, first published in Cologne in 1567 and, in a revised form, in 1572, in order to support his theory of the origin of Cambridge University.[17] The list of printed books includes such modern British writers as John Rastell, Richard Croke, Thomas Lanquet (in the augmented version in Cooper's chronicle), George Lily, John Leland and John Bale, who provided him with much information about scholars at both universities as well as with summaries of the writings of others he could not

[14]Ibid., 223–7.

[15]Mercuriale 1569; the Florence, L. S. Olschki, 2008 edition and translation of this work contains a major survey of Mercuriale's sources by Jean-Marie Agasse, 863–1118. See also Arcangeli and Nutton 2008; and Siraisi 2007, 42–55.

[16]Joannes Nauclerus (cited Caius, *De antiquitate*, 73) 1516; Sebastian Munster (cited, *De antiquitate*, 138), *Cosmographia*, various editions; Paolo Giovio and Carion, noted *De antiquitate*, 227; Albertus Crantzius (cited *De antiquitate*, 44, 95), *Chronica regnorum aquilonarium*, various editions; Christianus Massaeus (cited *De antiquitate*, 95, 145) *Chronicorum libri*, Antwerp, I. Crinitus, 1540; Lodovico Giucciardini (cited, *De antiquitate*, 95), probably *Description de tout le Pays Bas*, Antwerp, G. Sylvius, 1567; Achilles Pirmin Gasser (cited, *De antiquitate*, p. 129), *Historiarum et chronicorum …epitome*, various editions.

[17]Middendorpius 1567, or the revised edition, Cologne, G. Cholin, 1572 (cited Caius, *De antiquitate*, 16).

consult directly.[18] Leland and Bale in particular are often cited in the *De antiquitate*, particularly for earlier works no longer accessible.[19] Most striking of all is his use of a great variety of medieval chronicles, some of which he may have known at second hand from Polydore Vergil, John Rastell or John Stow's *Summary of English Chronicles*, published in 1565, but many of which he quotes directly and at length.[20]

Where he obtained all this information is not entirely clear, for, as Philip Grierson noted, few of these chronciles had yet appeared in printed form.[21] Certainly Caius had had access for some considerable time to the riches of his friend Archbishop Parker's famous library, his main source, as well as to a few manuscripts in the library of his old college and in those of others and in the university archives.[22] M. R. James suggested that he also used several of the manuscripts that later were bought by Sir Robert Cotton, and were then in private hands in or near London.[23] But the list of authors appears too long, and some of the citations too detailed, to be simply the result of a relatively swift exploration of Parker's collection for the purposes of furthering polemic, and even if his use of others' material largely explains the absence of medieval chroniclers from the 1573 probate Inventory of Caius' extensive library, a further problem remains. With the exception of Polydore Vergil, and possibly John Rastell, none of the modern authors listed by Caius figures in it either, not even Croke, Leland and Bale whom he cites often and with specific references.[24] Nor does it record any of the continental scholars whom he cites, sometimes precisely.[25] It seems unlikely that he did not

[18]Caius, *De antiquitate*, 227. For Rastell, see below, note 24; for Croke, Caius, *De antiquitate*, 106, citing Caius 1520. The revision of Thomas Lanquet's *Chronicle*, London, s.n., 1559, 1560 or 1565, is cited in *De antiquitate*, 106, 131. Caius seems to have cited Lily's *Descriptio Britanniae* (various printings), with its summary of Paolo Giovio, rather than his *History* because of its list of British scholars.

[19]Probably both from *The laboryouse Journey*, 1549, but he could also have known Bale's *Scriptorum illustrium*, various editions.

[20]For Stow, Caius, *De antiquitate*, 13, 38–9, citing the edition of 1566, 106, 203; cf. Hiatt 2004a, 45–56.

[21]Grierson 1978, 523. For the availability of chronicles in printed form, whether entire or summarised, see Gransden 1982, 476–9; Woolf 2000, 11–26; Hiatt 2004a, 45–55.

[22]Caius owned a parchment copy of Parker's edition of the St Albans chronicles, *Flores historiarum*, 1570, probably a presentation copy, Grierson, 518.

[23]See James' identifications in Caius, *Works*, viii–x. For Parker and his circle, and for London connections, McKisack 1971, 26–49, 69–74. Caius supplied Richard Grafton with information for his chronicle of 1569, *Correspondence of Mathew Parker, D.D.*, Cambridge, Cambridge University Press, *1853*, letter 266, 295, dated 1566, as well as borrowing from him a chronicle by Guido di Colonna.

[24]Ibid., 517, although this could be Rastell's *Expositiones terminorum legum Anglorum*, 1525 or later, rather than his *The Pastyme of the People*, 1530. Neither work is mentioned in the Caius College library catalogue of 1632.

[25]While one might imagine that his hunt for appropriate sources went back some time, even before 1558, his ability to summon up specific references must indicate that he took some quite detailed notes on what he had seen. He may have had a remarkable, if at times fallible, memory, but the precision of some of his citation is more likely to indicate notes, if not his own copy.

own his copy of at least some of these authors, for he was able to make speedy use of them, references and all, in his reply to Thomas Caius. Their absence from the inventory of his library may simply be because he had already disposed of them at the time of his death or because they were not in London when his other books were being listed for probate purposes.[26]

But this ostensible abundance of learning is not matched by any great critical faculty. His polemical zeal makes him a doughty assailant of the views of others, while at the same time obstructing serious investigation into the sources used to support his case. One example illustrates his strengths and weaknesses. Thomas Caius had poured scorn on the assertion by the Cambridge scholar John Herryson (d. 1473) in his chronicle that the Greek philosopher Anaxagoras was buried in Cambridge.[27] John Caius, who knew Herryson's work from a manuscript in the College Library, MS 249/277, defends his predecessor with a blizzard of learned authorities. Having earlier established to his own satisfaction that there were giants in the land when Cantaber and King Gurguntius ruled, he repeats an earlier chronology that placed the reign of Gurguntius, his purported founder of the university, in 487 B.C. (or on a more recent calculation 479 B.C). This dating, he argues, is entirely consistent with the dating of Anaxagoras and, indeed, Anaximander around 450 B.C. as given by the ancient writer Eusebius and confirmed by an array of modern chronographers. These may have quarrelled among themselves over a variety of biblical dates, but they are agreed on when the Greek philosophers lived. There may be a few discrepancies of 28 or 29 years, but that can easily be covered by the overlap of generations. Caius draws the logical conclusion: on these premisses, since the university is already in existence, there is nothing to say that Anaxagoras and Anaximander could not have been at Cambridge, although he refuses to commit himself as to whether Anaxagoras was actually buried there.[28]

This discussion of Anaxagoras in Cambridge is typical of the whole work. Caius is aware of differences of opinion between his sources, and he several times emphasises that accounts closest to the events they describe are likely to be far more accurate and trustworthy than the compilations of later authors who simply relied on them.[29] His experience with the manuscripts of Galen had also warned

[26]Grierson 1978 lists 142 items, not all of which are recorded in the inventory. The collection is less strictly medical than would have been the case a decade or so later, a function of the boom in medical publishing, which makes the absence of historical works, save for Livy, Polybius, Caesar, Marliani's *Antiquae Romae Topographia* (Rome, 1534 or 1544), and possibly Rastell, more surprising.

[27]The Middle Ages had seen a variety of resting places assigned to distinguished scholars of the past – both Aristotle and Galen were credited with tombs in Sicily. This story goes back to Cantelupe's *Historiola*, a copy of which was made by Herryson, Gonville and Caius College, MS 249/277, ff. 191–3 and annotated by Caius himself.

[28]Caius, *De antiquitate*, 144–146 noting the existence of a medieval house called 'the house of Anaxagoras', soon after called by the today more familiar name of that of Pythagoras, see Gray, *The School of Pythagoras, n.s. 4, 1938*, 38.

[29]Ibid., 174, 184, 191–2. His practice, when dealing with sources supporting his own case, was somewhat different.

him of the possibilities of scribal error, particularly in the transmission of unfamiliar names.[30] The fact that at least one of the authors he cites, Polydore Vergil, had cast doubt on Cantaber, Gurguntius and the like, was not sufficient to outweigh the evidence of what Caius took to be authoritative sources accepted by the university and the wider world, particularly when other recent investigations of the chronology of these events had not found, in his view, sufficient evidence to disprove them. He was far from alone in this belief, as his catalogue of modern scholars shows.[31] But this adherence to what to us are obvious fables is made more prominent by the polemical context of the whole work.[32] The refutation of the Oxford Caius matters far more than any sober account, and the repetition of the same arguments throughout the book only adds to the modern impression of Caius' naivety without making clear to the reader the amount of valuable information he himself had unearthed and made public.

His slightly later history of the university is a very different book, and deserves more attention than it has received. Hastings Rashdall, for example, merely notes that it was printed along with a revised version of the earlier polemic in 1574, while May McKisack misleadingly described it a purporting "to be a judicious summing up of the whole question from a historical standpoint".[33] Admittedly it begins with some 30 pages of erudite fiction, largely repeating the *De antiquitate*, but the rest of the book is of a very different type and quality. It is a survey of the university and town based on personal observation and familiarity with the archives in greater depth and at greater length than anything previously undertaken. Even the opening section draws on new personal and archival information; it includes an account of the Cambridge suburbs as well as a transcript of a letter of Philip II written from Windsor to the university in 1554, which Caius uses to support his theory of the foundation of the university by Cantaber.[34]

The preface explains why Caius was moved to collect the information included in the book. When in 1558 he made his first return to Cambridge after a long while away, he found everything had changed – the whole appearance, the people, their dress, their pronunciation, and the whole system of education. He no longer knew anyone, and had himself been forgotten. In the midst of all this change, and afraid that soon no trace of anything old would be left, he determined to write a history of the university both to preserve the past and to suggest a guide to the future.[35] Such

[30]Ibid., 29, 160.

[31]Ibid., 38. See further, below, note 52.

[32]Rashdall 1895, vol. 2, 544, remarks that "the exuberance of Cambridge imagination has, indeed, transcended that of our least scrupulous antiquaries", and, 543, implies that "antiquarian superstition" was a feature of all Cambridge histories down to the nineteenth century.

[33]Ibid., 533. For a comparison with the histories of universities by Jobst and Middendorp, see below, n. 59. McKisack May 1971, 71, a comment that neglects much of the work. Brooke, in his contribution to Morgan 2004, 457, notes in passing the difference between the two works, but does not amplify his judgment that the later one "has much sound learning in it, as well as occasional flights of fancy."

[34]Caius, *Historia*, 8–10, 12. On the Philip's letter and similar 'Catholic' entries, see below.

[35]Ibid., 3.

an intention is less overtly polemical than that of the *De antiquitate*, but, as we shall see, it also masks some striking criticisms.

Caius' dissatisfaction with the present resurfaces from time to time throughout the book. He laments the decision of the university in 1542 to adopt the new pronunciation of Greek and Latin, unintelligible to anyone who has not been educated in Cambridge.[36] But his greatest disdain is for the behaviour of modern students. In four pages of purple prose, he contrasts the experiences of his youth with the modern age. He paints a picture of a lost golden age when students spent their time in frugal study, properly dressed for lectures and services, packing the lecture rooms in their zeal for learning, and respectful of their elders. They raised their caps when they spied a don across the road, and graciously gave way if they met one on the path. They never wasted time wandering around the pubs and shops of Cambridge, drinking only on feast days, and then sparingly and in moderation. Sport formed merely a healthy break between periods of study, and, even then, students might use their leisure time to rehearse the tragedies and comedies for the Christmas season. University rules were obeyed, and colleges fined if they took back anyone expelled from the University for bad conduct. How different from the present age of luxury and licence, when students lavish money on fine clothes, ruffled shirts, slashed saffron breeches, fancy hats, dashing beards and the like![37]

Much of his account, however, is very different from these rhetorical effusions or from the imaginative early history (a belief in whose veracity he shared with almost all contemporaries). It is a detailed survey of all matters concerning the history of university and town based on a diligent search for information in earlier chroniclers and in the archives. Compared with the comprehensive list of worthies described in the *De antiquitate*, Caius gives relatively few names except when he has found them in the archives or wishes to flatter some of the leading members of the Court – and, in turn, their presence in his book, even if only recorded as in attendance on the Queen on her Cambridge visit, emphasises the wider support for the University.[38]

His more accurate sections begin in the thirteenth century when he can cite contemporary documents describing clashes between town and gown, as well as early evidence for university governance. He notes that, as in modern Italy, students at first lived in the houses of the citizens, but later began to take over empty properties or build their own halls.[39] There were 17 such halls in Caius' time, and he could report details of three defunct ones, as well as a few more names, leaving aside those that had once existed on the site of King's College. The various religious orders had also had their own hostels, most, like the halls, now subsumed into larger Colleges. These Colleges Caius enumerates in turn, listing their

[36]Ibid., 100.

[37]Ibid., 74–7. Compare Caius' denunciation of the lifestyle of the wealthy in Caius 1552, 19, 22 (*Works*, 19, 21), deploring the habit of bringing young children up on hot buttered toast.

[38]Ibid., 72. Cf. 106 for praise of Cecil.

[39]Ibid., 36, 64.

foundations and their heads. The first of the two books ends with a survey of the university buildings, including a catalogue of the holdings of the University Library, student life, relations with the town, and the various privileges of the University confirmed by both parliament and monarch.

The much shorter second book has the town of Cambridge as its main focus. Caius lists its churches and streets, but his view of the town's inhabitants is not complimentary. Its inhabitants are "mechanici" ("handworkers"), as different from the students as the sun and the moon.[40] The book ends with a survey of the officials of the University, including the preachers, and a description of some of the more unusual Cambridge traditions.

But behind the apparent accumulation of archival data one can also see a more subtle attack on contemporary Cambridge, and, more specifically, on the religious changes there in his lifetime. It is partly achieved through silence. Although Caius records the actions of "the most famous" King Henry VIII in abolishing all papal power and revoking all papal bulls while confirming the privileges therein given, there is no specific mention of any of the major changes in the University that had occurred during the 15 or so years of the work's preparation, and although due honour is paid to Queen Elizabeth and her leading ministers, there are striking references to her predecessors that would have raised more than an eyebrow in certain Cambridge circles.[41] The first document cited at length is a letter of 1554 of Philip II thanking the University for its loyal sentiments towards him on his entry into his new kingdom. Not only is the monarch given the full title of King of the Spains and described by Caius as "Most illustrious King", but the point that Caius wishes to prove by quoting the letter verbatim is not immediately obvious.[42] A similarly unnecessary reference to Philip occurs when he describes his own college as retaining the name of Gonville Hall "until the time of Philip and Mary rulers of England".[43] The select list of distinguished clerical Chancellors in Caius' lifetime mentions John Fisher, Stephen Gardiner and Reginald Pole, who is given his full title of Cardinal of the Holy Church and Archbishop of Canterbury.[44] Edward, Lord North, who retired from political life under Elizabeth, is described as a man "of notable piety" and cited alongside Thomas Lupset, the translator of Chrysostom, as a notable alumnus of Pembroke.[45] By contrast, a relatively minor change in the

[40]Ibid., 95.

[41]By the late 1560s, and for some time after, Caius and his College were viewed with suspicion by the university authorities as a nest of popery; Caius himself in 1572 had his rooms sacked, he alleges, at the behest of the Vice-Chancellor, the Master of Trinity and the Provost of King's, Brooke, 70–8.

[42]Ibid., 12, cf. 112.

[43]Ibid., 52. Brooke reminds me that the names and formal titles of both monarchs open the beautiful charter authorising the refoundation of Gonville Hall, see Caius, *Annals*, 43, 61.

[44]Ibid., 106. Thomas Cromwell is described as becoming Chancellor "after the death of the Bishop of Rochester", i.e. John Fisher.

[45]Ibid., 47. North's deathbed (re-)conversion to Catholicism was well known. Caius himself was a great admirer of Chrysostom's theology, but Lupset's friendship with Cardinal Pole will also have been remembered.

regulations for disputations under Edward VI is preceded by a parenthesis declaring disapprovingly that in his reign "innovation began to take over everything".[46] The praise duly given to Matthew Parker for his various benefactions also reflects his role in protecting Caius from some of the accusations of popery levelled against him, accusations that his history of the University would have done nothing to dispel. The repetition in the preface of the standard trope that history is more than an account of past deeds but should also be a guide to prudent and wise conduct in the future could appear to one who has read to the end of the book a manifesto for a return, not just of Henrician Anglicanism, but even of papal jurisdiction. None of the Catholic references detected in the text need be interpreted in that light, for Caius was quoting the official titles at the time, but cumulatively they indicate a definite anti-evangelical standpoint that was shared by few in Cambridge outside Gonville and Caius College.

The other two 'historical' works by John Caius are very different in character. His *Annals of the College of Physicians* is a record of the activities of the College from its foundation by Henry VIII until 1572, when it was agreed that he should be excused attendance at meetings because of his age and as an appreciation of the services he had rendered to the College. It was begun as a personal record in 1555, the year of his first presidency of the College, and is very uneven in its coverage of events. For the years when Caius was not a member or had little time to spend on College affairs, his account is brief and contains little more than a list of officers. But for others, and particularly for the years of his presidency, 1555–60, 1562–3 and 1571, it is extremely full. For these years Caius often cites documents and letters at considerable length, providing himself (and future historians) with detailed information on the struggles of the College to establish its authority over all medical practitioners in London and beyond, however humble or elevated they might be. There is also an implication that others should follow his example in seeking to maintain the traditions and role of the College as laid down by Thomas Linacre in 1518. This repository of information can thus serve as a weapon against all who might challenge the College's authority. Its value to the historian lies both in the information it contains and in what it tells us about Caius, the collector and creator of documents.

He is even more explicit about his aim in writing the *Annals* of his Cambridge College. In bringing together as much documentary evidence as he could he hoped thereby to preserve the memory of those who benefitted the College and to warn against the example of those who had damaged it.[47] It was a task that he insisted in his statutes should be continued each year by a specific fellow-registrar in a way that went beyond the mere keeping of accounts or the registration of property bought or sold.[48] Most of the information Caius himself included is of this sort,

[46]Ibid., 100. Note his association of Edward VI with the losses of his Cambridge College through currency devaluation in 1551. Similar losses in 1560 and 1561 are not attributed to Elizabeth, Caius, *Annals*, 81–2.

[47]Caius, 1.

[48]Cited by Venn, ibid., vi.

but he also found time to comment on fires, fortunately extinguished, that might have caused serious damage to the College fabric, as well as to provide small biographical details of College Fellows.[49] He had a precedent for this in the *Evidences* of Edmund Sherriff, Master 1472–1476/7, a volume which Caius rescued when it was in danger of falling apart and on which he drew for the first part of his *Annals*.[50]

John Caius, it must be admitted, was not a first-rate historian. Nonetheless he had a passion for the past and he hunted as diligently in the archives and on the shelves of Archbishop Parker's library for historical information as he did for readings of Galenic manuscripts in Italian repositories or in the margins of books owned and annotated by members of the London College of Physicians. Modern historians have, of course, added greatly to the archival record that he preserved, yet to contemporaries and for long after, his was a valuable contribution. He was a remarkably learned man, even if that learning often obtrudes into the narrative, and one should not underestimate his abilities as a writer. When he had a good story to tell, as with his accounts of the origins of the English Sweat, he could communicate his findings in a lively manner, even if his humanist Latin is more complex and more allusive than his English.[51] But it is also true that, particularly in the *De antiquitate*, his erudition and his polemical purpose together produce a disjointed and badly structured account whose faults have undoubtedly contributed to its subsequent neglect. There seems little point in studying a tract in which much is nonsense and where even the good things have been improved upon and surpassed.

But that dismissive verdict is to adopt a perspective that takes little notice of the context in which Caius was writing. While some Renaissance historians, and especially the most famous, are impressive forerunners of modernity, others, and particularly those who sought to go back beyond their own times, have a much greater kinship with their medieval predecessors. The stories of Gurguntius and Cantaber, to say nothing of Anaxagoras, the beautiful and learned Martia, and the erudite Druids carried off to adorn a Roman triumph, were not Caius' inventions, but those of Nicholas Cantelupe, d. 1441, and had been circulated by earlier chroniclers.[52] Some like Polydore Vergil, might demur at some parts of the story, which, like the tale that the university was refounded and given new buildings by King Octavius and his scholarch Maurice in A.D. 443, immediately appear implausible to modern eyes.[53] But such scepticism was rare in Caius' day, as Nancy Siraisi herself has shown. Two or three generations later, a sharper criticism was removing

[49]For fires, ibid., 111–2, 122.

[50]For this MS, Caius MS 706/692, conserved by Caius in 1564, with the addition of page numbers and an index.

[51]Caius, *The Swete*, 11, 16–7.

[52]Caius, *Historia*, 13–6. Putter 2003, 63–81, terms Cantelupe's procedure "conjectural emendation". Hiatt 1999, 2004a, 70–101.

[53]Caius, *Historia*, 26. The Oxford legend of its foundation by King Arthur had a somewhat longer life, in part because of its comparative sobriety.

much of this fictional history, but such a development had not yet occurred widely in Elizabethan England.[54] A belief in the existence of giants was well-nigh universal; after all, they figured prominently in the Old Testament as well as in the legends of Greece and Rome. Students of anatomy, like Caius, were only too familiar with the argument that the Galenic body was different from that dissected by Andreas Vesalius, because in Roman times the bodies of the great were much larger and much more heroic than their modern counterparts.[55] The fact that these stories went back centuries and were copied in medieval manuscripts was a further reason for giving credence to them, in accordance with the principle that the earlier the source, the more likely its truth. Cantelupe's authority would also have been increased by the fact that his *Historiola* had been copied out by an earlier Fellow of Caius, John Herryson, and, still more, by its presence in the Black Book of the University, kept in the University Archives, a guarantee from the university itself of its veracity.

In his uncritical reliance on earlier chroniclers, Caius was little different from the great majority of historians of the period. Giucciardini or Sarpi might compose modern-sounding histories of their own day, but those who ventured into the remoter past were still entrapped in a late-medieval world of imaginative reconstructions.[56] To label them antiquarians, rather than historians, is to impose too sharp a distinction between the two.

It also underplays Caius' achievement in his *History*. If his preface is to be believed, he began collecting information in 1558 and continued writing from then on. For the early history he could rely on Cantelupe and the various Cambridge copies of his *historiola*, a work written imaginatively to show the greater antiquity of Cambridge compared with Oxford. But the title Caius gives to his book shows that he was contemplating something bigger and, in its day, very unusual. It was to be a history of an institution, not like Paolo Giovio's *Elogia* a series of biographies of university worthies, nor, like Leland and, in particular, John Bale, a list of writers who had connections with the university.[57] These sources were quarried by Caius, but his focus was different, and new. At least at the time of inception, he had no examples of others writing the history of their own university to draw on. He mentions the two earliest studies of European universities, Wolfgang Jobst's survey of the early history of universities and, Jacob Middendorp's *De celebrioribus academiis*, first published in Cologne in 1567 and in a much enlarged edition there in 1572, and uses them to confirm the superior antiquity of Cambridge

[54]Siraisi 2007, 127–33. One might also suppose that the refutation of this story by Thomas Caius (like the later refutation of it in Brian Twyne's *De antiquitate Oxoniae*, Oxford, 1608) could be viewed from Cambridge as little more than sour grapes and bad history; John Caius' own demolition of Thomas Caius' claims for Oxford was regarded by Oxonians in the same light.

[55]Caius, *De antiquitate* 16. Siraisi 2007, 35–42; Piggott 1989, 48–9. Caius, a small man, supports the possibility of giants with a reminiscence of his time in Germany when he had seen huge German oxen, reaching up to the third floor of the houses, *De antiquitate*, 96.

[56]Burrow 2009, 278–9, 308–9, 314–5, showing the coexistence of different approaches to the writing of history.

[57]Cf. McKisack, 1–25.

University.[58] But neither author offers a description of a university, its institutions and its surroundings that would have served as a model for Caius' H*istory*.[59] Jobst's is a very slender booklet, concerned solely with origins, while Middendorp, although more extensive, is equally at pains to create a chronology of universities rather than to describe their activities or settings, and his work may not have been available when Caius began his project. Here, if at all, Caius comes close to some of the Elizabethan antiquaries and topographers whose work culminates in Stowe and Camden. But he had no history of Cambridgeshire to follow, and none of the Elizabethan topographers appears in the probate list of his books. His London links may have introduced him to antiquarian circles there, but all the indications are that both the idea and the organisation of his *History* were his own. If so, he deserves greater credit for his independence than is usually given him.[60] His pioneering status has been obscured both by the notoriety of the *De antiquitate* and by the much more detailed and accurate histories of later writers from the nineteenth century onwards.

History mattered to Caius. He was deeply interested in the history of the places and institutions with which he was involved, from his planned history of Norwich in his early manhood to the *Annals* of his two Colleges. He took seriously the canonical injunctions about "time that eateth all things", and the sentiments in the preface to the *History* were no mere clichés to him.[61] His pride in his institutions led him to use history as a way of defending them against the aspersions of others, and of preserving as far as possible a memory of the past to instruct and warn future readers. But throughout, explicitly as well as implicitly, there is a nostalgia for the past and, even more, an eagerness to justify his own position. The *Annals* of the College of Physician present detailed records of his years as president, where he strove, and at times succeeded, in establishing the authority of the College over other purveyors of healing. His sober account of the history of Cambridge University can be read both as a lament for a past golden age and as a criticism of his own day. Other references to his own activities are more subtle. His laudatory references to the visit of Queen Elizabeth to Cambridge in 1564 do not mention the leading role that he played in the proceedings, although it would have been known to many

[58]Caius, *De antiquitate*, pp. 38, 79, citing W. Justus 1556, and 39, citing Middendorp, above, n. 17.

[59]Middendorp's later edition, Cologne, G. Cholin, 1594, 627–9, in turn cites Caius on Cambridge Colleges, but seems to have misinterpreted a letter of the Caian John Hatcher and transferred the "recently founded and well endowed" College of Physicians from London to Cambridge. Jobst, following Bale, sig. A.4 r-v, attributes the foundation of Cambridge to Sigebert, King of East Anglia, although noting that others placed it even earlier. Both authors place its origins before those of Oxford.

[60]Rowse 1953, 49–86, discusses the Elizabethan topographers, but the great flowering of antiquarian studies in London post-dates Caius' death, see Harris 2004, 27–36. McKisack, 127–54, discusses the local historians and topographers, but none, save possibly Leland, is likely to have influenced Caius' projects. Although he was a considerable land-owner, he did not share in the gentry culture described by Broadway 2006.

[61]Caius, *Historia*, 3, a sentiment that went back to the Preface to Livy's *History of Rome*.

of his readers.[62] But to suggest that Caius was only interested in justifying his own position in an uncertain present is also unfair to him. It cannot explain the efforts that he, and a variety of secretaries, employed in reading and copying manuscripts of medieval historians or the volumes of chronographers. The past was important to Caius, and not only because in many respects he had experienced the destruction of things he held dear, in religion, morals, learning and even the pronunciation of Latin. His gifts to his College and to the College of Physician show a belief in the importance of symbols as a way of linking the present and future to what had gone before in an institution. He was aware of the fragility of memory, and also of manuscripts, and was convinced of the importance of preserving, and continuing, a record of the past for future consultation. His use of history, for all its flaws, should be viewed in the light of this not ignoble aim.[63]

Bibliography

Arcangeli, Alessandro, and Vivian Nutton, eds. 2008. *Girolamo Mercuriale. Medicina e cultura nell'Europa del Cinquecento*. Florence: L. S. Olschki.

Broadway. 2006. *'No Historie So Meete': Gentry Culture and the Development of Local History in Elizabethan and Early Stuart England*. Manchester: Manchester University Press.

Brooke, Christopher. 1985. *A History of Gonville and Caius College*. Woodbridge: The Boydell Press.

Brooke, Christopher, Damian Leader, Victor Morgan, and Peter Searby, eds. 2004. *A History of the University of Cambridge*, vol. 2, 1546–1750. Cambridge: Cambridge University Press.

Burrow, John. 2009. *A History of Histories. Epics, Chronicles and Inquiries from Herodotus and Thucydides to the Twentieth Century*. Harmondsworth: Penguin.

Caius, John. 1568. *De antiquitate Cantabrigiensis Academiae*. London: H. Bynneman.

———. 1574. *De antiquitate Cantabrigiensis Academiae*. London: John Daye.

———. 1912. *The Works of John Caius, M.D. Second Founder of Gonville and Caius College and Master of the College 1559–1573*, ed. E.S. Roberts. Cambridge: Cambridge University Press.

Caius MS. 706/692.

Clark, Sir George. 1964. *A History of the Royal College of Physicians of London*, vol. 1. Oxford: Clarendon Press.

Flores historiarum per Mathaeum Westmonasterium collecti, praecipue de rebus Britannicis ab exordio mundi usque ad annum Domini 1307. London: Thomas Marsh, 1570.

Grafton, Richard. 1853. *Correspondence of Mathew Parker, D.D.* Cambridge: Cambridge University Press.

Gransden, Antonia. 1982. *Historical Writing in England, II. C. 1307 to the Early Sixteenth Century*. London: Routledge.

Gray, J.M. 1938. *The School of Pythagoras (Merton Hall)*. Cambridge: Cambridge Antiquarian Society, Quarto Publications, n.s. 4.

Grierson, Philip. 1978. John Caius' Library. In *Biographical History of Gonville and Caius College*, vol. vii. Cambridge: Cambridge University Press.

[62] Caius, *De antiquitate*, 5–6, 76; *Historia*, 72.

[63] I am grateful to Christopher Brooke for his advice and encouragement to a non-Caian over many years.

Gundert, Beate. 2006. Zu den Quellen der Basler Galen-Ausgabe (1538). In *Ärzte und ihre Interpreten. Medizinische Fachtexte der Antike als Forschungsgegenstand der Klassischen Philologie*, ed. C. Brockmann, C.W. Brunschön, and C.W. Müller, 87–106. Munich: K.G. Saur Verlag.

Harris, Oliver. 2004. John Stow and the Elizabethan Antiquarian Network. In *John Stow (1525–1605) and the Making of the English Past*, ed. Ian Gadd and Alexandra Gillespie, 27–36. London: British Library.

Hiatt, Alfred. 1999. Forgery at the University of Cambridge. *New Medieval Literatures* 3: 95–118.

———. 2004a. Stow, Grafton, and 15th-Century Historiography. In *John Stow (1525–1605) and the Making of the English Past*, ed. Ian Gadd and Alexandra Gillespie, 45–56. London: British Library.

———. 2004b. *The Making of Medieval Forgeries. False Documents in Fifteenth-Century England*. London: British Library.

Justus, W. 1556. *Academiarum, et praestantium quarundam scholarum fundatarum secundum aetatis seriem*. Frankfurt an der Oder: J. Eichhorn.

McKisack, May. 1971. *Medieval History in the Tudor Age*. Oxford: Clarendon Press.

Mercuriale, Girolamo. 1569. *De arte gymnastica*. Venice: Giunta.

———. 2008. *De arte gymnastica*, ed. and trans. Jean-Marie Agasse. Florence: L. S. Olschki.

Middendorpius, Jacobus. 1567. *De celebrioribus academiis*. Cologne: P. Horst.

———. 1594. *De celebrioribus academiis*. Cologne: G. Cholin.

Nauclerus, Joannes. 1516. *Memorabilium omnis aetatis et omnium gentium chronici commentarii*. Tubingen: T. Anshelm.

Nutton, Vivian. 1987. John Caius and the Manuscripts of Galen. In *Proceedings of the Cambridge Philological Society*, Suppl. 13. Cambridge; Cambridge Philological Society.

———. 2004. Caius, John (1510–1573). In *Oxford Dictionary of National Biography*. Oxford University Press. http://www.oxforddnb.com/view/article/4351.

Piggott, Stuart. 1989. *Ancient Britons and the Antiquarian Imagination. Ideas from the Renaissance to the Regency*. London: Thames and Hudson.

Putter, A.D. 2003. King Arthur at Oxbridge: Nicholas Cantelupe, Geoffrey of Monmouth, and Cambridge's Arthurian Foundation Myth. *Medium Aevum* 72: 63–81.

Rashdall, H. 1895. *The Universities of Europe in the Middle Ages*, vol 2. Oxford: Clarendon Press.

Rowse, A.L. 1953. *The England of Elizabeth*. London: Macmillan.

Siraisi, Nancy. 2007. *History, Medicine and the Traditions of Renaissance Learning*. Michigan: University of Michigan Press.

———. 2013. *Communities of Learned Experience: Epistolary Medicine in the Renaissance*. Maryland: Johns Hopkins University Press.

Venn, John, ed. 1904. *The Annals of Gonville and Caius College by John Caius*. Cambridge: Cambridge Antiquarian Society.

———. 1912. John Caius. In *The Works of John Caius, M.D., Second Founder of Gonville and Caius College and Master of the College 1559–1573*, ed. E.S. Roberts, 1–78. Cambridge: Cambridge University Press.

Woolf, Daniel R. 2000. *Reading History in Early Modern England*. Cambridge: Cambridge University Press.

Chapter 5
A Medical Man Among Ecclesiastical Historians: John Caius, Matthew Parker and the History of Cambridge University

Anthony Grafton

John Caius is no longer a household name, except in a few households in East Anglia. Yet he was in many ways a characteristic and dominating figure of a particular moment in the 1560s and 1570s. For a few years, British courtiers, churchmen and country aristocrats—as well as successful medical men like Caius—shared a particular late humanist culture. They believed in the power and utility of ancient and medieval texts. These common assumptions kept them engaged in the scholarly study of the past long after their formal studies were over, and inspired them to nurse what were sometimes exaggerated hopes for the power of education. Many of them took a special interest in pragmatic political history, which they saw as a guide to public life.[1] Caius was also a historian, like so many other medical men.[2] But as we will see, he practiced a particular kind of history—one normally focused on the history of the church and produced more often by groups than by individuals.

The academic life began for Caius when he came as an undergraduate to Gonville Hall, a small establishment in Cambridge that was, in the 1530s, as much a monastic community as a college. Evidently he did not make much of an impression at first, at least to judge from the variety of ways in which the bursar spelled his name in the account books. But brilliance and mastery of Greek and Hebrew soon set him apart, first at Cambridge, where he graduated at the head of the list of BAs in 1532–33, and then, in 1539, at Padua, where he studied medicine and stayed to teach. Caius won a European reputation as a Galenic scholar and academic physician. He produced some impressive critical editions before working his way home to England, library by library, collating manuscripts as he went. From

[1]Jardine and Grafton 1990; Todd 1987; Womersley 1991, 1992; Kewes 2011; Cox Jensen 2012.
[2]See the classic work of Siraisi 2007.

A. Grafton (✉)
Princeton University, Princeton, NJ, USA
e-mail: grafton@princeton.edu

G. Manning, C. Klestinec (eds.), *Professors, Physicians and Practices in the History of Medicine*, Archimedes 50, DOI 10.1007/978-3-319-56514-9_5

1547 on he held anatomical demonstrations and practiced medicine in London. After 10 years or so, he had made his pile, and in 1557 he offered to refound his college. Caius's proposal found acceptance, he himself was elected master, and he devoted himself to equipping what was now Gonville and Caius College with new statutes and new buildings.[3]

Like many other brilliant academics—Richard Bentley comes to mind—Caius had no great diplomatic gifts. He laid out his refounded college as a humanistic machine for character formation, carefully designed to stamp its inmates with good morals as well as good learning. Students entered it through a "Gate of Humility"; passed through a "Gate of Virtue" every day during their years in college; and finally left for the Examination Schools through a "Gate of Honour." Yet somehow the conditioning did not stick.[4] "Young men now-a-days," Caius wrote in 1567 to his friend Matthew Parker, the Archbishop of Canterbury, "be so negligent that they care for nothing."[5] In his *History of Cambridge University*, written at the end of his life, he inserted a long passage in which he mourning for the austere Cambridge of his youth: "no fancy caps were worn at disputations, no clothing of uncertain identity, no ruffled shirts, no round caps, no barbered frivolity, no bearded vanity, no jollity, no arms, no dice, no dances, all of which the laws of the university then prohibited, no haughtiness in dress and bearing, in which many shine with a borrowed light, like the moon."[6] Many other Cambridge men—Thomas Smith, for example—welcomed the transformation of the university from religious community to training ground for young aristocrats and gentlemen. Caius did not.

When Caius applied mild disciplinary measures—such as beating refractory BAs and putting them in the stocks—they resisted. In fact, they went so far as to accuse their master, to Parker, of being an atheist and a crypto-Catholic. The archbishop believed the fault lay with both sides and urged compromise. In the end, Caius resigned as Master.[7] He would, one feels, have happily spent his last years telling the young to stay off his lawn—except, of course, that even the most

---

[3]Venn 1912; and Nutton, "Caius, John" in *Oxford Dictionary of National Biography* (http://www.oxforddnb.com.ezproxy.princeton.edu/view/article/4351?docPos=1; consulted May 19, 2014).

[4]For this program, which was completed only after Caius died, see Nickson 2005; and Binski 2013.

[5]John Caius to Matthew Parker, April 8, 1567; Parker 1853, 299.

[6]Caius, *Historiae Cantebrigiensis Academiae ab urbe condita liber primus* [–*secundus*], in *Works*, 76–77: "Quas ob res eos tum temporis omnes fovebant, omnes amplexabantur, ut quos virtus, & eruditio commendabant populo, non ut hodie insolentia alienabat, non luxurians mensa, non vestis, non cubiculum, non intumescentia crocotillis crusculis faemoralia, non inter disputandum galeri, non ambiguae vestes, non crispatae camisiae, non rotundi pilei, non capiti pressi, non tonsa levitas, non barbata vanitas, non lascivia, non arma, non alea, non choreae, universitatis legibus iam olim prohibitae, non fastus denique vestis atque vitae, in quo multi ita splendent luce aliena, ut solet luna."

[7]Venn, "John Caius," in *Works*, 22–27.

disrespectful of them would never have dared to step onto it. Instead, he left the college and spent most of his last year or two in London.[8]

In 1568, Caius published the least successful of his works—the one in which even his most enthusiastic modern admirers have been unable to find much to praise. The historian of Gonville and Caius college, for example, calls it "the largest and most calamitous of its author's works: a farrago of invention and credulity."[9] *On the antiquity of Cambridge University* is a very polemical little book.[10] It sprang from a strange but recognizable form of academic politics. Elizabeth loved pageantry, both universities provided it, and when she deigned to visit either of them, it tried to impress the queen with its superior antiquity. In 1564, when Elizabeth processed to Cambridge, a local orator argued that his university was the oldest. In 1566 she graced the other place with her presence. Thomas Caius, fellow and warden of All Souls and Register of the University of Oxford, composed an "Assertion of the Antiquity of Academy of Oxford." He presented this to the queen when she visited his university. He had not meant the work for general circulation. But Elizabeth's court was a hive of information masters, above all William Cecil and the spymaster Francis Walsingham. And they were Cambridge men. As John Strype told the story in the early eighteenth century, "This MS. as it seems by the Secretaries means, a *Cambridge* man, coming into the hands of the Archbishop, a *Cambridge* Man also, was transcribed, and communicated by him unto another *Caius*, and a learned *Antiquarian* of Cambridge; the Archbishop exhorting him to consider well the Book, and to vindicate his University."[11] So Caius was taking part in a confrontation, one university against the other, when he sent his little book *On the Antiquity of Cambridge* to be printed in London, together with the work by Thomas Caius that he meant to refute. It was the pedant's version of the Boat Race, and as in the Boat Race, one side had to lose.

The controversy went on for another 50 years, I am happy to say, as generations of scholars engaged in the quintessential scholarly activity, belaboring one another with slapsticks and bladders. But the arguments involved were relatively simple. Oxford—so Thomas Caius and others maintained—was clearly the older of the two universities. The city, after all, had been founded by good King Mempricius, a century after Brutus the Trojan came to Britain. It was known from its handsome setting as "Beaumont." Later on, the Greek scholars who had accompanied Brutus to England and settled in Cricklade (Greeklade), migrated to Beaumont not long before the Saxons settled in England. But it was really King Alfred who created the university, in 873, when St Grimbald became the first chancellor.

Cambridge was also clearly the older of the two universities. It was founded—as Caius pointed out—by Cantaber the Spaniard. He belonged to a group of Spanish

[8]For a vigorous account of Caius and his college, see Brooke 1985 (repr. with corrections, 1996), 55–78. See also Venn 1897–1998, vol. 3, 30–63.

[9]Brooke 1985, 75.

[10]Caius 1568.

[11]Strype 1711, vol. 3, ch. 18, 257.

exiles whom King Gurguntius Brabtruc encountered, sailing in Scottish waters. Cantaber married the king's daughter, and founded a city on the river Cant, which was named after him. His son Grantinus bridged the river, and provided names for both Grantchester and Cantbridge or Cambridge. Cantaber gathered scholars around him. In the course of time, both Anaximander and Anaxagoras visited his city. Local scholars helped convert good king Lucius of Britain to Christianity. King Arthur granted a charter in 531. When King Alfred founded Oxford, the Cambridge historians made clear, he peopled it with the descendants of a band of Cambridge scholars who had migrated to Cricklade before his time.[12]

The details of this debate need not detain us. The central narratives rested, as Alfred Hiatt has recently made clear, on documents created in the late fourteenth and fifteenth centuries.[13] True, the critical weapons that these scholars wielded against one another could be quite sharp. Caius made clear more than once that when the other Caius cited three modern sources, he was really only citing one, since "all of them are following Polydore Vergil as their source."[14] He also showed that his opponent cited texts in deceptively incomplete forms.[15]

But their positive arguments—and the foundations these rested on—tended to be much wobblier. Like most Renaissance scholars concerned with questions of distant origins, Caius and Caius drew their evidence indiscriminately from sources of very different kinds and qualities. Caius even quoted the text of Berosus forged by Annius of Viterbo—though he suggested that the term "giant," applied by Annius to the men who lived just before and after the Flood, should not be taken literally, but read as a statement that they were aborigines, born of the earth (γηγενεῖς).[16] Etymology served them as a philological oyster knife. Inside the name of any city or institution, the name of its founder was locked up. Etymology opened the shell and revealed, for example, that Cantaber and Grantinus lurked inside Cambridge, waiting to be discovered. Any reference in a source, however vague, could be pressed into service to prove any argument, however specific. If Bede connected Alfred to scholars in East Anglia, then he must—obviously—have had Cambridge in mind, even though Alfred's kingdom had not extended so far.[17]

[12]See in general, Parker 1885, 5–62; Gabriel 1988, I, 601–626.

[13]Hiatt 2004, ch. 4.

[14]Caius, *De antiquitate*, in *Works*, 26: "Nam postquam ab uno atque altero recentiori auspicatus fueris, ut Polydoro, Baleo, atque Lilio (quorum oscitante uno oscitat & alter, nam unum Polydorum authorem sequuntur omnes) cum locus sit introducendi veteres scriptores, nullum prorsus introducis, sed in alium locum differs ... quasi trium istorum testimonia unum non essent, sed plura, uno Polydoro authore, pluribus qui eum authorem sequuti sunt, referentibus."

[15]Ibid., 133: "Sed ad rhetoricam tuam, quae etsi dicat, nude & ieiune proferenda adversarij verba, non tamen dicit decerpenda duo aut tria vocabula testimonij ut causam adiuves: (incivile enim est arripere historiae particulam, & totam ex ea causam iudicare) nec ita implicanda scriptorum testimonia ut imponas."

[16]Ibid., 14–16. For Annius's giants, see Stephens 1989.

[17]For further complaints about the forms of criticism used in this debate, see Parker 1885; and Kendrick 1950, 76–77.

What interests us here is not the content of Caius's text but its character. His short treatise is stuffed with information of a kind not found in the rest of his works. Quotations, not from Galen or other Greek writers but from documents preserved in the Cambridge archives and the Tower of London and in the Latin and Anglo-Saxon historians of the Middle Ages, cram its pages. A dense bibliography lists Caius's sources.[18] Most of these works were unpublished in 1568—as most of them still were in 1574, when the posthumous second edition of the work appeared. And a great many of them—as M.R. James pointed out a century ago—were available in a single collection: the massive manuscript library at Lambeth Palace in London, which Parker and his secretaries created, and part of which is now to be found in the Parker Library at Corpus Christi College Cambridge. The Parker Library was larger then than it is now: it included the Cotton MSS that Caius also used.[19] Strype's conclusion is lapidary: "from the first, to the last, the Archbishop's Influence and Assistance, ran through this curious Work."[20]

This formulation—though basically accurate—remains very general. In fact, the evidence allows us to establish the nature and genesis of Caius's book in detail—and to make clear that it was not a word of occasion, tossed off by its author at the suggestion of his friend the archbishop, but the result of intensive collaboration and consultation. Four sets of evidence will bring us closer and closer to Parker's own enterprise and Caius's place in it.

Parker practiced ecclesiastical scholarship, and from antiquity on, those who have played this game, with its emphasis on finding, cataloging and quoting original documents, have organized teams of helpers.[21] Parker was no exception. Emulating—and surpassing—the Magdeburg Centuriators, whose enterprise did much to inspire his, he built a group of secretaries, scribes, illuminators and printers, who worked together to organize and study his manuscripts and to produce printed books from them.[22] In 1572, John Day printed a massive study of the origins of the British church and the lives of the archbishop of Canterbury, down to Matthew Parker himself. Entitled *De antiquitate Britannicae ecclesiae*, the book showed that the British church—especially in its Anglo-Saxon form, before the mission of

---

[18]Caius, *Works*, [223]–227: "Catalogus scriptorum, quibus usus est duobus hisce libris Londinensis. Historici nostri antiqui & scripti."

[19]Ibid., viii–x, after the Preface, where a note by M.R. James appears, identifying as many of the sources Caius used as possible. He points out Caius's use "of Parker's collection before it was bequeathed to Corpus Christi College" and of "many of the MSS, which Sir Robert Cotton (who seems to have begun to collect books about 1588) afterwards secured" (x). He does not note that the Cotton MSS, or many of them, would have been at Lambeth Palace. See also Grierson, "Appendix IV: John Caius's Library," in *Biographical History of Gonville and Caius College*, ed. Venn et al., VII, 509–535, at 523. For the growth and use of Parker's collection see Graham and Watson 1998; Graham 2006, II. The portion of it that Parker left to Corpus Christi College has been digitized and made available as The Parker Library on the Web (http://parkerweb.stanford.edu/parker/actions/page.do?forward=home, consulted on May 20, 2014). See also Budny 1997.

[20]Strype 1711, 257.

[21]See Grafton 2012, 3–26.

[22]This point is best made by McMahon 2013.

Augustine of Canterbury and the Norman Conquest corrupted it—had practiced a relatively pure form of Christianity. Look the title up in any catalogue, and you will find its authorship ascribed to Parker.[23] But it's a wise book that knows its author.

A copy of this book, heavily annotated by Parker's principal secretary, John Joscelyn, his son, John Parker, and others, is now in Lambeth Palace Library. On the title page Joscelyn, writing after Parker's death, claimed authorship: "This Historie was collected & penned by John Joscelyn ... by ye appointment & oversight of Matthwe Parker Archbp. of Cant. ye saide John being intertained in ye said Archb: howse, as one of his Antiquaries."[24] The evidence of Joscelyn's notebooks in the BL and Lambeth confirms that he did the main work of excerpting sources and composing narratives.[25]

But the bibliographical evidence makes clear that he did not do all the work. The *De antiquitate* includes, along with many other original sources, a text of the Accord of Winchester—an agreement made in 1072, which established the primacy of the archbishop of Canterbury in the English church. A British Library copy of the *De antiquitate* —described in notes on its fly-leaves as a set of bound proofs—contains not only the text of the Accord, but also the entire list of those who signed it, with crosses like those that marked their names in the original document.[26] But in the copies of the book that were actually given to Queen Elizabeth, William Cecil and others, the list of signatories was pruned back. The reader learned only that "it is clear from the archives" that many bishops and abbots signed the Accord.[27] The change must reflect discussion—presumably, a discussion between Parker, making a final decision about the form his book should take, and Joscelyn, his hard-pressed content provider.

The Lambeth manuscript also contains a second printed version of the life of Augustine of Canterbury, the first archbishop. This differs from the one that appears in normal copies of the first edition, both in content—it is much more oriented to the world outside England—and in form—it has sets of extended notes in the margins. A Parkerian annotator identified its author as a colleague: "Thes. 24 pages of Augustins life, were thus begun, by George Acworth, d. of the Law at ye appointment of Matthew Parker Archb. of Cant. & the lives of all ye Archb. shold have in this course bene perfected wt a generall storie. but deth prevented it."[28] Other documents—including everything from original charters with seals to printed broadsides—are tipped into this complicated but fascinating copy of the book.

[23] Parker 1572.

[24] Lambeth Palace Library MS 959, 36 recto.

[25] See e.g., London, British Library, MS Cotton Vitellius D. VII and MS Cotton Vitellius E XIV (the latter contains earlier drafts, the former later ones).

[26] London, British Library C.24.b.6, 2d pagination, 94–96.

[27] Parker, ibid., 2nd pagination, 95: "cum multis alijs Episcopis et Abbatibus, ut in Archivis patet."

[28] Lambeth Palace Library MS 959, 18 recto. On Acworth see Fritze, "Acworth, George," in *Oxford Dictionary of National Biography* (http://www.oxforddnb.com.ezproxy.princeton.edu/view/article/78, consulted May 20, 2014).

Many notes were added by members of Parker's team.[29] Not only was the original *De antiquitate* the product of collaboration, then: long after Parker's death, the loyal members of his household were still at work on a collaborative revision.[30]

It seems clear that the members of Parker's team regarded Caius and his project as connected organically to their own efforts. Matthew Parker printed the second, posthumous edition of *On the antiquity of Cambridge*, and he himself distributed presentation copies to the good and the great.[31] His son John went on doing the same long after his father's death. He had a copy with a special illuminated title page made up for King James I, who remarked, "What shall I do with this book? Give me rather Caius *De Canibus*."[32] Lambeth manuscript 959 includes Parker's printed work on Cambridge, which lists contemporary senior members of the university and describes his building projects at the university. Marginal notes add material from Caius's book.[33] Most suggestive of all, Parker had the printer add a cautionary note to the verso of the title page of some copies of the second edition. It explained that Caius had been less interesting in carrying on a polemic with his namesake in Oxford than in using a wide range of "ancient monuments" to establish Cambridge's privileges—a typical Parkerian effort to damp down controversy and emphasize the importance of primary source research, a value that Parker and his men shared with Caius, though also one that understates Caius's powerful

[29]Most striking—and most revealing of the labyrinthine quality of this printed book which has morphed into a manuscript—are two marginal notes in which members of the Parker circle identify beyond question the handwriting of a colleague: Lambeth Palace Library MS 959, 132 recto: "manus Domini Yale, ni fallor. autographum Yalei iam prae manibus habeo." "valde dubito. Manus enim Johannis Parker est perquam similis."

[30]See Knight 2013, 40–51.

[31]Plomer 1926, 252–268. London, British Llibrary C.32.h.15.(1.) is the Arundel/Lumley copy. A note on the title page reads: "Ex dono Mathei Cantuariensis Archiepiscopi."

[32]London, British Library C.24.a.27. (1.) is a presentation copy from John Parker to King James VI and I, with a bespoke colored title page. On the verso of the first fly-leaf appears the following: "Excellentissimo Principi Iacobo Angliae Scotiae / Franciae et Hiberniae Regi dignissimo. / (Matthei dudum Archiepiscopi filius.) / Subditus humilimus / Johannes Parker / hunc." James's response is quoted by Brooke, 75.

[33]Lambeth Palace Library MS 959, 359 recto: "Johannes Lydgatus, Galfridi Chauceri discipulus author est tempore Gurguntij Regis Britanni qui regnavit anno mundi 4317. Cantabrum Regis Hispaniae filium et Bartholum regis Hiberniae fratrem Cantabrigiam super Cantam fluvium condidisse, nomenque Cantabrigiae dedisse: anno mundi 4346. A transmigratione babylonica anno 538. eumque Athenis edoctum inde Philosophos advocasse et Cantabr. docendi gratia collocasse & ab alijs initiijs ad suam Bedae et Alfredi memoriam primae scholae et universitatis nomine Cantabr. claruisse. Johannes Caius"; 374 verso: "Singulares patronos et restitutores habuit Cantabrigia: et habet hodie quoque multos. Habet enim tot ex multis paucos referam.) lucens ille et pulcherrimus orbis literarum et virtutum Cantabrigia prae caeteros, tres summos et primarios viros, tanquam tres stellas radiantes, de quibus multum sane gloriatur. Reverendissimum Matthaeum Parker Cantuar. Archiepm. et totius Angliae primatem: D. Nicolaum Bacon equestris ordinis virum, summi Cancellarij locum tenentem et Magni Sigilli custodem. Et Guliel. Cecilium equitem auratum, summum Angliae Thesaurarium, regiae maiestati a Consiliis, atque Cantabrigiensis Academiae Cancellarium summum. Qui ut eodem tempore Cantabrigiae omnes studuerunt etc. ut Caius de antiquitate Canteb. Academiae. p[a]. 129 et 130."

sense of competition with Oxford.[34] So far, in other words, the evidence supports the view that *De antiquitate* was in part a Parkerian collaborative effort.

By contrast, Caius's book lacks one form of evidence that one would expect to find there—and that would, if present, have confirmed that he compiled it on his own. A library rat, Caius liked to brag about his adventures among the bookworms. He took clear pride in telling his readers, in his 1570 *De libris propriis*, about the contrasting conditions that awaited scholars at the library of the Medici, which was open to all, thanks to the generosity of Cosimo de' Medici, and the library in Urbino, which was very hard to access.[35] He not only explained that he had created an apparatus for future work on the text of Galen, but surveyed a whole series of Italian libraries where he had collated his manuscripts.[36] More important still, as a philologist Caius knew that location and provenance mattered to experts on manuscripts. In his collations of manuscripts of Galen, he usually noted the owners of the manuscripts he studied and often dated his working sessions with them—vital information for later editors, rarely supplied by scholars at this time (Parker and his secretaries seem almost never to have done so).[37]

Caius thought it important that England develop research libraries. In his posthumously published *History of Cambridge*, he excoriated the inhabitants of Oxford for neglecting Duke Humfrey's library (he clearly had fellow-feeling for another patron spurned).[38] Then he drew up a detailed list of the books in Cambridge University Library.[39] He insisted again and again, in his polemic against Thomas Caius, that his arguments deserved belief because he had used very few recent texts, "just as I used few printed texts, but used all the other manuscripts,

[34]"Non tam solicitus fuit Caius noster cum adversario suo de utriusque Academiae antiquitate in hoc opere contendere, quam quae ex varijs antiquis monumentis de statu, privilegijs, dignitate, ac praerogativa Cantebrigiae ipse collegisset, edere ac in lucem proferre. In quo eum maxime elaborasse facile erit sano ac prudenti lectori deprehendere."

[35]Caius 1570, *Works*, 86–87, contrasting "Bibliotheca publica illustrissimi principis Cosmi Medices, quae omnibus literarum studiosis principis humanitate atque gratia patet" (86) with the library at Urbino ("usque adeo difficilis accessus est in Bibliothecam ejus viri") (87).

[36]Ibid., 100–102.

[37]Nutton 1987; Berlier 2011, 1–14 (which argues that Caius probably stole at least portions of one of the manuscripts that he saw in Italy).

[38]Caius, *Historia*, in *Works*, 68: "Ita utraque Oxoniensium Bibliotheca cum alijs perquisitis in illas scholas Theologicas translata, a nobilibus viris ornata quondam fuit. Quae iam vereor ne una cum Patronorum memoria deleta pene atque consumpta sit. Tam paucis annis gratitudinem extinguit negligentia, & benemeritorum oblivionem parit. Proinde admonendi sunt utriusque universitatis studentes, ut diligenter conservandis his quibus affecti sunt beneficijs, colendaque fraequenter Patronorum memoria a supina illa negligentia se prorsus vendicent atque seiungant. Eo enim modo Patronos novos indies conciliabunt, & quae profutura sibi sunt, acquirent."

[39]Ibid., 68–71: "Hi autem veteres libri in Cantebrigiensi Bibliotheca iam supersunt." At [115–116] thete is a "Note on pp. 68–71 by Dr. M.R. James, Provost of King's College," which identifies many of the books and manuscripts listed by Caius.

which are my principal delight, as exemplars of pure and venerable antiquity."[40] Yet nowhere in his works in defense of Cambridge did he so much as mention Parker's collection—much less make clear that he had done the vast majority of his research there.[41] One simple explanation for this lacuna would be that Parker and his men did a good deal of the research for him.

A third document of another kind allows us to come closer still: to stand behind Caius at his desk and watch him work. On April 8th, 1567, he wrote to Parker, sending him the handwritten draft of *De antiquitate* with the request that he assess it. "I wholly commit it to your grace's pleasure, and trust no man shall see it until I hear further of your grace's pleasure." Caius made clear that the text was not complete: "View it again I would before it should be printed, for that many things be roughly left for want of leisure, and haste to satisfy your grace." He asked Parker to correct it, and gave detailed procedural instructions: "Because all things should be the readier to your grace I have put to every *pagina* his number. If anything your grace will note, the number is ready to tell the place. If anything your grace will have altered, note it *seorsum* for avoiding diversity of styles." Caius even cautioned Parker not to let Joscelyn "shew it to everybody, and give out copies *ante maturitatem*, and do little good in it himself."[42] This was clearly no pro forma letter, but a precise request for editorial help, as scholars since Strype have recognized.

Still, two points in the letter have generally escaped notice. Caius tells Parker: "Some things that your grace thought best should be put out, were by the writer put in before I was aware, and therefore remain, but so that what your grace will have done with them shall be done."[43] This sentence clearly refers to a previous stage of editorial discussion. Caius had already shown Parker a first draft, and Parker had

[40]Ibid., 184: "Etenim si quis volet superiorum temporum omnia comprehendere, ex ipso fonte petat, unde ad alios rivus dimanavit. Quapropter neque ego certe usus essem recentium authoritate, nisi tibi in hac controversia hos placere ex usu animadvertam, sic ut non alij aeque. Usus autem sum paucissimis, ut & impressis paucis, coeteris omnibus scriptis, quibus delector maxime, ut incorruptae & venerandae vetustatis exemplaribus." Note also his description of the value of Asser as a contemporary witness to Alfred's doings, ibid., 74: "Quid multis? Fidem & authoritatem maiorem semper affert vetustas in omnibus controversijs, ut quae, res ut erant viderat, aut illis quam proxime accesserat. Posteriores igitur examinabo ad primos illos incorruptae vetustatis scriptores, qui fide supereminent omnes, quod aut illis diebus vixerant quibus haec gesta sunt, aut his non longe aberant, aut quam proxime (ut dixi) accesserant, ut ex illis haurire possent. Inter quos primus primaeque fidei Asser seu Asserus est, oculatus & auritus testis, qui ex intimis Aluredi familiaribus fuit, qui in eius Aula vixit, res eius & domesticas & forenses novit, atque etiam cum doctis regiae familiae viris consuetudinem habuit, omniaque in vita & in morte diligenter observavit, ut solent qui historias veras scribere decreverunt." It was of course Parker who arranged for the publication of Asser's work on Alfred.

[41]Though Caius's bibliography contains substantive comments on some of the works it lists, it says nothing about their provenance or location. See e.g. *Works*, 226: "Antonini Augusti itinerarium, in quod vir magnae diligentiae, & praestabilis nostri temporis Antiquarius Robertus Talbotus scripsit commentarios, satis certe luculentos atque elaboratos."

[42]Parker, *Correspondence*, 299.

[43]Ibid.

already suggested revisions, at an earlier stage. Apparently, Caius had not managed to carry all of them out. No doubt remains: the book was a joint project. The other point is even simpler. Caius dated his letter not from London but from Cambridge.[44] But the research base for a large part of his work was in Lambeth, on the south bank of the Thames. For at least part of the period when the text was taking shape, Caius was in one place and the manuscripts he relied on in another. Caius deployed the evidence of medieval texts as if he were using a salt-shaker rather than a sharp nib, as many modern readers have complained. It seems likely that Parker and his secretaries stuffed the text with at least some of these inapposite citations.

The three sorts of evidence that we have examined so far all reveal material flowing from Parker and his men to Caius. But Parker trusted and looked up to Caius, and the traffic moved in both directions. It seems likely that Caius helped Parker bring a new and more sympathetic perspective to bear on the Anglo-Norman historians whom he and his secretaries published. Parker consistently held that the church had been purest at its beginnings, and that its corruption accelerated after the Norman Conquest. But in his 1570–71 edition of the thirteenth-century Benedictine chronicler Matthew Paris, Parker made clear his liking for that stout medieval critic of the papacy. More striking still, he suggested that Matthew's chronicle formed part of a longer tradition of histories commissioned by the kings of England, which were composed and preserved at the Monastery of St Albans.[45]

To support this argument, Parker drew on the preface to the British history of the Bellunese humanist Ludovico da Ponte, perhaps better known by his Latin name, Ponticus Virunius.[46] Learned historians, Parker claimed, had recorded and then archived contemporary history:

> It was our law that the monasteries and ecclesiastical colleges, especially Saint Albans, should be a sort of common treasury, where all the historical events of this kingdom that were worth remembering were recorded. This was also noted by Ponticus Virunius, in his British history, in which he attests that it was the custom of western rulers to have scholars with them who could master their excellent sayings and deeds exactly and by heart. But they did not want to publish these splendid deeds abroad, while they or their sons lived. These outstanding examples were carefully preserved in the registers of the kings, so that posterity might be shaped by them to imitate the virtues of their ancestors.[47]

[44]Ibid., 300.

[45]Parker, ed. 1570–71. † iii verso – † iiii recto: "Sane studiosius eum hanc provinciam in se suscepisse credibile est, quod lege et communi decreto cautum erat apud nos, Monasteria et Collegia ecclesiastica, in primis vero et prae caeteris Albanense Coenobium, quasi communem thesaurum et receptaculum debere esse, ubi reponerentur ac fidelissime reservarentur omnia historica gesta huius Regni, et quaecunque memoria ac fama digna essent."

[46]On Da Ponte see Ricciardi 1986 (http://www.treccani.it/enciclopedia/ludovico-da-ponte_%28Dizionario-Biografico%29/; consulted May 19, 2014).

[47]Parker, ed. 1570, loc. cit.: "quod et notavit PONTICVS VIRVNIVS. in historia sua Britannica, in qua hoc testatum relinquitur, morem et consuetudinem Occidentalium Principum fuisse, semper apud se domi habere tales eruditos et doctos viros, qui sua et suorum dicta et facta egregia, vere possent et memoriter statim expedire: nolebant tamen haec sua magnifica et heroica gesta in publicum prodire et evulgari, quamdiu aut ipsi aut ipsorum filii viverent. Semper vero haec illustria et nobilia exempla in Principum regestis diligentius custodiebantur, ut hinc insequens posteritas ad maiorum suorum excellentes virtutes et imitationem facilius informarentur."

He then added what he called a "more weighty testimony" to the same effect from one of the medieval historical texts in his library, the *Scotichronicon*.[48]

The general description of western historians that Parker quoted did appear in the prooemium to Da Ponte's little history. The Italian historian, trying to shore up the credibility of his sources, had evoked "... the custom of western kings, of having with them men who would record their deeds, with particular accuracy, but not to reveal them in their lifetimes or those of their sons, for it would be shameful to state what they had not been able to achieve with such great authority. They kept them for posterity in the royal archives."[49] Parker did not distort Da Ponte's views.

Still, he may not have known them directly. For neither the reference to Virunius nor the effort to connect it to the Saint Albans chronicle tradition reflected new research or thinking. A sequence of arguments strikingly similar to Parker's appeared previously in Caius's book of 1568. First he conjured up the Saint Albans chronicles, presenting them as a sort of national archive: "Edward I ordered the monastery at Rochester to record the events of his times, as they happened, and he chose that monastery, with many others, and especially St Albans, as if for a treasury of memorable things as Matthew of Westminster and the author of the Rochester chronicle write ..."[50] Then, on the bottom of the same page and on the

[48]Ibid.: "Pro graviori testimonio huius antiqui et perveteris moris in hac Anglia nostra (quod his nostris diebus valde optabile esset, ut servaretur accuratius) memoriae mandatum est ab illo qui scripsit SCOTICHRONICON, libro. 16. cap. 39. *Statutum est (inquit) convenienter in Anglia (ut audivi) quod unumquodque Monasterium a Regibus fundatum, haberet de ipso loco suum certum scribam, vel scriptorem, quo omnia notabilia tempore Regis, saltem in Regno, vel e vicinis contingentia, secundum quod veritas facti se haberet, cum data annotarentur: ad proximum generale Concilium, post obitum Regis, omnes illi Chronographi convenirent, et sua vere dicta, sive scripta, in medium producerent, et delectis a Concilio sagacioribus, et in talibus peritis et expertis, scripta examinarent, et diligenti habita collatione, de congestis summarium extraherent, et Chronicam compingerent, ac in Coenobicis Archivis librariorum, pro authenticis Chronicis, quibus fides daretur, scripta reponerent, ne temporum labilitate memoriae gestorum in Regno deperirent.* Haec ille. Hic mos continuabatur et studiose observabatur a pluribus Coenobiis, sed maxime ab illo Monacho Albanensi THOMA WALSINGHAM, qui in summam et compendium congesserat omnia memoratu digna, quae gerebantur ab ultimo anno Henrici tertii, ad annum primum Henrici sexti: cuius historia (si hic labor noster, quod speramus, satis tibi probabitur, amice Lector) a nobis posthac fortasse in ordinem rectius composita adducetur, et prelo etiam in tuum commodum commitetur." The passage from the Scotichronicon is from Corpus Christi College Cambridge MS 171B, folio 353 recto. The first words of the portion quoted by Parker are underlined in red chalk.

[49]Ludovico da Ponte 1534, C1 recto: "Ad manus meas historiae regum Britannorum nobilissimae supra cladem Trioianorum pervenere, quas esse verissimas arguebat regum Occidentalium consuetudo semper secum habere, qui eorum gesta notarent veritate praecipua, sed nec viventibus ipsis, nec filijs aperire, obprobrium vero fore, scilicet attribuere, quae ipsi in tanto Imperio facere non potuissent, eas deinde in regalibus archivis in posteros custodire ..."

[50]Caius 1568, 239: "imperavit coenobio Roffensi, ut in commentarios referrent res gestas sui temporis ut acciderint, idque coenobium cum multis alijs monast. & praecipue S. Albani, in hoc delegit, tanquam in Thesaurum et custodiam rerum memorabilium, uti Matth. Westm. & is qui Roffensem historiam aedidit, scribunt."

next one, he drew on Da Ponte to set this particular form of history-writing into a larger context.

> For it was once the custom of kings in the western regions, to have with them men who would accurately record the events of each year as they happened. But they did this without making them public in the time of the king or his sons, but passed the histories that were composed into the royal archives, where they were preserved for posterity, as Virumnus Ponticus explains in his history of the Britons.[51]

The passages are not identical, and Caius did not cite the *Scotichronicon*, with which Parker developed his vision of the St Albans tradition. But the similarities are striking. To draw on Da Ponte was, in this context, quite peculiar. He was best known not as a historian at all, but as a witty Greek teacher who enlivened his lessons about grammar at the school in Reggio Emilia with many jokes. In his lectures, he kept the students awake by describing masculine declensions as monasteries that had expelled women, and feminine ones as nunneries.[52] It seems far likelier that Caius, the sophisticate and onetime Italian traveler who loved the British history, as Da Ponte did, brought this foreign work into the discussion, than that Parker did.[53]

Caius, moreover, had reasons of his own for insisting on the value of Matthew Paris and his ilk. In general, he felt a more sympathetic interest in the Norman church and its institutions than Parker did. Parker insisted that all ministers conform in the use of vestments, and instructed colleges to melt down plate that seemed irremediably Catholic in its decoration. By contrast Caius referred to himself, in the first, anonymous edition of his work on the Antiquity of Cambridge, as an antiquary.[54] Like the seventeenth-century antiquaries so vividly evoked by Alexandra Walsham, he seems to have felt a considerable interest and affection for material remains that more strenuous Protestants would have wished not to collect but to burn.[55] According to the dissident members of Caius College, the Master "mainteyneth wythin his colledge copes vestments albes suinches sensors crosses tapers also all kinde of Masse bookes Porteses pies grales processionalls wth all massinge abominations and termeth them the college treasure."[56] The Vice-Chancellor, the Master of Trinity and the Provost of Kings demanded an inventory

[51]Ibid., 239–240: "Etenim Regibus occidentis mundi partis consuetum fuit olim, apud se habere eos, qui res eorum ut erant gesta, annis singulis bona fide scriberent. Sic tamen ut neque regis aetate, nec filiorum eius facerent publicas, sed ita concinnatas historias, in archiva referrent regia, ubi ad posteros reservarentur: uti Virumnus Ponticus in historia Britannorum refert."

[52]Grafton and Jardine 1986, 107–109.

[53]For Caius's interest in Italianate forms of symbolism and sculpture see Fox 1986, 46–56, and Radcliffe 1987, 121–126.

[54]Caius 1568, e.g. 5, 14 (Caius keeps up a feeble pretense that the "antiquary" and he himself are not the same person).

[55]Brooke, 72–73; Walsham 2011; and Aston 1973, 231–255.

[56]Lambeth College Library MS 720, quoted by Venn, "John Caius," 26. Caius was also accused of papism in a memorandum of 1572 or later, written by a critical member or members of the Royal College of Physicians, of which he was elected President nine times: Clark and Cooke 1964–72, I, 127–130.

of the "treasure" from the college, and on December 13, 1572, all of these items that could be found were confiscated and burnt.[57] It is certainly possible that Parker's close friend Caius nourished secret Catholic beliefs, as some have suggested.[58] But it also seems not at all impossible that the old doctor, nostalgic for the monastic colleges of his youth, looked for evidence to convince Parker that medieval monastic chronicles deserved credence.

It is a truth universally acknowledged that collaboration in the humanities is a recent phenomenon. Humanists, we know, have traditionally modeled their lives on the tragic figure in Dürer's *Melencolia* I: sunk in Saturnian misery and isolation, weighed down by their learning, they have labored endlessly in solitary cells. In the study of nature, by contrast, collaboration played a central role in practice as early as the late sixteenth and seventeenth centuries. The alchemical laboratories, astronomical observatories and scientific societies that took shape across Europe in the late Renaissance and after employed numerous workers and often performed operations that required many hands to work together. In the nineteenth and twentieth century, scientific work and authorship were very largely collaborative. Humanists, by contrast, have only begun in the recent age of European and national funding councils and pressure for outside funds to organize large-scale research projects.

Or so we are often told. These generalizations are as untrue as they are widespread, and it is well past time that they were laid to rest. Caius was only one of many distinguished humanists who, during the sixteenth, seventeenth and eighteenth centuries, participated in collaborative scholarly enterprises that were ecclesiastical in their inspiration and their goals. By working with Matthew Parker, Caius gained access to the scholarly resources that he needed and to the expert editorial advice that he wanted. His books on Cambridge were not only inspired, but also shaped, by the help that Parker and his antiquaries provided. Yet Caius did not passively accept what the members of Parker's team offered him: in fact, it seems, he helped to shape Parker's editorial program. The story of these exchanges is only a little twig from the great oak of Nancy Siraisi's work on medicine and history in early modern Europe. But it helps to show that our understanding of humanistic scholarship has often been impoverished in the past, by our failure to realize how often the goals it aimed at and the practices it adopted were ecclesiastical in their origins.

[57]Brooke, 76–77. For Caius's own account of this event see Caius 1904, 185.

[58]Informed opinions on this subject have differed. John Venn 1923, 55, thought that Caius "never ceased to be at heart a decided Roman Catholic." Brooke disagrees, as does Grierson, "Caius's Library," 524–525.

Bibliography

Aston, Margaret. 1973. English Ruins and English History: The Dissolution and the Sense of the Past. *Journal of the Warburg and Courtauld Institutes* 36: 231–255.

Berlier, Stéphane. 2011. John Caius et les manuscrits de Galien. *Revue d'Histoire des Textes*, new ser., 6: 1–14.

Binski, Paul. 2013. Humfrey Lovell and the Gates of Gonville and Caius College: A Note on the Sources. *Journal of the British Archaeological Association* 166: 179–188.

Brooke, Christopher. 1985, repr, 1996. *A History of Gonville and Caius College*. Woodbridge/Dover: Boydell.

Budny, Mildred. 1997. *Insular, Anglo-Saxon, and Early Anglo-Norman Manuscript Art at Corpus Christi College, Cambridge: An Illustrated Catalogue*, 2 vols. Kalamazoo: Medieval Institute Publications.

Caius, John. 1568. *De antiquitate Cantabrigiensis academiæ libri duo*. London: Bynneman.

———. 1570. *De libris propriis*. Seres.

———. 1904. *The Annals of Gonville and Caius College*, ed. John Venn. Cambridge: Printed for the Cambridge Antiquarian Society.

———. 1912. *Works*, ed. E.S. Roberts. Cambridge: University Press.

Clark, G.N., and A.M. Cooke. 1964–72. *A History of the Royal College of Physicians of London*, 3 vols. Oxford: Clarendon Press.

Cox Jensen, Freyja. 2012. *Reading the Roman Republic in Early Modern England*. Leiden: Brill.

Da Ponte, Ludovico. 1534. *Britannicae historiae libri sex*. Augsburg: Alexander Weyssenhorn.

Fox, Paul. 1986. On the Symbolism of the Arms of John Caius and of the College Caduceus. *The Caian* 46–56.

Fritze, Ronald. Acworth, George. In *Oxford Dictionary of National Biography*. http://www.oxforddnb.com.ezproxy.princeton.edu/view/article/78.

Gabriel, Astrik. 1988. *Translatio studii*. Spurious Dates of Foundation of Some Early Universities. *Fälschungen im Mittelalter*, 6 vols, I: 601–626. Hanover: Hahn.

Grafton, Anthony. 2012. Church History in Early Modern Europe: Tradition and Innovation. In *Sacred History: Uses of the Christian Past in the Renaissance World*, ed. Katherine Van Liere, Simon Ditchfield, and Howard Louthan, 3–26. Oxford: Oxford University Press.

Grafton, Anthony, and Lisa Jardine. 1986. *From Humanism to the Humanities*. London/Cambridge: Duckworth/Harvard University Press.

Graham, Timothy. 2006. Matthew Parker's Manuscripts: An Elizabethan Library and Its Uses. In *The Cambridge History of Libraries in Britain and Ireland*, 322–341. Cambridge: Cambridge University Press.

Graham, Timothy, and Andrew Watson. 1998. *The Recovery of the Past in Early Elizabethan England: Documents by John Bale and John Joscelyn from the Circle of Matthew Parker*. Cambridge: Published for the Cambridge Bibliographical Society by Cambridge University Library.

Grierson, Philip. 1897–. Appendix IV: John Caius's Library. In *Biographical History of Gonville and Caius College*, 8 vols, ed. Venn et al., 7: 509–535. Cambridge: Cambridge University Press.

Hiatt, Alfred. 2004. *The Making of Medieval Forgeries: False Documents in Fifteenth-Century England*. London: British Library.

Jardine, Lisa, and Anthony Grafton. 1990. 'Studied for Action': How Gabriel Harvey Read His Livy. *Past and Present* 129: 30–78.

Kendrick, T.D. 1950. *British Antiquity*. London: Methuen.

Kewes, Paulina. 2011. Henry Savile's Tacitus and the Politics of Roman History in Late Elizabethan England. *Huntington Library Quarterly* 74: 515–551.

Knight, Jeffrey Todd. 2013. *Bound to Read: Compilations, Collections and the Making of Renaissance Literature*. Philadelphia: University of Pennsylvania Press.

Lambeth College Library MS 720.

Lambeth Palace Library MS 959.
London, British Library, MS Cotton Vitellius D. VII and MS Cotton Vitellius E XIV.
London, British Library C.24.b.6.
London, British Llibrary C.32.h.15.(1.)
London, British Library C.24.a.27. (1.)
McMahon, Madeleine. 2013. '*Ani one example of the primitiue churche': Church History and Confessional Identity in Sixteenth-Century England.* BA Thesis, Princeton University.
Nickson, Tom. 2005. Moral Edification at Gonville and Caius College, Cambridge. *Architectural History* 48: 49–68.
Nutton, Vivian. 1987. *John Caius and the Eton Galen.* Cambridge: Cambridge Philological Society.
———. Caius, John. In *Oxford Dictionary of National Biography.* http://www.oxforddnb.com.ezproxy.princeton.edu/view/article/4351?docPos=1.
Parker, Matthew, ed. 1570–71. *Matthaei Paris, Monachi Albanensis, Angli, Historia Maior, a Guilielmo Conquaestore, ad ultimum annum Henrici tertii. Cum indice locupletissimo.* London: Reginald Wolf.
———. 1572. *De antiquitate Britannicae ecclesiae & privilegiis ecclesiae Cantuariensis, cum archiepiscopis eiusdem.* London: John Day.
———. 1853. *Correspondence*, ed. John Bruce and Thomas Thomason Perowne. Cambridge: At the University Press.
Parker, James. 1885. *The Early History of Oxford, 727–1100.* Oxford: Printed for the Oxford Historical Society at the Clarendon Press.
Plomer, Henry. 1926. The 1574 Edition of Dr. John Caius's *De antiquitate Cantebrigiensis academiae libri duo. The Library*, new ser. VII, 3: 252–268.
Radcliffe, Anthony. 1987. John Caius and Paduan Humanist Symbolism. *The Caian* 121-126.
Ricciardi, Roberto. 1986. Da Ponte, Ludovico (Ponticus Virunius). *Dizionario biografico degli italiani* 32. http://www.treccani.it/enciclopedia/ludovico-da-ponte_%28Dizionario-Biografico%29/.
Siraisi, Nancy. 2007. *History, Medicine, and the Traditions of Renaissance Learning.* Ann Arbor: University of Michigan Press.
Stephens, Walter. 1989. *Giants in Those Days: Folklore, Ancient History, and Nationalism.* Nebraska: University of Nebraska Press.
Strype, John. 1711. *The Life and Acts of Matthew Parker.* London: John Wyat.
The Parker Library on the Web. http://parkerweb.stanford.edu/parker/actions/page.do?forward=home.
Todd, Margo. 1987. *Christian Humanism and the Puritan Social Order.* Cambridge: Cambridge University Press.
Venn, John. 1897–1998. John Caius. In *A Biographical History of Gonville and Caius College*, 8 vols, ed. John Venn et al. Cambridge: At the University Press.
———. 1912. John Caius. In *Works*, ed. E.S. Roberts. Cambridge: University Press.
———. 1923. *John Caius.* Cambridge: Reprinted by the Cambridge University Press for Gonville and Caius College.
Walsham, Alexandra. 2011. *The Reformation of the Landscape: Religion, Identity, and Memory in Early Modern Britain and Ireland.* Oxford: Oxford University Press.
Womersley, David. 1991. Sir Henry Savile's Translation of Tacitus and the Political Interpretation of Elizabethan Texts. *Review of English Studies* 42: 313–342.
———. 1992. Sir John Hayward's Tacitism. *Renaissance Studies* 6: 46–59.

Chapter 6
A Medical Collection Anatomized: The *Catalogus bibliothecae Hieremiae Martii* (1572)

Ian Maclean

In 2008, an article by Giles Mandelbrote appeared entitled 'The first printed library catalogue? A German Doctor's library of the sixteenth century and its place in the history of the distribution of books by catalogue.'[1] The evidence for this very rare volume being a sales catalogue lies principally in its *mise-en-page*. The catalogue has two vertical lines at the outside margin of every page, which could accommodate prices estimated and received. There are certainly other lists at this time which use this typographical layout to record prices[2]; but it might also be the case that the publication is a public expression of pride or vanity. In either case, the collection itself is of great interest, and worthy of close scrutiny. Its study fits in well with Nancy Siraisi's two impressive recent books on the wider scholarly interests of doctors, and as such has a natural place in this festschrift.[3]

The catalogue begins with a brief preface:

> You will find in this catalogue, good reader, an abundance of the most select books of all kinds, most especially relating to the medical art. Among them many are new and elegantly bound; some others also are not to be found anywhere else today. I wanted to point this out to you, in case you might think that this library has been brought together by me without discrimination. Only someone putting something of this sort to the test will come to understand properly what this has really cost me.[4]

---

[1]Mandelbrote 2008.

[2]There is an example in Belper 2011, 9–21.

[3]Siraisi 2007, 2013.

[4]'Invenies, optime Lector, in hoc catalogo librorum selectissimorum omnis generis copiam, praecipue vero in arte Medica. Inter quos plerique novi et eleganter sunt compacti, nonnulli etiam hodie nullibi extant. Quod te monere volui, ne existimes hanc Bibliothecam absque iudicio a me fuisse collectam. Quanto vero mihi constiterit, is demum optime intelligerit, qui huius rei periculum fecerit.'

I. Maclean (✉)
University of Oxford and Fellow of All Souls College, Oxford, UK
e-mail: ian.maclean@all-souls.ox.ac.uk

G. Manning, C. Klestinec (eds.), *Professors, Physicians and Practices in the History of Medicine*, Archimedes 50, DOI 10.1007/978-3-319-56514-9_6

In his short preface to the reader, Martius sets out a number of claims about the catalogue which follows. According to him it is comprehensive, but selective, in respect of its medical holdings ('librorum selectissimorum omnis generis copiam, praecipue in arte medica'); it contains many up-to-date items ('plerique novi'), and a number of very rare (presumably out-of-print) books ('nonnulli etiam hodie nullibi extant'); some books have luxury bindings ('eleganter compacti'); the contents were chosen with care ('non absque iudicio'), and it is has cost its owner much to assemble them (he does not specify whether this is in terms of money of effort: 'quanto vero mihi constiterit'). He suggests that the collection covers all the salient parts of what would be expected from a medical library at this time, and implies that what is absent from the collection is the result of conscious exclusion on his part. He does not specify what sort of medical library he has in mind, although we may deduce from the reference here to 'ars medica' and to himself on the title-page as 'doctor medicus augustanus' that it is one to be associated with the learned medical profession, and possibly with Augsburg also. 'Augustanus' on the title-page might simply be a conventional descriptive personal adjective, but I suspect that it is more than this. A designation of place or nation is often used at this time as an 'epitheton ornans', that is, a rhetorical gesture to express local pride; it is pertinent to note that many of the entries in the catalogue retain the indication of the authors' place of birth.[5] From the contents of the collection, we may infer also that the 'medical art' is conceived of more broadly than the discipline associated with university-trained Latinate physicians, for there are works in German, Italian and French, works written for a non-professional readership, and works covering subjects which are normally thought to fall outside the remit of learned physicians: alchemy, metallurgy and surgery being some of these.

I shall set out in this essay to evaluate Martius's claims about the medical component of the collection – that it is comprehensive, well-chosen, up-to-date and includes rare items – by relating it first to his medical training and early career, comparing it next with bibliographies available to Martius containing sections on medicine, and thereafter with libraries having similar contents owned by his contemporaries. I shall take an analytical look at the collection itself from the point of view of subject, region, and date, and consider how this collection relates to Martius's home town of Augsburg and his employment there in the Blatterhaus (hospital for syphilics), to his academic peregrination up to 1563, to the book market of his time and the way it operated, and to the efficiency (or otherwise) of the diffusion of books throughout Europe.

[5] See Maclean 2009, 372–9.

6.1 Martius's Academic Career

Martius includes an account of his training in the opening epistle of his edition of the ancient Greek physician Nonnus Theophanes, dedicated to the Town Council of Augsburg in whose employ he then found himself, which appeared in 1568.[6] As one would expect, he evinces strong loyalty to his patrons, his tutors, and to those who impressed him during his formative years. Through the financial support he received from members of the Fugger family (Anton, and later his children Marcus and Hans), he was able to gain experience in Germany, France and Italy, which gave him a European perspective on his chosen profession. After an excellent humanist education at the protestant Gymnasium of St Anna in Augsburg, where he collaborated with his teacher Hieronymus Wolf (1516–80) in the edition of medieval Greek chronicles,[7] he went on to the staunchly Catholic local university Ingolstadt, whose medical faculty was small but distinguished.[8] The renowned botanist and physician Leonhard Fuchs had taught there up to 1535, before moving to Tübingen, where his protestantism did not cause the problems it had in Ingolstadt. The senior professor there in Martius's time was Johannes Agricola Ammonius (1496–1570), who had a longstanding connection with Martius's patron Anton Fugger.[9] Agricola Ammonius was a very distinguished linguist and medical writer, who had fostered botanical studies and given preference to Greek over Arab medicine.[10] Martius possessed at least two of his books,[11] but he does not mention him by name in his autobiographical essay; he declares rather that his tutor was the Bavarian Lorenz Gryll, a client of another member of the Fugger family and pupil of Johannes Agricola Ammonius, who taught at Ingolstadt from January 1556 until his premature death in 1558.[12] Gryll seems to have been a model for Martius. He chose in his inaugural lecture to give an account of his own academic peregrinations and encounters with physicians in Vienna, Ingolstadt, Tübingen, Padua, Ferrara, Bologna, Florence, Rome, Montpellier, Toulouse, Paris, England, the Low

[6]Theophanes 1568, 4v ff. The book contains poems in support of Martius by Achilles Pirmin Gasser and Hieronymus Wolf.

[7]These MSS, as well as the MS of Nonnus Theophanes, were in the possession of Anton Fugger; the edition appeared in 1557. Both Martius and Wolf were paid by Fugger for this work: see Lehmann 1956–60, i.31–3, 37. On the Gymnasium of St Anna, see Müller 2010.

[8]Rotmar 1571; Liess 1984; Nauck 1956. According to Nauck, there were three professors of medicine in Ingolstadt between 1554 and 1567. The curriculum, which is unusually ambitious (it embraces almost the whole corpus of Galenic and Hippocratic texts, as well as clinical precepting and botanical studies) is dated by Nauck to 1556, but Nutton 1989, 421, has shown convincingly that it was more likely to have been introduced after 1562. See also Nauck 1954, 181, which records 19 matriculations in the medical faculty at Ingolstadt between 1550 and 1560.

[9]See Ammonius 1539, dedication, which is translated and commented in Hieronymus 2005.

[10]See Bietenholz and Deutscher 1985–7, s.v. Ammonius.

[11]Ammonius 1534, 1537.

[12]According to Gryll's editor Adam Landanus; see Gryll 1566, γ1r; but Gryll himself names Johann Vietmüller as his tutor: *Oratio*, f.2r.

Countries and Germany. This allowed Gryll to impress on his student audience that 'a solid general knowledge of medicine is to be gained not only from books and study, but from conversations with leading doctors, travel, and first-hand experiences': advice which Martius seems to have taken to heart.[13] The final part of his lecture was devoted to a strong defence of the value of botanical studies.

It is unlikely that Martius would have encountered Johannes Boschius [Lonaeus van der Bosch] (d. 1578?), who came from Louvain to Ingolstadt in 1556 or (more probably) 1558, but it is pertinent to mention here the lecture he gave after his arrival 'on the best doctor and on medical writers'.[14] After a summary history of medicine from the earliest times, Boschius affirms his adherence to humanist medicine and his hostility towards the Arabic tradition; among recent doctors, he praises Giovanni da Vigo (1450?–1525) of Rome, Girolamo Fracastoro (1478–1553) and Giambattista da Monte (1498–1551) of Padua, Giovanni Argenterio (1513–72) then of Naples, Jean Fernel (1497–1558) and Jacques Dubois (1478–1555) of Paris, and the celebrated anatomist Andreas Vesalius (1514–64). Boschius also referred to new illnesses (the great pox and the English sweat) and new remedies (China Root, Sarsaparilla and guaiac, the wood of which the Fuggers held the monopoly, used in the treatment of the great pox). If indeed Martius had cognizance of the contents of this lecture (which was not published until 1571), it would seem that he took its recommendations to heart, in terms of both of subject matter and of authors.

Martius spent only 1 year in Ingolstadt; he alludes to problems he encountered there ('I thought it was not a place for me to linger in, for a variety of reasons'[15]), which may well have been of the same confessional nature as those which drove away his fellow protestant Leonhard Fuchs. Gryll urged him to go to Montpellier, which he was able to do with his patron's financial support.[16] He encountered all the members of the distinguished medical faculty during his time there, and praises the location as especially good for the study of plants and fish. He singles out Guillaume Rondelet for special commendation, as Gryll had done.[17] After his 3 years in Montpellier, his patron summoned him back to Augsburg to begin his medical practice. Anton Fugger died shortly thereafter, in September 1560. On the advice of

[13]Ibid., f. 12r: 'solidam universae medicinae cognitionem non solum ex libris et privatis cogitationibus, sed etiam multorum excellentium virorum conversatione, peregrinationibus variis, et rerum omnium αὐτοψία et crebra inspectione comparari.'

[14]Rotmar 1571, 268–276; see also the World Biographical Information System (http://db.saur.de/WBIS) s.v. Boschius, where the various entries in biographical dictionaries disagree about the date of Boschius's arrival in Ingolstadt.

[15]Theophanes 1568, 5r: 'cum putarem multis de causis mihi non diutius Ingolstadii manendum esse.'

[16]Ibid.: 'suadente ita praeceptore et annuente maecenate.'

[17]Ibid. At Montpellier, all students were assigned to a 'parrain', that is, one of the four regius professors who undertook to look after the student's personal and academic needs: see Dulieu 1975–9, ii.71. It is likely that Martius was assigned to Rondelet. See also Gryll, *Oratio*, f. 6v (Gryll stayed in Rondelet's house).

an Augsburg patrician (Johann Baptist Haintzel[18]), and with the support of Anton's sons, he set off again, travelling first to Padua where he spent 6 months and sat at the feet of one of the most distinguished medical faculties in Europe. Two years in Florence followed, one of which was spent working in Cosimo de' Medici's hospital. A stay in Rome brought yet more encounters with prominent physicians and naturalists, bringing the total of scholars named to more than 20; some of their works he seems to have purchased on his travels, others more probably later.[19] On his return to Augsburg (probably before 1566), he obtained a medical post in the Blatterhaus through the good offices of the town doctor and bibliophile Achilles Pirmin Gasser (1505–77).[20] Gasser's protestant convictions, his own medical training, career and writings, and his close association with Konrad Gessner, to whom Martius was known as a successful empirical botanist, predisposed Martius to friendship with Gasser, and made him susceptible to his intellectual influence.[21]

Martius's publishing career as an author in his own right had begun in 1563 with a funeral oration for Cardinal Giovanni de' Medici.[22] After his return to Augsburg, he produced his edition of Nonnus Theophanes, which a prominent Strasbourg printer of scholarly works, Josias Rihel, printed in 1568, probably at his own expense.[23] Martius's only other publications produced outside Augsburg were his translation into Latin of Jacques Grévin's book on poisons, printed by the prominent Antwerp publisher Christophe Plantin in 1571, and Africo Clemente's book on agriculture for another Strasbourg publisher, Bernhard Jobin, which appeared in 1580. Martius's local productions were all translations into German of works designed to benefit the urban literate community: books of secrets, remedies and gynecology by Nostradamus, Falloppio, Nicolaus de Metri (fl. 1488) and Giovanni Marinelli, financed by the local bookseller Georg Willer.[24]

---

[18](1524–81); A protestant member of the Geheimer Rat, and an Oberkirchen- und schulpfleger. See *Augsburger Stadtlexikon* 2009; Burmeister 1970, i.167.

[19]Theophanes 1568, 5v: the most relevant names are Fracanzanus, Fallopius, Landus, Capivaccius, Salvianus, and Eustachius. It is interesting to note that he did not possess the important ichthyological and anatomical works of the last two named authors.

[20]On Gasser and his relationship with Martius, see Burmeister 1970, i.193–4, 154–5.

[21]Ibid. Gasser was also in contact with the botanist Hieronymus Bock, (ibid, i.51), whose works Martius possessed. Another of his friends was Giulio Alexandrino (ibid., i.185), who wrote against one of Martius's favoured modern physicians, Giovanni Argenterio (the *Antargenterica pro Galeno* of 1552).

[22]Ibid., 146–152.

[23]It appears to be protected by Rihel's imperial privilege 'cum gratia et privilegio Caesareo ad annos octo', which was accorded to him for publications he himself financed, although the Nonnius is not mentioned in the list submitted on 30 March 1568: see *Die kaiserlichen Druckprivilegien im Haus- Hof- und Staatsarchiv Wien* 2008, 443 (59.17). There exists, however, a 1559 draft of an eight-year privilege which seems to have been used to protect Rihel's editions of Johannes Sturm and Philippe de Comines in 1566, and which may have been used again for the Nonnus edition (ibid., 442 [59.11]).

[24]See Mandelbrote 2008, 300–1.

6.2 Guides to Medical Publications Up to 1572

It is pertinent to consider next the bibliographies available to Martius, who seems to have been an avid collector of this genre of writing. The earliest printed catalogue devoted exclusively to medical authors was that by the Lyonnais humanist, physician and historian Symphorien Champier (1471–1538/9). Although Martius possessed a number of other works by this polymath, he does not list the *De medicine claris scriptoribus [...] tractatus* (Lyon, 1506) in his own catalogue. He does however possess copies of the next medical bibliographies in date to appear, by the Strasbourg physician and reformer Otto Brunfels (1488–1534), the German humanist poet Helius Eobanus Hessus (1488–1540), the French scholar and linguist Robertus Constantinus (1530–1605), and the father of systematic bibliography, Konrad Gessner (1516–65). As in the case of Champier, Martius possesses a number of Brunfels's works (both medical and theological), and it is reasonable to infer that he approved of his writings in both of these disciplines. Brunfels's *Catalogus illustrium medicorum* (1530) is less sketchy than Champier's work, and contains both a survey of the productions of various nations and a rudimentary subject index. Eobanus's *Chorus illustrium medicorum* (Strasbourg, 1530) is a celebratory poem which adds little if anything to the other writings. Constantinus's *Nomenclator insignium scriptorum, quorum libri extant vel manuscripti, vel impressi: ex Bibliothecis Galliae et Angliae: Indexque totius Bibliothecae atque Pandectarum doctissimi atque ingeniosissimi viri C. Gesneri* (1555) was compiled with the help of a catalogue in the form of tables produced around 1548 by William Burnell, a Royal auditor in the Court of Augmentations in England, which Constantinus supplemented with new publications that appeared after 1548, and the contents of the French Royal Library at Fontainebleau. The sections on medicine and alchemy cover 20 octavo pages. The work pays very explicit tribute to the Swiss physician Konrad Gessner, who lists many doctors in his ground-breaking *Bibliotheca universalis* (1545), in which living and dead authors of all disciplines and ancient languages are listed by first name. Gessner's project was deliberately unselective, at least initially, unlike that of Martius. The Swiss bibliographer set out to produce a subject catalogue of his listed authors in a work of 1548–9 entitled *Pandectae*, but, as he confessed on the title-page of its second volume, he regretfully deferred to some future date his systematic bibliography of medicine.[25] Two elements of this missing bibliography were to appear, one on surgery (Gessner's *De chirurgia scriptores optimi quique veteres et recentiores*, 1555), the other on botany (published by Gessner in Hieronymus Tragus [Bock], *De stirpium historia*, 1552); both of these are to be found in Martius's collection. Martius lists neither Gessner's *Bibliotheca universalis*, nor its continuation by Conrad Lycosthenes in 1555, nor the *Pandectae*, but it is very likely that he had access to them: there were copies in

[25]Gessner 1549, titlepage, 'ad lectorem': '[...] medicinales etiam partitiones quae solae restant, alias seorsim, si Deus annuerit, locupletissimas edituri.'

the Fugger libraries in Augsburg.[26] The *Pandectae*, it is true, does not have a separate medical section, but its various sections are each preceded by a dedication to a leading European printer-publisher, together with the catalogue of his publications organised by subject: this would have provided Martius with bibliographical evidence about the medical books available from these printing houses.

A final subject-specific bibliographical resource was made available to him from 1564 onwards: this is the composite catalogue of the Frankfurt Book Fair compiled by Georg Willer (and later other Augsburg booksellers) at the Fair, and published in his home town. This contains a section on *libri medici*, and another on *libri philosophici* where works on natural history and natural philosophy are listed.[27] From the juxtaposition of books in Martius's catalogue, it seems certain that Martius used this resource, and from the fact that he does not possess copies of all advertised books (some but not all of which he might already have possessed), it is safe to conclude that he was indeed exercising discrimination by declining to purchase all of the books on offer. It is easier to demonstrate that Martius made use of the Book Fair Catalogues than of the bibliographies of Brunfels, Gessner, and Constantinus,[28] but the fact that he possessed copies of them suggests that he was keen to secure an overview of works on the medical art which had already been published for some time. Martius must have had access to other sources of information about new publications, too, for his purchases in the years 1564–1572 include a number of items not listed in the Fair Catalogues. It is not easy to establish what these sources were, but it is likely that they included the *nomenclaturae* or broadsheet inventories of stock which were produced by publishers, and found their way to the book fairs even if the publishers themselves did not.[29] It may well be that Willer brought a number of these back to Augsburg.

6.3 Martius's Medical Collection Anatomised

Martius's catalogue, whether advertising books for sale or his pride in his collection, does not tell us whether the books listed there constitute the whole of his collection or not. A surprising omission is his own translation into Latin of Jacques Grévin's *Deux livres des vénins* which appeared in 1571 in Antwerp from the presses of Christophe Plantin; the original text by Grévin is however listed. My comments on notable absentees from the collection and on significant

[26]Lehmann 1956–60, ii.341, 344, 552.

[27]In Autumn 1572, the heading reads 'medici et chemici libri'.

[28]For medical books alone, the statistics of Martius's purchases at the Fair are the following: Autumn 1564: 6; Spring 1565: 12; Autumn 1565: 5; Autumn 1566: 5; Autumn 1567: 5; Spring 1568: 7; Autumn 1568:2; Spring 1569: 5; Autumn 1569: 2; Spring 1570: 1; Autumn 1570: 4; Spring 1571:7; Autumn 1571: 7; Spring 1572: 9; Autumn 1572: 3. Another available bibliography, this time of ancient writers on medicine, was Neander 1565: see Maclean 2009, 62–3n.

[29]See Richter 1974, 183–229; Engelsing 1969; Wittmann 1984.

concentrations of material are therefore to be treated with caution; they are warranted only by the claim that the catalogue contains only 'libri selectissimi'.

It is clear that Martius understands by 'ars medica' something beyond the university conception of the subject. His collection includes vernacular books, books on natural history and metallurgy, secrets literature, popular guides to regimen and to survival in times of plague, as well as to the texts of medieval masters, humanist medicine, modern polemical writings, the works of doctors with heterodox opinions, and more besides. We may break down this broadly medical field in various ways: by subject; by date of composition (ancient, medieval, modern medicine); by date of publication; by language; by region of authorship or publication (France, Italy, Switzerland, Germany, Iberia). In order to characterize Martius's collection, I shall concentrate on the first of these categories. It would be fastidious to comment on every text in his collection, especially as the more general works often contain sections on topics (the great pox, or plague, or secrets, for example) on which Martius collected monographs. I have chosen under each of the headings below a representative sample to indicate the flavour of Martius's holdings. I have also indicated under each heading what is absent but found in the other collections and bibliographies which I have consulted. The categories which I have used indicate coherent areas in which Martius was interested, but even here the existence of overlapping bodies of knowledge (botany and *materia medica*, for instance) threatens to undermine any attempt at systematization. Moreover, the division of the discipline of medicine into *theoria* and *practica* became blurred during the sixteenth century. In traditional terms, *theoria* involved a general understanding of the arts course and the physiological principles specific to medicine, while *practica* constituted a body of information of proven medical usefulness in the management of illness (pathology and therapy) and health (regimen). *Practica* was conceived of as inferior to the more theoretical parts of medicine such as physiology, pathology and semiotic (concerned with signs and cures), which had greater dignity as they dealt with the principles of the art and with causal analysis; this hierarchy was reflected in the relative remuneration received by professors of theory and practice, at least until the end of the sixteenth century. But pathology, semiotic and anatomy came increasingly to be seen to as belonging to both theory and practice, and innovative medical teachers such as Giambattista da Monte of Padua (1498–1551) stressed their interdependence.[30] It would seem that Martius went even further than this in collecting secrets literature in the vernacular and promoting medical knowledge of various kinds in his own language.

As I have suggested, private medical libraries possessed a core group of texts in many of the areas even before the time that comprehensive medical bibliographies appear, such as those by Israel Spach (1591), Paschalis Gallus (1590), Georg Schenck (1609), and finally Joannes Antonides van der Linden (1637).[31] By

[30]Bylebyl 1991; Maclean 2001, 68–9.

[31]Gallus (Paschal Le Coq) 1590; Spachius 1591; Schenckius 1609; Antonides van der Linden 1637.

1650, collectors wishing to possess copies of these core texts could aspire to do so either through their being republished speculatively, or through back-purchase. It is not entirely clear how the European market for second-hand books operated, nor how physicians came to hear about newly emerging innovative medical authors before the establishment of the book fair catalogues. Travels such as those undertaken by Gryll and Martius, and correspondence with colleagues were two means of transmitting such knowledge; it may be significant that one of the genres regularly purchased by Martius is precisely collections of letters written by physicians in the emerging medical republic of letters.[32]

Medieval Authorities and University Texts Martius himself belongs to the generation of doctors who were exposed to the fruits of medical humanism when they undertook their university studies. This does not mean that the medieval syllabus had disappeared altogether[33]; in his collection are some of the basic teaching manuals and authoritative texts used in that period (two copies of the *Articella* – a collection of Galenic and Hippocratic texts with other materials relating to diagnosis and prognosis; Avicenna's massive *Canon*; Averroes's *Colliget*; Pietro d'Abano's *Conciliator*).

Medieval Practica The majority of Martius's books (and his own translations) deal with practical medicine, whether ancient, medieval or modern. I have mentioned the trend towards the revaluation of *practica*; this included the medieval texts of this genre. Two humanist doctors who commended medieval writers and arranged for them to be printed for the first time are Symphorien Champier of Lyon and Otto Brunfels of Strasbourg. Martius possesses the latter's compendium of medieval masters, in which he asserts their superiority over the ancients; and nearly all of those mentioned by Brunfels are represented in his catalogue.[34] A very common arrangement of material, borrowed from the Arabic physicians Rhases's *Liber ad Almansorem* and Avicenna's *Canon*, fen I.4, discusses illnesses and therapy from head to toe.[35] Two of Martius's authors are worthy of mention here. Michaele Savonarola's work provides a good example of the scope of the genre: it covers

[32]See Maclean 2008; Siraisi 2013. He also possessed eight other letter collections, mainly by humanists (Guillaume Budé, Johannes Reuchlin, Paulus Manutius, Etienne Dolet, Christophe Longueil, Bartolomeo Ricci).

[33]See Siraisi 1987.

[34]Brunfels 1533, *5v–6r; see also Maclean 2009, 93–7.

[35]There are two broad groups of medieval writers in this field, separated by the year 1400: the earlier group includes the twelfth-century Salernitan doctor Nicolaus Praepositus, Gilbert the Englishman (fl. 1250), Dino del Garbo (d. 1280), Taddeo Alderotti (1223–1303), Arnau de Vilanova (1238?–1311), Bernard of Gordon (1283–1320), Niccolò Bertrucci (d. 1347), Gentile da Foligno (d. 1348), Pietro Torrigiano (1270–1350), Marsilio de Santasofia (d. 1405), Tommaso del Garbo (d. 1370), John of Gaddesden (1280?–1361), Jean de Tournemire (1329–1396), and Jacopo da Forlì (d. 1413). Of these, only the last three are not represented in Martius's catalogue. The second group all belong to the fifteenth-century, and are predominantly Italians: Ugo Benzi (1376–1448), Antonio Guainerio (d. 1440), Bartolommeo de Montagnana (d. 1460), Michele Savonarola (d. 1461), Giovanni Arcolani (c. 1390–1458), Gianmatteo Ferrari de Gradi (d. 1472), and Marco Gatinaria (d. 1496); of these, only Ugo Benzi is missing from the catalogue.

general pathology, therapy, some surgery, dietetics, material medica, prognosis and nosology, and was produced for practising doctors as well as those being trained in universities. Bartolommeo de Montagnana's work is a collection of *consilia* which as a genre was to flourish and mutate in the sixteenth century.

Medical Humanism and Humanist Commentaries In Martius's catalogue there are the major editions in Latin and Greek of the newly recovered texts of high and late antiquity, by Galen, Hippocrates, Paul of Aegina, Aetius, Celsus, Dioscorides, Oribasius, Alexander of Tralles and others. Martius was himself the editor of an ancient Greek medical author, as has been noted. He seems also to have collected modern commentaries, especially of Hippocrates, including those by Antonio Musa Brasavola, Girolamo Cardano, Andrea Laguna, Leonardo Jacchini, Cristofo da Vega, and Thomas Rodriguez da Veiga.

Humanist Polemic Martius possesses the most important polemical writings which promote the new humanist medical learning at the expense of the Arabic Middle Ages, most notably the Florentine Academy's declaration of hostility of 1534 (the *Opuscula: adversus Avicennam et medicos neotericos qui Galeni disciplina neglecta, barbaros colunt*). But there were other important strands in medical polemic, beginning with Niccolò Leoniceno's attack on Pliny of 1492. Martius's collection includes this, as well as Fuchs's *Errata recentiorum medicorum* (1530), Andreas Laguna's attack on Cornarius (1543), and a number of the polemical texts discussing botany of the 1550s, involving Fuchs, Matthioli, Gessner, Guillandinus, and others.

Innovative Textbooks There are several important humanist textbooks in the catalogue: Fernel's *Universa medicina*, Fuchs's *Isagoge*, Cornarius's *Medicina sive medicus*, and *De rectis medicinae studiis amplectendis*,Wecker's *Syntaxes*, which sets the medical art out in dichotomized tables, and the *De methodo* by Hieronymus Capivaccius, whom he met in Padua. But this does not seem to be an area in which Martius showed much interest.

Sixteenth-Century Practica (1) Nosology, new illnesses, and new cures

I come now to the areas in which Martius was most active as a purchaser. Many (indeed, probably the majority of) titles concern therapeutics (*de morbis curandis*), medicines (*de compositione medicamentorum, de simplicibus*), and dosage; these being areas of developing knowledge and practice. There are also significant thematic clusters. The outbreak of the great pox in the 1490s provoked a re-examination of the Galenic doctrine of illness, and the pursuit of a cure.[36] At the same time, the emergence of new epidemic diseases such as the English Sweat gave impetus to the investigation of the nature and prevention of plague. The most famous innovator in these areas was Girolamo Fracastoro, whose *Syphilis* of 1536 and *De contagionibus* of 1538 have their place in Martius's collection. In the catalogue, there are nearly 40 separately published plague pamphlets, a text in

[36]See Arrizabalaga et al. 1997.

German on the English Sweat, and about 20 texts on the great pox (including a two-volume folio collection of treatises published in 1566). Martius was of course professionally involved with the great pox at the Blatterhaus, as well as through his connection with the Fuggers and their monopoly of guaiac wood.[37] Related to this section of his collection are the texts which deal with new cures, such as the china root, guaiac wood, and new procedures in phlebotomy.[38] Surprisingly, Pirmin Gasser's *Unterricht wieder die Pestilentz* of 1565 is not to be found in the catalogue, but it does contain the plague tract of another of Anton Fugger's clients, Heinrich Stromer von Auerbach (1476–1542) of Leipzig. Two further areas are worthy of mention: Martius seems to have collected works relating to local dispensaries and apothecaries, which around this time were coming under the control of municipal colleges of physicians; and he evinces a clear interest also in texts about poisons, one of which he was himself to translate.

(2) Anatomy

By making Vesalius's *Humani corporis fabrica* the first entry in his catalogue, Martius is declaring his allegiance to the new anatomy, which he reinforces by his possession of texts by Falloppio and Dryander. There are also some late medieval textbooks here, by Zerbi and Curtius (a commentary on Mondino's much earlier work). At the same time, Martius seems to have collected works by the humanist Parisian doctor Jacques Dubois, who opposed the vogue of anatomical illustration, while at the same time as promoting anatomy itself.[39]

(3) Hippocratism; innovative and heterodox medicine

Authors associated with the rise of Hippocratism (that is, the preference for Hippocratic texts over those of Galen, and an approach to medical knowledge through individual cases rather than precepts) are well represented in this catalogue.[40] Among the innovative medical writers whose work is collected by Martius are Jean Fernel, Jérôme Monteux (1480?–after 1558), Giovanni Argenterio, and Girolamo Cardano (nearly all of whose works are present here).[41] Remarkably, Martius possesses one text (the *Antoniana Margarita* of 1554) by a radical Spanish writer, Gomez Pereira, although it is not the one in which his anti-Galenism is most clearly expressed.[42] The catalogue includes Galenic diehards as well, notably Donato Antonio Altomare (who declared that he 'would prefer to be wrong with Galen than right with his detractors') and Jacques Dubois.[43] In that sense, this is a balanced collection.

---

[37]See Pagel 1958, 24–5; Stein 2008, 128–144; 2006; and 2009.

[38]On the debates about phlebotomy, see Saunders and O'Malley 1948.

[39]See Cunningham 1997; Kellett 1961.

[40]See Lonie 1985; and Nutton 1989.

[41]See Sherrington 1946; Siraisi 1990, 1997.

[42]The more radical work is Pereira 1558.

[43]Altomare 1574, *2r: '[hoc opus] si qui pertinacia adducti aut malevolentia potius suffusi audeant contradicere sciant non mihi sed Galeno se contradicturos, quocum errasse malo, quam ad aliorum mentem recte sapere'; for the use of this topos, see Maclean 2001, 192n.

(4) New genres: *mirabilia, observationes, epistolae*

The genre of *consilia* (advice given to patients or colleagues on specific cases) predates the new forms of writing about extraordinary medical events; Martius owned at least two fifteenth-century collections of consilia, by Montagnana and de Gradi. Antonio Benivieni's *Libellus de abditis nonnullis ac mirandis morborum et sanationum causis*, which first appeared in 1509 and was reprinted in 1529, launched a new genre devoted to miraculous or strange illnesses and cures attested by the doctor who had witnessed them at first hand.[44] The most impressive collection of these was composed by Amatus Lusitanus between 1551 and his death in 1568; both his work and that of Benivieni are in the catalogue, as well as the work by the town doctor of Arles, François Valleriola, who was particularly commended by Lorenz Gryll.[45] Martius possessed also other popular works on nature's *mirabilia*, notably those by Levinus Lemnius and Giambattista della Porta, whose work on natural magic (first edition 1558) is not however to be found in the catalogue.[46] Martius acquired also a considerable collection of medical letters, including the composite volume of five authors (Manardo, Massa, Mundella, Theodosio and Lange) which appeared in 1556.

(5) Regimen

Martius possessed copies of the classic medieval texts on diet and hygiene, by the Salernitan school and Maimonides (Rabbi Moses), as well as some modern writings by Paracelsus, Gratarolo and others; but not the popular defence of the sober life by Alvise Cornaro (the *Trattato della vita sobria*, 1558).

(6) Gynecology and obstetrics

There was a resurgence of interest in these topics in the sixteenth century, which saw the republication of Moschion and 'Trotula' as well as new tracts. These were published as a composite volume entitled *Gynaicea* in 1566, which Martius possessed. He also owned a number of vernacular works in this field, and was to translate one in 1571 from Italian into German: Giovanni Marinello's *Le medicine partimenti alle infermità delle donne* of 1563.

(7) Urology

One would have expected to see a copy of Joseph Struthius's radical new work on the pulse, which appeared at Padua with the title *Ars sphygmica* in 1555 and was immediately very popular[47]; but it is absent. Martius seems to have taken a deep interest instead in urology; 11 monographs on the subject are listed.

Astrological and Divinatory Medicine; Oneirocritic; Crises and Decretorial Days Achilles Pirmin Gasser was very interested in iatromathematics, that is, the relationship

[44]See Pomata 2010.

[45]Gryll, *Oratio*, f. 5v.

[46]See Eamon 1996, 194–233.

[47]According to Jöcher 1750–1, iv.892 (quoted by Durling 1961, 237), 800 copies were sold in a single day; but Zedler, *Universallexikon*, s.v. Struthius, (which may be Jöcher's source) gives the number as 80.

between astrology and medicine (which concerns also phlebotomy through the association of planets and part of the body).[48] Martius has a small collection on this topic, which includes some vernacular texts, several works by Cardano, the one empirical approach to astrological predictions, by Thomas Bodier, which appeared in 1555.[49]

Natural History, Balneology Around the core concerns of the medical art lies a penumbra of subjects which were given new impetus by empirical and other advances. Martius's collection contains a number of books on metallurgy, most of the humanist commentaries and modern works on zoology (including those of Scaliger, Gessner and Wotton), as well as Albert the Great's *De animalibus*; Rondelet's and Salviani's works on fishes are not however present. Martius possesses the composite volume of balneology which appeared in 1553.

Botany and Materia Medica (Simples) I have already noted that Martius was distinguished for his botanical learning, so it is not surprising that the works connected with this subject form the most striking component of the collection. This was an area fraught with controversy over the nature of illustrations, the identification of plants, and plagiarism; Martius possesses works from all sides of the debates, in both Latin and German. The collection also contains many pharmacological works on simples and their preparation.

Surgery, Military Medicine Although the distinction between learned (internal) and artisanal (external) medicine, which is also that between medicine licensed by universities or colleges of physicians and surgery licensed by guilds, is upheld in many parts of Europe, it is not universal. Bologna, one of the most prestigious of all medical universities, offered a doctorate in medicine and surgery; there were French medieval masters of the latter (notably Guy de Chauliac and Jacques Depars) who were respected by university-trained physicians; many medieval masters of *practica* include discussion of surgical procedures.[50] Martius's most recent acquisition in this area was Amboisé Paré's *Chirurgie* of 1572; he also possessed one book in Italian by Rostini, revised by Leonardo Fioravanti, one of his favoured authors, two German-language books on surgery by Hieronymus Braunschweig and Lanfranc, as well as a number of books on wound therapy in that language, including that of Paracelsus. Among the more important treatises in Latin to be found here are Vidus Vidius's book on Hippocratic surgery, Vigo's *Practica chirurgica* which Boschius had recommended, the Gessner anthology of 1555, the *Chirugia magna* published under Vesalius's name by Prospero Borgaruccio in 1568, and Antoine Chaumette's *Enchiridion chirurgicum* of 1564.

48 Burmeister 1970, i.57–61.

49 On Bodier, see Maclean 2001, 198n.

50 Malgaigne 1861, cxi–xx, names Galeatus de Santa Sofia, Guiainerio, Savonarola, Arculani, Montagnana, Matteo de Gradi, and Gatenaria as significant figures before Paré.

He possessed also two innovative works dedicated to wounds caused by firearms (by Rota and Botello).

Paracelsus Through his connection with the Fugger family and with Achilles Pirmin Gasser, one would have expected Martius to have been hostile to Paracelsus and his followers.[51] There is some evidence *e nihilo* that he chose not to purchase Paracelsian books (especially the Latin translations of his works) which were advertised in considerable numbers (over 40 titles) in the Frankfurt Book Fair Catalogues between 1564 and 1572. But Martius's interest in wound therapy, balneology, regimen, new remedies and plague tracts led him to acquire certain of Paracelsus's works, often, it seems, in early editions; and he bought at least one of the treatises written to mediate Paracelsian medical theories to learned physicians (Petrus Severinus's *Idea medicinae philosophicae, continens totius doctrinae Paracelsicae, Hippocraticae et Galenicae* of 1571). As in the case of Hippocratism and its opponents, this may be a sign of eclecticism, or fair-mindedness, on Martius's part.

Vernacular Medicine Secrets Throughout the Middle Ages, various versions of secrets literature – notably the pseudo-Aristotelian *Secreta secretorum* and the Pseudo-Albertine *Secreta* – were widely disseminated through scribal transmission, and came to be printed in due course. A second wave of such publication began in 1555, with the appearance of the *Secreti* of Alexis of Piedmont. These books are explicitly empirical: that is, they do not attempt to give a causal explanation for the remedies and recipes they offer.[52] Very soon notable exploiters of the medical market such as Girolamo Ruscelli and Leonardo Fioravanti produced their own versions, and others came to be associated with natural philosophers such as Giambattista della Porta, Girolamo Cardano and with distinguished learned physicians such as Gabriele Falloppio. Martius possessed many of these, and even translated the spurious 'Falloppio' volume into German.[53] The popularity of such texts was partly due to the pre-existing German genre of Kunstbücher, of which Martius had several examples. Secrets literature contains recipes of various kinds, which may be linked to the writings of apothecaries, of which there are examples in the catalogue. Some authors, notably Walther Hermann Ryff (d. c. 1562) and Georgius Pictorius (c. 1500–69), made their publishing careers through vernacular transmission of medical and natural-philosophical texts; these are also well represented in the collection.

[51]Burmeister 1970, i.54.

[52]Cf. the title of the 1564 French translation of Alessio Piemontese (Girolamo Ruscelli): *Empirie, ou les secrets d'Alexis*. See Eamon 1996, 986–104, 168–93.

[53]By the time of his translation, Martius could have learnt from Andrea Marcolini, the editor of another Falloppio book in his possession, the *De medicatis aquis* of 1564, that the secrets text was not the work of Falloppio, but he makes no reference to this fact.

6.4 Geographical Distribution by Author and Place of Printing; Market Zones

As is clear from the annotations on Achilles Pirmin Gasser's books now in the Vatican Library, Augsburg was a particularly favoured place for the availability and rapid acquisition of books. The presence of enterprising booksellers such as Georg Willer and the Portenbachs ensured that works advertised at the Frankfurt Book Fair were quickly made available in impressive numbers; the surviving records of the Plantin-Moretus Museum in Antwerp show Willer to be one of the largest clients of this publishing house, and it is clear that they were operating on a similar scale with other printer-publishers of learned texts such as the Wechels in Frankfurt and their colleagues in Lyon, Strasbourg, Basle, Zürich and Venice.[54] A surprising exception to this is Paris. Very few printers from that city chose to attend the Frankfurt Fair even before the St Bartholomew's day massacre of 1572, and it is to be surmised that their own immediate market zone for clients was sufficiently profitable for them not to be concerned about diffusion through the fairs.[55] Martius certainly collected works by Parisian medical authors, but it is notable that he often possesses them in Venetian editions. Even before 1559, when the Spanish King Philip II put in place restrictive measures to control the import and export of books, very few books from Iberia were made available in the international markets[56]; as in the case of Parisian authors, Spanish and Portuguese doctors were mainly known through editions of their works published in Antwerp and Venice, which Martius seems to have collected. There is one exception to this rule of foreign publication, which probably came about through the publicity given to it by Konrad Gesner in the *Bibliotheca universalis* of 1545. It is the work composed by Antonio Ludovicus (Luís) with the intriguing title *De occultibus proprietatibus* (Lisbon, 1540). Achilles Pirmin Gasser bought a copy of this lavishly produced book in 1552, and Martius also possessed it.

As one would expect, Martius collected books published in Augsburg on his chosen subjects, those by Hebenstreit and Stengel, for example), although strangely there are none here by Achilles Pirmin Gasser or Adolf Occo. He also has books from the places of his early academic career: Ingolstadt and Montpellier, whose physicians published most of their work in Lyon. Other works he may have acquired on his travels in Italy: Curtius's *Dosandi methodus*, Argenterio's *Varia opera* and his *Consultandi ratio*, Landus's *De origine pestis*, da Monte's *Quaestio quomodo medicamentum dicatur aequale vel inaequale*, his *De excrementis* and his *In nonum librum Rhasis explicatio*. There are also some publications from parts of Italy not served by the Book Fairs, including Perugia, Brissino, Rome, and Florence. He obviously went out of his way also to purchase incunables and early

[54] Plantin-Moretus Museum, Antwerp, Archives, 962ff. (Cahiers de Francfort). Willer is often the first and largest entry.

[55] The evidence for this is *ex nihilo*: see Fabian 1972–2001.

[56] See Maclean 2009, 386, 390–6.

printed books, although it is not possible to know how and where he acquired them. One of the motives for this might have been Martius's desire to collect practical medicine by medieval masters, of which he has an impressive number.

6.5 Conclusion: Martius's *iudicium* in the Light of Contemporary Private Libraries

I should like to stress again that it may be incautious to draw inferences from this collection of books as it is possible that it does not constitute the whole of Martius's library. The analysis by subject in which I engaged above does however suggest some reasons why Martius made the claims he did about the catalogue. There is a clear relationship to his academic training, in that he owned the writings of three of the professors of Ingolstadt, two from Montpellier (a centre renowned for its strong emphasis on *practica* and its botanical and zoological studies), and he seems to have collected certain authors with whom he came into contact while in Italy. He did not, however, acquire much in the way of theoretical medicine from Padua, which specialised in text books in this area. His interest in bibliographies was certainly one he shared with several bibliophiles in Augsburg; the medical finding lists he acquired mainly favoured *practica* and humanistic medicine, both of which are very well represented in the catalogue. His recourse to the Frankfurt Book Fair Catalogues supports his claim that his collection is up-to-date. It is less easy to establish however that this is a select library in the sense that Martius has brought together texts from the whole spectrum of the medical art. It does not contain the latest versions of Paracelsian writings, there is not very much evidence of an interest in chemical medicine, and writings on the theory and precepts of the discipline are not well represented; on the other hand, it does contain polemical writings in various areas from both sides of the debate, and is impressive in its representation of medieval masters as well as humanist editions of ancient texts. It also does not eschew secrets literature and the newly emerging investigations into natural history and the marvels and aberrations of nature. These features give some weight to its claim to be comprehensive. The declaration that it contains very rare books can be upheld largely through the presence of tomes of medieval masters published before 1520. Another feature to be noted, which is reflected also in Martius's own career as a translator, is his commitment to works in the vernacular, and to the making accessible of the core texts of the learned profession to the medical non-latinate laity. This would associate him with a group of writers, including Walter Hermann Ryff and later Jakob Horst, who have a programmatic vision of the importance of promoting vernacular culture in the German lands.[57]

Martius was of course not alone in collecting medical and other works; it is pertinent here to mention five medical libraries or more general collections brought

[57]See Eamon, 96–102; and Maclean 2006, 73–4.

together by contemporaries of Martius of which fairly detailed records survive. All the collections mentioned below have many features in common, and there is a high degree of overlap: some books, such as Vesalius's *Humani corporis fabrica*, Matthioli's various works on Dioscorides, or Fernel's *Universa medicina*, are clearly near-universal desiderata, and collectors of medical books would also have hoped to secure copies of the most innovative or controversial material, such as the Paracelsian canon, and works by Cardano, Argenterio and others. But rather than claim that they point to the existence of an ideal medical library, it would be more prudent to suggest that they form a polythetic class, in which the absence of any one element from a given collection is not in itself crucial, and does not disqualify a collection from being seen as up-to-date or comprehensive. The ideological and intellectual bias of the collectors (towards or against Paracelsus, or chemical medicine, or *theoria*, for example) is, I believe, not always to be seen as a negative factor, as it relates to the specific function of the library, and it is the way in which the collectors' *iudicium* is expressed.

Fairly detailed accounts of the five collections which I shall use as points of comparison have survived. The first two are local. The extensive Augsburg library of Ulrich Fugger (1526–84), the brother of Martius's patron Anton, was sold to the Count Palatinate and transported in 1567 to Heidelberg; it contains (with a great deal of other material) about 140 medical and botanical books in larger formats; the transport manifest survives in the Vatican Library.[58] This specifies format, but only in rare cases date of publication. The collection, like that of Martius, contains multiple copies of some books, but this seems to be the result of purchasing policy, whereas most of Martius's duplication of titles seems to result from his purchase of works which had already been bound together with other texts, some of which he may already have possessed.[59] Ulrich Fugger was a collector on a grand scale, and the only member of his family in this time to embrace protestantism. There are a great many theological and polemical writings in this collection which do not concern us here. For our purposes, Ulrich Fugger's library is enlightening in that his acquisition of expensive works on anatomy, of the herbals of Fuchs, Bock, and others, and of the lavishly illustrated natural histories of animals and fishes including those of Gessner and Rondelet, shows how the medical book market shades into that aimed more generally at bibliophiles. Martius's possible reference to the costs he incurred in bringing together his collection may be motivated by his possession of anatomical studies, herbals and natural histories.

The second Augsburg collection is that of Martius's friend and colleague Achilles Pirmin Gasser which contained some 900 medical items.[60] This, like

[58]Lehmann, *Fuggerbibliotheken*, ii.149–453.

[59]Some of these books are presumably the ones he refers to as 'eleganter compacti'; a serious accumulator of books – even a rich magnate such as Duke Julius of Braunschweig – usually looked on fine bindings as otiose and unnecessarily expensive, and it is reasonable to suppose that Martius would have adopted the same attitude towards the unbound purchases he made. See Graefe 1989.

[60]Burmeister 1970, i.53–61, 121–9.

that of Ulrich Fugger, is not a medical library as such, but reflects rather the acquisitions policy of a bibliophile in the general area of medical, natural-historical, zoological and botanical books. Some indications of the contents of Achilles Pirmin Gasser's library, which was bought by Ulrich Fugger in 1583 and eventually found its way, via Heidelberg, to the Vatican Library, where a catalogue of its contents still exists.[61] An idea of the library can be derived from Enrico Stevenson's annotated *Inventario dei libro stampati palatino-vaticani*, and Giovanni Mazzini's supplement to this list[62]; Karl Heinz Burmeister's biographical monograph on Pirmin Gasser of 1970 makes useful reference to this catalogue, and offers some comments on Gasser's own medical interests. According to Burmeister, he embraced traditional Arabic and Galenic medicine, and, unlike Gryll and possibly also Martius, thought that the art was to be acquired not empirically but through theory learnt from authoritative books. He possessed defences of Arabic medicine not to be found in Martius's catalogue, as well as other polemical writings, showed great interest in botany (Hieronymus Bock was one of his friends) and iatromathematics, but rejected Paracelsian medicine. This gives the impression of a somewhat conservative stance, an impression reinforced by his possession of medieval masters, but belied by the books he acquired by Champier, Fracastoro, Gessner, Fuchs, Thriverus and Cornarius. Like Martius, he had a large collection of tracts on the great pox and the plague. He also collected the new editions of physicians of high and late antiquity.[63] One fact which emerges from the annotations he made on acquiring his books is the speed of their transmission, even to Lindau where he lived until 1546, but a fortiori to Augsburg.[64] Martius could not have been better located in Europe for the acquisition of a library from both Southern and Northern Europe.

The third library, also near-contemporary, consists of the 1422 books owned by the Wittenberg Professor Caspar Peucer (1525–1602), of which 421 are medical.[65] Most of these were acquired before 1574, when Peucer was imprisoned until 1586 for crypto-Calvinism.[66] These are classified by format, as (for the most part) is the catalogue of Martius. Peucer possessed a fine collection of *practica*, a very similar range of innovative writers to that of Martius (Fracastoro, Argenterio, Cardano, Amatus, Valleriola), works on balneology, tracts on the great pox, the plague and other new illnesses, pharmacology, and botany. His anatomical holdings are stronger, as are those on zoology (he has both Saviani and Rondelet on fishes, neither of which have a place in Martius's catalogue), and he owned a copy of Struthius's work on the pulse, which, as I have noted, was absent from Martius's collection. Peucer was the author of a work on divination (which Martius possessed), and there

[61] Ibid., i.122.

[62] Stevenson 2010; Mazzini 1953.

[63] Burmeister 1970, i.54–6.

[64] These annotations are transcribed in Stevenson and Mazzini.

[65] See Kolb 1976.

[66] Hillebrand 1966: s.v. Peucer.

are, as one would expect, more works on this topic than are to be found in Martius's catalogue; he also evinced an interest in acquiring books about chemical medicine. But there is very little secrets literature here, and no Paracelsus at all.

A fourth collection is the later library catalogue of the Paduan professor Girolamo Mercuriale (1530–1606), in which contains books with imprint dates up to 1587.[67] Of the 600 or so books listed, about 420 fall in the period covered by Martius's catalogue. Mercuriale was known to Martius; like Peucer, the Paduan professor possessed Martius's edition of Nonnus; in both cases, these could have been presentation copies. Mercuriale has many more works produced in Italian towns and cities which did not advertise at the Book Fairs (including Rome, Perugia, Brissino, Naples and Ravenna) than Martius. Once Mercuriale began acquiring the works of a given author, he seems to have been more persistent in accumulating his other works than Martius, who only purchased works by Cardano in this way. Mercuriale owned a number of Paracelsus's works, in spite of his presence on the Roman Index. His library contains an impressive number of French and German works printed in those countries, and gives the impression of being more eclectic and comprehensive than Martius's collection.

The fifth library is that of Thomas Lorkyn (1528?–91), Professor of Physic at Cambridge, which was listed in the Donors' Book of Cambridge University Library in 1591, and consists of 281 items, which coincide significantly with those of Martius.[68] As in the case of Peucer and Mercuriale, he owned a copy of Martius's Nonnus, but there is no indication that this was a presentation copy. Like Martius, he possessed a number of incunables of medieval masters of *practica* (most of which were gifts made to him by tutors and colleagues), much less botany, more surgery, and a near-complete collection of Paracelsian tracts that had been translated into Latin and advertised at the Book Fairs. There are fewer collections of medical letters, and, unsurprisingly, a few local publications (including one by Lorkyn himself) which did not find their way to the continent.[69] Struthius's *Ars sphygmica* is there, together with one or two works on *theoria* not found in Martius's catalogue; Lorkyn's professorial duties would account for these. But in its mixture of the ancient, the medieval and the modern, and its emphasis on *practica*, Lorkyn's library is kindred to that of Martius.

It is pertinent finally to comment on the numbers of books in the various formats in each collection, as this indicates something about the market for which they were produced, and the choices made by the collector. The smallest format – 16mo, usually reserved for the student market – is represented in all these collections. Octavo, 12mo and 16mo were formats more popular with physicians than with lawyers, as doctors were consulted not just in their houses but also at the bedside of their patients. If they wanted to transport a manual of practical medicine with them,

[67] See Agasse 2002–3.

[68] See Sayle 1921. Lorkyn's books can be identified from books still in Cambridge University Library; Mercuriale's catalogue gives the place and date of publication in nearly all cases.

[69] On the poor exposure of English books in Europe, see Maclean 2009, 291–351.

it had to be in a small format. By the 1520s, such books began to be produced for them and for their students.[70] Before that, it seems that the needs of the medical profession were mainly served by the two larger formats (called 'libri da banco' by Armando Petrucci[71]) which would have stayed in their owners' houses or have been accommodated on university lecterns, just as they had been in the age of manuscript books. Peucer, Mercuriale and Lorkyn have more octavo books that Martius, and fewer folios and quartos; this in part arises from Martius's higher percentage of medical books published before 1520. It is noteworthy that the three medical libraries all contain bibliographical guides, and evidence that the catalogues of the Frankfurt Book Fair were used as vehicles for purchasing, even in faraway England and in Catholic Italy which officially was hostile to the circulation of such publications as they contained references to much heretical material and to the works of authors who had been placed on the Index.

The *ars medica* section of Martius's catalogue, which may or may not be his whole library, is most similar to Lorkyn's library: in comparison to those of Gasser, Peucer, and Mercuriale, it has a somewhat modest look. The implicit claim for comprehensiveness and rarity in the preface appears therefore somewhat inflated. Unlike the collection of Peucer, which is marked by its holdings in chemistry and divination, and that of Mercuriale, with its broader sweep of interests and geographical origins, its emphasis lies in its holdings in *practica* and botany, and in this it resembles Gasser's collection as far as this can be deduced from Stevenson's catalogue, without, however, a strong holding in iatromathematics. One may therefore conclude that Martius's *iudicium* is one which characterised a practitioner of somewhat limited means with a professional commitment to nosology, pharmacology and therapeutics, a strong interest in botanical study, both for its own sake and as a source of *materia medica*, and an open mind with regard to the writings of Paracelsus and empirics outside the learned profession.

Bibliography

Agasse, Jean-Michel. 2002–3. La bibliothèque d'un médecin humaniste: l'*Index Librorum* de Girolamo Mercuriale. *Les Cahiers de l'Humanisme* 3–4: 201–254.

Agricola, Johann. 1539. *Medicinae herbariae libri duo*. Basel: Bartholomaeus Westhimer.

Altomare, Antonio. 1574. *Omnia opera*, ed. Giovanni Altomare. Venice: For Jacobus Anielus de Maria.

Ammonius. 1534. *Scolia in methodum threrapeuticam Galeni*. Augsburg: Philippus Ulhardus.

———. 1537. *Commentarii in Galeni lib. 6 de locis affectis*. Nuremberg: Johannes Petreius.

Antonides van der Linden, Johannes. 1637. *De scriptis medicis*. Amsterdam: Johannes Blaeu.

Arrizabalaga, Jon. 1998. *The Articella in the Early Press, c. 1476–1534*. Cambridge/Barcelona: Wellcome Unit for the History of Medicine, CSIC Barcelona.

[70] Arrizabalaga 1998, 36.

[71] Petrucci 1969, 297–8.

Arrizabalaga, Jon, John Henderson, and Roger K. French, eds. 1997. *The Great Pox: The French Disease in Europe*. New Haven: Yale University Press.
Augsburger Stadtlexikon. Augsburg: Wissner-Verlag, 2009.
Belper, Jill. 2011. Inventare lesen. Quellen zu Fürstinnenbibliotheken des 17. Jahrhunderts und ihrer Erschliessung. In *Öffentlichkeit. Bibliotheken zwischen Barock und Aufklärung*, ed. Claudia Brinker-von der Heyde and Jürgen Wolf, 9–21. Kassel: University Press.
Bietenholz, Peter G., and Thomas B. Deutscher, eds. 1985. *Contemporaries of Erasmus*. Toronto/Buffalo/London: Toronto University Press.
Brunfels, Otto. 1533. *Neotericorum aliquot medicorum in medicam practicam introductiones*. Strasbourg: Joannes Albertus.
Burmeister, Karl Heinz. 1970. *Achilles Pirmin Gasser*. Wiesbaden: Pressler.
Bylebyl, Jerome J. 1991. Teaching *Methodus medendi* in the Renaissance. In *Galen's Method of Healing*, ed. Fridolf Kudlein and Richard J. Durling, 157–189. Leiden: Brill.
Cunningham, Andrew. 1997. *The Anatomical Renaissance: The Resurrection of the Anatomical Projects of the Ancients*. Aldershot/Brookfield: Scholar Press.
Dulieu, Louis. 1975–9. *La médicine à Montpellier*. Avignon: les Presses universitaires.
Durling, Richard J. 1961. A Chronological Census of Renaissance Editions and Translations from Galen. *Journal of the Warburg and Courtauld Institutes* 24: 230–305.
Eamon, William. 1996. *Science and the Secrets of Nature: Books of Secrets in Medieval and Early Modern Culture*. Princeton: Princeton University Press.
Engelsing, R. 1969. Deutsche Verlegerplakate des 17. Jahrhunderts. *Archiv für Geschichte des Buchwesens* ix: 217–338.
Fabian, Bernhard. 1972–2001. *Die Messkataloge des sechzehnten Jahrhunderts*. Hildesheim/New York: Olms.
Gallus, Paschalis. 1590. *Bibliotheca medica, sive catalogus illorum, qui ex professo artem medicam in hunc usque annum scriptis illustrarunt*. Basel: Conrad Waldkirch.
Gesner, Conrad. 1549. *Partitiones theologicae, Pandectarum universalium liber ultimus*. Zürich: Christoph Froschauer.
Graefe, Christa. 1989. *Staatsklugheit und Frömmigkeit: Herzog Julius zu Braunschweig-Lüneburg*. Weinheim: VCH Acta Humaniora.
Gryll, Lorenz. 1566. *De sapore dulci et amaro libri duo [...] accessit in fine oratio Laurentii Grylli de peregrinatione studii medicinali ergo suscepta*. Prague: Georg Melantrichius ab Aventino.
Hieronymus, Frank. 2005. *Theophrast und Galen – Celsus und Paracelsus*. Basel: Universitätsbibliothek.
Hillebrand, Hans J., ed. 1966. *The Oxford Encyclopaedia of the Reformation*. Oxford: Oxford University Press.
Jöcher, Christian Gottleib. 1750. *Allgemeines Gelehrtenlexikon*. Leipzig: J.F. Gleditsch.
Kellett, C.E. 1961. Sylvius and the Reform of Anatomy. *Medical History* 5: 101–116.
Kolb, Robert. 1976. *Caspar Peucer's Library: Portrait of a Wittenberg Professor of the Mid-Sixteenth Century*. St. Louis: Center for Reformation Research.
Koppitz, Hans-Joachim, ed. 2008. *Die kaiserlichen Druckprivilegien im Haus- Hof- und Staatsarchiv Wien*. Wiesbaden: Harrassowitz.
Lehmann, Paul. 1956–60. *Eine Geschichte der alten Fuggerbibliotheken*. Tübingen: Mohr.
Liess, Leonore. 1984. *Geschichte der medizinischen Fakultät in Ingolstadt von 1472 bis 1600*. Gräfelfing: Demeter.
Lonie, Iain M. 1985. The 'Paris Hippocratics': Teaching and Research in Paris in the Second Half of the Sixteenth Century. In *The Medical Renaissance of the Sixteenth Century*, ed. Andrew Wear, R.K. French, and Iain M. Lonie, 155–172. Cambridge: Cambridge University Press.
Maclean, Ian. 2001. *Logic, Signs and Nature in the Renaissance: The Case of Learned Medicine*. Cambridge: Cambridge University Press.
———. 2006. *Le monde et les hommes selon les médecins de la Renaissance*. Paris: CNRS Editions.

———. 2008. The Medical Republic of Letters Before the Thirty Years War. *Intellectual History Review* 18: 15–30.

———. 2009. *Learning and the Market Place: Essays in the History of the Early Modern Book*. Leiden/Boston: Brill.

Malgaigne, Jean François. 1861. *Histoire de la chirurgie en occident depuis le VIe jusqu'au XVIe siècle, et histoire de la vie et des travaux d'Ambroise Paré*. Paris: Baillière.

Mandelbrote, Giles. 2008. The First Printed Library Catalogue? A German Doctor's Library of the Sixteenth Century and Its Place in the History of the Distribution of Books by Catalogue. In *Le biblioteche private come paradigma bibliografico. Atti del convegno internazionale*, ed. Fiametta Sabba, 295–311. Rome: Tempio di Adriano.

Mazzini, G. 1953. *Libri stampati Palatini Vaticani 2670–2830*. Città del Vaticano: Tipografia Poliglotto Vaticano.

Müller, Gernot Michael, ed. 2010. *Humanismus und Renaissance in Augsburg*. Berlin: de Gruyter.

Nauck, E.Th. 1954. Die Zahl der Medizinstudenten deutscher Hochschulen im 14 bis 18. Jahrhundert. *Sudhoffs Archiv* 38: 175–186.

———. 1956. Der Ingolstädter medizinische Lehrplan aus der Mitte des 16. Jahrhunderts. *Sudhoffs Archiv* 40: 1–15.

Neander, Michael. 1565. *Graecae linguae erotemata ab autore recognita*. Basel: Joannes Oporinus.

Nutton, Vivian. 1989. Hippocrates in the Renaissance. *Sudhoffs Archiv Beiheft 27*: 420–439.

Pagel, Walter. 1958. *Paracelsus*. Basel/New York: Karger.

Pereira, Gomez. 1558. *Novae veraeque medicinae, experimentis et evidentibus rationaus comprobatae prima pars*. Medina del Campo: Franciscus a Canto.

Petrucci, Armando. 1969. Alle origini del libro moderno. Libri da banco, libri da bisaccia, libretti da mano. *Italia medioevale e umanistica* 12: 295–313.

Piemontese, Alessio. 1564. *Empirie, ou les secrets d'Alexis*. Lyon: par Guillaume Rouille.

Plantin-Moretus Museum, Antwerp, Archives: Cahiers de Francfort, 962ff.

Pomata, Gianna. 2010. Sharing Cases: The 'Observationes' in Early Modern Medicine. *Early Science and Medicine* 15: 193–236.

Richter, G. 1974. Bibliographische Beitrage zur Geschichte buchhändlerischer Kataloge im 16. und 17 Jahrhundert. In *Beiträge zur Geschichte des Buches und seiner Funktion in der Gesellschaft. Festschrift fur Hans Widman*. Stuttgart: Hiersemann.

Rotmar, Valentin, ed. 1571. *Tomus primus orationum ingolstadiensium*. Ingolstadt: Weissenhorn.

Sabba, Fiametta, ed. 2008. *Le biblioteche private come paradigmo bibliografico. Atti del convegno internationale, Roma 2008*. Rome: Bulzoni.

Saunders, J.B., and C. Donald O'Malley, trans. and eds. 1948. *Andreas Vesalius Bruxellensis, the Bloodletting Letter of 1539: An Annotated Translation and Study of the Evolution of Vesalius's Scientific Development*. London: Heinemann.

Sayle, Charles. 1921. The Library of Thomas Lorkyn. *Annals of Medical History* 3 (4): 310–323.

Schenckius, Johannes Georgius. 1609. *Biblia iatrica*. Frankfurt: Joannes Spiessius for Antonius Hummius.

Sherrington, Charles. 1946. *The Endeavour of Jean Fernel*. Cambridge: Cambridge University Press.

Siraisi, Nancy. 1987. *Avicenna in Renaissance Italy: The Canon and the Medical Teaching in Italian Universities After 1500*. Princeton: Princeton University Press.

———. 1990. Giovanni Argenterio and Sixteenth-Century Medical Innovation, Between Princely Patronage and Academic Controversy. *Osiris*, 2nd series, 6: 161–180.

———. 1997. *The Clock and the Mirror: Girolamo Cardano and Renaissance Medicine*. Princeton: Princeton University Press.

———. 2007. *History, Medicine and the Traditions of Renaissance Learning*. Ann Arbor: University of Michigan Press.

———. 2013. *Communities of Learned Experience: Epistolary Medicine in the Renaissance*. Baltimore: Johns Hopkins University Press.

Spachius, Israel. 1591. *Nomenclator medicorum*. Frankfurt: Martin Lechler for Nicolaus Bassaeus.

Stein, Claudia. 2006. The Meaning of Signs: Diagnosing the French Pox in Early Modern Augsburg. *Bulletin of the History of Medicine* 80: 617–647.

———. 2008. Blatterhäuser als frühmoderne Kliniken? In *Dienst am Kranken - Krankenversorgung zwischen Caritas, Medizin und Ökonomie*, ed. Gerhard Aumüller et al., 128–144. Marburg: Elwert.

———. 2009. *Negotiating the Pox in Early Modern Germany*. Aldershot: Ashgate.

Stevenson, Enrico. 2010. *Inventario dei libri stampatai Palatino-Vaticani*. Rome: BiblioBazaar.

Theophanes, Nonnus. 1568. *De omnium particularium morborum curatione, sic ut febris quoque et tumorum praeter naturam complectatur, liber editus et conversus per Hieremiam Martium*. Strasbourg: Iosias Rihel.

Wittmann, Reinhart, ed. 1984. *Bücherkataloge als buchgeschichtliche Quellen in der fruhen Neuzeit*. Wiesbaden: Harrassowitz.

World Biographical Information System. http://db.saur.de/WBIS.

Zedler. 1742. *Universallexikon*, 46 vols. Leipzig.

Part III
Medicine, Collections and Their Illustrations

Chapter 7
The Death of a Naturalist: Knowledge and Community in Late Renaissance Italy

Paula Findlen

In the spring of 1556 the sixteenth-century community of scholars engaged with natural history, one of the most exciting and rapidly evolving sciences of the century, lost one of its most important and beloved members. On May 4th of that year Luca Ghini (ca. 1490–1556), who had been the first physician to teach medicinal simples at the University of Bologna, died, leaving behind a widow and young children (Fig. 7.1)[1]. News of his death was not entirely unexpected since Ghini had been ailing for several months. They mourned the man and they commiserated with his family. Most importantly, they wondered how they could possibly replace this living repository of all the important and hard-won botanical knowledge.[2] Ghini played a singularly authoritative role in establishing the significance of botany to sixteenth-century medicine. His death marked a turning point for this community.

A preliminary version of this essay was originally presented at a session in honor of Nancy Siraisi at the Renaissance Society of America (Toronto, 2003), the Medieval and Early Modern Seminar at the University of Virginia, and the Renaissance Studies workshop at Indiana University. Thanks to Ann Blair, Nico Bertoloni Meli, Bill Newman, Duane Osheim, Anne Jacobsen Schutte, and other participants for their comments, to Giuseppe Olmi and Lucia Tongiorgi Tomasi (whose superb work on the Italian Renaissance naturalists has guided my own studies for many years), to fellow travelers in Renaissance natural history, Karen Reeds, Brian Ogilvie, Sachiko Kusukawa, Alix Cooper, Hannah Marcus, and Mackenzie Cooley, and to Gideon Manning and Cynthia Klestinec for providing this opportunity to bring this research to conclusion. Finally, to Nancy Siraisi whose intellectual generosity, friendship, and unparalleled knowledge of the history of medicine in medieval and Renaissance Italy inspired this essay.

[1]Sabbatini 1926.

[2]On the renaissance of botany, a good starting point is Arber 1986; Greene 1983; Pavord 2005; and Ogilvie 2006.

P. Findlen (✉)
Stanford University, Stanford, CA, USA
e-mail: pfindlen@stanford.edu

G. Manning, C. Klestinec (eds.), *Professors, Physicians and Practices in the History of Medicine*, Archimedes 50, DOI 10.1007/978-3-319-56514-9_7

Fig. 7.1 Posthumous portrait of Luca Ghini (*Source*: Anonimo Emiliano, ritratto di Luca Ghini secolo XVIII, Iconoteca degli Illustri Imolesi, Museo di San Domenico, Imola. Archivio Fotografico, Musei Civici di Imola)

Less than a year before his death, Ghini returned to Bologna after a celebrated career in Pisa where he founded the first university botanical garden in 1543 under the patronage of Cosimo I de' Medici.[3] He hoped to resume teaching the science of medical botany that he originally pioneered in Bologna (1534–44), transplanting the rich program of teaching and research that he developed in Pisa (1544–1555) to the university where he began his career. Ghini's Bolognese colleagues joyfully celebrated his return, knowing that his lectures would attract students to their ancient and venerable medical faculty. They hoped that his international reputation as a judicious observer of nature would rejuvenate Bologna's reputation as a leading center for natural history; perhaps his presence might even encourage the city to fund a public botanical garden in imitation of the ones in Pisa and Padua. However, it soon became clear that Ghini had come home to die.[4]

During the last winter of his life close associates anxiously awaited bulletins from Bologna about Ghini's health and prayed for his recovery. From Naples on 9 February 1556, the physician Bartolomeo Maranta (1500–71) wrote to Ghini's younger colleague in Bologna, Ulisse Aldrovandi (1522–1605), expressing his concern that the 66 year old naturalist probably would not survive this latest illness.

[3]The priority of the Pisa botanical garden over Padua and Florence, both founded in 1545, has been long established; see Chiarugi 1953; and Garbari et al. 1991. Documents reveal that Ghini began works towards this garden in 1543, even before starting his position in 1544. On the Padua garden, see Azzi Visentini 1984; and Minelli 1995.

[4]For a fuller account of Ghini's life, see Mazzini 1949; Galassi et al. 1992; Meschino 1999; Engelhardt 1995 and 2011. The only substantial studies of Ghini in English are Keller 1970; and Greene 1983, vol. 2, 702–722.

He began his letter by denying the very possibility of Ghini's mortality. "This week I am not writing to our Messer Luca because I am very skeptical about his death, which God would not permit."[5] Unable to engage his beloved master about the ordinary matters that animated their correspondence – the exchange of specimens and drawings, persistent problems of botanical identification, the uses of plants, animals, and minerals in the Renaissance pharmacopeia, and their judgment of recent books in relation to the writings of ancient authorities – Maranta chose not to send a letter directly.[6] Instead, he willed Ghini to remain alive, hoping that they might resume their conversation about nature.

Ghini's foundational role in the emergence of natural history within the medical curriculum of the Italian universities during the 1530s becomes especially apparent when we read the correspondence of this era.[7] He belonged to the generation of physicians that increasingly valued travel, observation, and collecting, and considered long-distance communication to be essential to the pursuit of knowledge. Renaissance physicians frequently studied, taught, and practiced medicine in multiple locations. As their sense of community became more expansive, conversations about books, manuscripts, specimens, images, and experience multiplied. The most active participants in the learned world of Renaissance medicine considered themselves to be citizens of a medical republic of letters.[8] While Brian Ogilvie has rightfully defined the world of Renaissance natural history as a vast network with many nodes, Ghini's centrality to this community became especially apparent at the moment of his death.[9]

During the previous decade the Milanese physician Girolamo Cardano (1501–76) composed *De consolatione* (1542), a learned guide on managing grief and misfortune inspired by Cicero.[10] It was written shortly before the untimely death of the German physician Valerius Cordus (1515–44), who succumbed to malarial fever in Rome at age 29 after herborizing with his companions between the Appenines and the Pontine marshes in the pestilential heat of summer. "His death was a great loss to medicine and to his friends in Germany," lamented Johannes Crato. The Swiss naturalist Conrad Gesner (1516–65) published a number of his unfinished botanical works so that Renaissance naturalists might benefit from his careful and systematic descriptions of plants and their medicinal properties.[11]

---

[5]De Toni 1911–12: 1521 (9 February 1556).

[6]On the role of letters in Renaissance natural history, a good starting point is Olmi 1991 and Findlen 1991. Sixteenth-century scientific correspondence has recently been a subject of renewed interest; see Ogilvie 2006, esp. 14, 38, 46, 55, 80–82, 211; Egmond 2007; Delisle 2004 and 2008.

[7]Schmitt 1976 and 1984, XV, 297–336; Palmer 1985; Olmi 1992; Findlen 1994; and the literature on botanical gardens in note 3.

[8]Maclean 2008; Siraisi 2013 and Marcus 2016. For a parallel discussion of the idea of a botanical republic, see Findlen 1999.

[9]Ogilvie 2006, 82.

[10]McClure 1991, esp. 161–162. On Cardano's medical activities, see Siraisi 1997.

[11]Cordus 1561, which includes his *Historia stirpium et Sylva*; quote from Crato's letter to Gesner, sig.b2v–b3r. See Arber 1986, 74–76; Greene 1983, vol. 2: 368–415; and Ogilvie 2006, 135–136, 145–148 (translation of Crato on p. 145).

Cordus died on the job, so to speak, too soon to have an immediate impact on his contemporaries who only fully understood the import of his work in its posthumous publication. By contrast, Ghini taught and researched for several decades. He trained several generations of students and generously shared what he knew with other colleagues. At the time of his death, the majority of university professors who taught medicinal simples in Italy, as the science of botany was then called, and the prefects of the botanical gardens in Pisa, Florence, and Padua were his disciples. The extraordinary attention given to his death allows us to explore this crucial moment in the development of a new and exciting science.

In February 1556, Maranta acknowledged the gravity of Ghini's illness, observing that his mentor was no longer writing his own letters. He offered Aldrovandi a heartfelt confession that anticipated his imminent sense of loss. "I am almost without hope about his health. If it occurs, I will be the most aggrieved man in the world because I am certain to lose another father and perhaps the best one."[12] When confirmation of Ghini's death finally reached Naples in early summer, Maranta wrote again to Aldrovandi to express his "grief and displeasure," leaving us with an indelible image of Ghini's impact on the naturalists closest to him: "Certainly if I were to say that I was more displeased at his death than I was at that of my father, I would not be lying."[13] He immediately made plans to honor and preserve Ghini's formidable knowledge of plants. Maranta's *Methodi cognoscendorum simplicium libri tres* (1559), a book he planned to submit to Ghini's judgment, became a funerary monument to his beloved master. By contrast, Ghini's modest family tomb at Santa Maria dei Servi in Bologna no longer survives.[14]

The strong sense of intellectual kinship expressed by Maranta in conversation with Aldrovandi reminds us that Ghini actively shaped a "school" while also inspiring a broader scholarly community to think about how to continue to improve natural history in the sixteenth century.[15] He created a well-defined intellectual family which derived its character, in part, from its relation to Ghini's actual family. Maranta shared the pain of losing "my preceptor Luca Ghini" with other disciples, including Aldrovandi who studied unofficially with Ghini between Bologna and Pisa. They declared their allegiance to Ghini by calling him "our common friend and father."[16] All of them keenly felt the absence of Italy's leading botanist.

In the months after Ghini's death, letters arrived in Bologna from virtually every corner of Italy, and beyond, as Ghini's students and colleagues consider how best to commemorate him. In Naples the great bibliophile and amateur botanist Giovan Vincenzo Pinelli (1535–1601), whose private garden had benefited from Ghini's efforts to increase the stock of available plants in Italy, consoled Maranta with the

[12]De Toni 1911–12, 1521 (9 February 1556).

[13]Ibid., 1522 (21 June 1556).

[14]Meschini 1999, 769.

[15]Fantuzzi 1784, 135.

[16]Maranta 1559, sig. *3r, 21, 27, 86, 182; De Toni 1924–25, 618 (12 December 1556).

"good memory of Messer Luca."[17] From Padua, Gabriele Falloppio (1523–62), long remembered for his investigations of female anatomy (and equally forgotten for his efforts to teach botany to medical students for a few years in the 1550s), expressed his sorrow about "the loss of such a man" who had been a dear colleague in Pisa. He worried about the fate of Ghini's family, even going so far as to take Ghini's eldest son Galeazzo under his tutelage, once he was old enough to be away from his mother, while also working with the imperial physician Pier Andrea Mattioli (1500–1577) to provide dowries for Ghini's daughters.[18] The rituals of mourning surrounding Ghini's death underscored the extent to which the sixteenth-century republic of letters conceived of itself as a family of knowledge, whose intellectual offspring not only preserved the memory of their academic father but honored him by taking care of his actual progeny.

The outpouring of grief at Ghini's death stands in stark contrast to the virtual absence of specific knowledge that we have today regarding the nature of his contributions to medicine and natural history. Why exactly was he so widely admired? Ghini published nothing under his own name in his own lifetime, and very little appeared posthumously.[19] His fabled herbarium has long since disappeared; few letters or manuscripts survive in his own hand. His lectures on syphilis and botany, and certain medical recipes (such as the *elixir vitae* associated with his name whose recipe could still be found in the grand ducal pharmacy in Tuscany in the early seventeenth century) survived because physicians saved copies; some eventually appeared in print. Ghini's written botanical opinions, known as the *Placiti*, also circulated but, all in all, this is a slim legacy.[20] Fundamentally, Ghini disappeared because he did not become an author of an important printed book in an age of great illustrated anatomies and natural histories.

Ghini deserves further study precisely because he is the antithesis of contemporaries such as Andreas Vesalius (1514–64) and Leonhart Fuchs (1501–66), who have been indelibly associated with big, beautifully illustrated, densely written books that transformed respectively the sciences of anatomy and botany.[21] He

[17]Biblioteca Universitaria, Bologna (hereafter BUB), *Aldrovandi*, ms.38^{2}, vol. 1, c.66 (1 October 1556).

[18]BUB, *Aldrovandi*, ms.38^{2}, vol. 1, c.43r (29 May 1556), c. ## (16 March 1557); Di Pietro 1970: 27–28, 34.

[19]For his posthumously published work, see Ghini 1589. His recipe for *acqua vitae* appears in Wecker 1617, 441, and was republished in Cortesi 1629, 84; see Sabbatini 1927a.

[20]Ghini's cure for Cosimo I's kidney ailments (*renella*) was preserved in the grand ducal archive and rediscovered by the Roman medical professor Pompeo Caino, when he came to Florence in 1609 to treat an ailing Ferdinando I and his son, Cosimo II. Caino copied *Ad Arennias Lucae Ghini pro Magna Duce Hetruriae Syrupus* into a blank page in one of the medical books in his library; see Sabbatini 1922–23. There are references to Ghini's *Experimenta in praxi*, which do not seem to have made it into print; Velschius 1676. I will discuss Ghini's *Placiti* and his lectures copied by Aldrovandi throughout this essay; for another example of how Ghini's work survived, see Biblioteca Riccardiana, Florence, ms. 806, fasc. 3: *Simplicium medicamentorum nomina ... a ... Luca Ghino Pisis* (April 1553).

[21]Carlino 1999; Kusukawa 2012.

also took a different path than Mattioli, whose commentary on Dioscorides was as close to a bestseller as any Renaissance scientific book.[22] Ghini challenges us to understand Renaissance natural history at the moment of its institutionalization, when many of its constituent features were not yet fully in place and when many of the leading figures had just begun to develop the techniques for generating knowledge that became a standard part of this field of inquiry.

Studying Ghini immerses us in the materials of sixteenth-century scholarship – notes, images, specimens, lectures, commentaries, and of course letters. The explosion of scholarly materials was yet another artifact of what Ann Blair describes as the first age of information overload in world of print.[23] Ghini's seemingly ephemeral legacy raises interesting questions about the making of knowledge because how he practiced natural history mattered much more than what he published. Fundamentally, this essay reconstructs the community that Ghini formed in order to understand why his contributions to natural history were so important to his heirs that they sought to preserve both his material and immaterial legacy.[24]

7.1 Becoming a Renaissance Naturalist

To the extent that we can trace Ghini directly, he exists in the institutional records of the two Italian universities – Bologna and Pisa – where he became the first professor to teach "medicinal simples," the nascent field of botany. Under the patronage of Cosimo I, he not only inaugurated the botanical garden in Pisa in 1543, the first such laboratory for the study of nature in any Italian university, but also the one in Florence in 1544, and eventually inspired the one in Bologna founded in 1568. During the decades in which he practiced medicine and encouraged colleagues and institutions to care about the study of plants and minerals, Ghini was a reliable and seemingly inexhaustible source of useful information, a precise observer of thousands of specimens, and an attentive reader of and contributor to the growing books written about natural history.

Ghini belonged to the second generation of Italian Renaissance naturalists. Probably trained by the Ferrarese humanist physician Niccolò Leoniceno (1428–1524), he took his medical degree in Bologna in 1527.[25] He immediately began teaching practical medicine and married a physician's daughter, Gentile Sarti, a year later. His interest in botany emerged in this context, nurtured in the

[22]Ferri 1997; Findlen 1999; and Fausti 2004.

[23]Blair 2010.

[24]On the relationship between material and immaterial inheritance, see Levi 1988; and Findlen 2004.

[25]Bronzino 1962, 21.

private gardens of the city including his father-in-law's.[26] Throughout his career Ghini was a well-regarded practicing physician who served as town physician in Fano (1536–39) and later personal physician to Duke Cosimo I (1544–55). His medical activities supported his passion for natural history since his university stipend was modest and erratic.[27] Even as he championed the importance of studying nature for its own sake, Ghini remained committed to understanding the medical utility of plants, animals, and minerals. He was serious about trying to find more effective treatments for syphilis, for instance, and earned a reputation as a physician who tested the medicinal properties of plants and minerals to improve medical practice.[28]

What did Ghini learn from masters such as Leoniceno? Leoniceno's *De Plinii et plurium aliorum medicorum in medicina erroribus* (1492) emerged as part of a vociferous debate about the errors in Pliny the Elder's *Natural History*.[29] Comparing this popular Roman encyclopedia to a number of its Greek sources – Theophrastus's philosophical writings on plants, Dioscorides's *De materia medica*, and Galen's pharmacological works – many discrepancies emerged.[30] Some were evidently the result of incomprehension and mistranslation, others reminded Pliny's Renaissance readers that he was more of a compiler than an observer. Ghini was heir to a previous generation that rediscovered ancient writings on natural history, in part, because they were more readily available in newly printed editions. Renaissance naturalists used their philological and observational skills to rectify problems that they found in an expanding ancient corpus. Leoniceno played a prominent role in these debates, arguing for the importance of correlating words with things. Observation increasingly mattered. Such practices inspired a new attentiveness to the role of seeing and describing in the writings of the ancients. As a young medical student, Ghini absorbed lessons from masters such as Leoniceno whose personal library contained an impressive collection of Greek medical writings.[31] And yet Ghini's learning seems far less bookish than this earlier generation. He wore his learning lightly and did not privilege it.

By the early 1530s Ghini began to advocate for teaching medical botany at the University of Bologna. His first opportunity came in 1534 when he was permitted to teach "simples" without any additional remuneration. This was an exciting moment

[26]Recalling the Bologna gardens that inspired his own interest in natural history, Aldrovandi described going "ad hortos quosdam Bononiae, ut S[anc]ti Salvatoris, et Poetarum, et Lucae Ghini in domo de Sartis." BUB, *Aldrovandi*, ms. 136, XXIV, c. 29r. Also discussed in De Toni 1910–11, 818.

[27]Sabbatini 1922–23, 244–245.

[28]Sabbatini 1927b; Bernabeo 1992. For the general context, see Arrizabalaga et al. 1997.

[29]Castiglioni 1953; Nauert 1979; Ferrari 1990; and Ogilvie 2006. For the intellectual climate in which these debates occurred, see Nutton 1997.

[30]Schmitt 1980; Stannard 1966; and Riddle 1980.

[31]Mugnai Carrara 1991. On the new emphasis on observation in the late fifteenth and sixteenth centuries, see the essays by Katharine Park and Gianna Pomata in Daston and Lunbeck 2011, 15–80.

for the subject. A series of influential publications had begun to define the intellectual agenda of this field. Jean Ruel's 1516 Latin commentary on Dioscorides provided medical botanists with a new scholarly edition of the single most important book for their subject that advocated the value of *autopsia*, or seeing for oneself. With the appearance of his *Herbarum vivae eicones* (1530) the German botanist Otto Brunfels created a new standard for how to illustrate plants. The year that Ghini first began to teach, Euricus Cordus, another student of Leoniceno, published his botanical dialogues. The *Botanologicon* (1534) effectively captured in print the conversations of Renaissance naturalists in nature, puzzling over the relationship between ancient descriptions and modern plants, and searching for reliable informants.[32]

In 1537 Ghini was lecturing on Galen's considerable corpus of pharmacological writings, though struggling to earn a decent stipend. A new spate of publications such as Antonio Musa Brasavola's *Examen omnium simplicium medicamentorum* (1536) increased the number of known plants almost fourfold – from Dioscorides's approximately 550 Mediterranean plants to around 2000.[33] Ghini incorporated the lessons of these modern authors into his approach to research and teaching, creating an exciting agenda for the inclusion of natural history in the medical curriculum. His growing ranks of admirers found his lectures a refreshing expansion of the traditional curriculum in practical medicine. The Florentine humanist Benedetto Varchi fondly remembered Ghini as a striking exception to the overly theoretical tendencies of most natural philosophers "to always believe and not ever prove everything that one finds written in good authors, especially in Aristotle." By contrast, he warmly praised "messer Luca Ghini, a most singular physician and botanist, who not only had great knowledge but also practice in all kinds of minerals, or so it seemed to me when I heard him publicly in the Studium of Bologna."[34] Ghini embodied a new breed of physician who advocated learning from experience, initially as an antidote to problems and lacunae in ancient natural history and ultimately as a technique in its own right, capable of generating new knowledge.[35]

While Varchi understandably took note of Ghini's alchemical interests, the majority of his students recognized that his greatest passion was plants. Luigi Anguillara (1512–70), who also studied with Ghini in Bologna, assisted him in Pisa, and subsequently became the first prefect of the botanical garden in Padua in 1546, 1 year after its foundation, later recalled Ghini's experimental approach to plant identification as one of the reasons his identifications were authoritative.[36] At the height of his fame, Ghini's reputation in navigating thorny issues of plant identification was virtually unparalleled. He had much to offer medical students

[32]Arber 1986, esp. 55–64; Ogilvie 2006, 126–139; and Kusukawa 2012, 21–22.

[33]Engelhardt 1995, 12.

[34]Varchi 1827, 37.

[35]Pomata and Siraisi 2005.

[36]Anguillara 1561, 126–128.

interested in the nature of the ingredients that constituted the Renaissance pharmacopeia, while also enticing a handful of dedicated students to follow his lead in considering botany a subject worth pursuing beyond its medical utility.

In 1539 the Senate of Bologna finally offered Ghini a 5 year contract and a raise specifically to teach medicinal simples. In acknowledgment of Ghini's arguments that the best material for this subject lay beyond Galen, they permitted him to teach "out of other ancient and recent medical books," incorporating this expanding canon into the new curriculum.[37] The English naturalist William Turner (1508–68) who studied in Ferrara and Bologna, later recalled how "Lucas the reader of Dioscorides in Bonony" showed him plants such as wild parsley that he could find neither in England nor in Northern Europe. There was no plant too ordinary to escape his attention. Like Maranta, Turner proudly called Ghini "my master."[38] Other foreigners interested in natural history began to make the pilgrimage to Bologna. Among them was Valerius Cordus who visited in 1542, with his Dutch colleague Pieter van Foreest who would accompany him to Rome. Van Foreest later recalled Ghini's lectures on Dioscrides and demonstrations of plants, and encouraged his colleagues in Leiden to found a botanical garden.[39] Ghini's reputation was growing.

During the 1530s Ghini began to collect and press dried plants. Anna Pavord rightfully compares the pressed plant to the articulated skeleton.[40] Admiring Ghini's method, Gesner eventually gave it a name by calling the book of plants a *herbarium*.[41] The herbarium was an instrument that arose from new habits of observation. As the most influential teacher of his generation, Ghini taught others how to create herbaria. He laid the foundation for the organization of the earliest examples; four of his disciples, Gherardo Cibo, Michele Merini, Andrea Cesalpino, and Aldrovandi, created some of the earliest surviving herbaria, containing the occasional reference to Ghini's role in their own decisions about plant identifications.[42] We do not know how much of this empirical work Ghini did before moving to Pisa in 1544, where these practices evolved further. "I'm not finding more than three hundred kinds of dried herbs," Ghini wrote regretfully to Aldrovandi in October 1553 in response to his request for samples, "though I could have sworn that I had over six hundred of them, but I don't know where they are."[43] Certainly,

[37]Archivio di Stato, Bologna, *Senato. Partitorum* (1533–42), cc. 103v–105r; also transcribed in Sabbatini 1926, 32: "Item d. M. Luca Ghino nominatum in quinquennium proximum simplicis medicinae tradendae et monstrandae munus, eamque ex Galeni et aliorum medicorum veterum, et recentiorum libris in publicis scholis interpretandum."

[38]Engelhardt 2011, 241; Turner 1888, 22.

[39]Meyer et al. 1999, 14; Santing 2010, 154–156.

[40]Pavord 2005, 206.

[41]Galassi 1992, 199.

[42]Arber 1986, 138–143; Ogilvie 2006, 280–281n35; and especially Camus 1895; and Tongiorgi Tomasi 1992. Much ink has been spilled debating who pioneered this technique of preserving plants, but there is virtually universal agreement that Ghini popularized the herbarium.

[43]De Toni 1905, 11 (16 October 1553).

the profusion of plants in the Medici botanical gardens gave him a far greater stock of materials to collect, distribute, and preserve.

Many years later, Mattias de Lobel and Pierre Pena recalled visiting Ghini's legendary school in Pisa where they praised him for not only taking notes but drying and depicting plants.[44] As colors faded from the pressed specimen, the painted simulacrum became a tangible reminder of what the living plant had once been. The difficulty of preserving living color as long as possible represented a technical challenge that Ghini and his disciples sought to address. Aldrovandi later boasted that he was "the first in Europe to discover the way to dry green plants between paper strips, reducing them to the form and shape so that they seemed depicted, dried and glued forever to the papers, which is very useful to those who cultivate these sciences who otherwise forget them because of the difficulty of this arrangement."[45] He dated his innovation to 1553, 2 years after he initiated a herbarium inspired by his work with Ghini. This was one of many instances in which Ghini's disciples demonstrated how they improved on what they had learned from their master who institutionalized these practices.

7.2 Creating a Garden in Pisa

Ghini's crucial role in the formation of the Italian Renaissance community of naturalists crystallized with his move to Pisa. In 1543 the University of Pisa reopened after being closed for 8 years. Its restoration was one of Cosimo I's important projects designed to publicize the activities of the newly reconfigured Medici state. The duke wanted to attract recognizable talent as well as support the teaching of new subjects. Cosimo I's decision to offer Ghini a well remunerated position to teach medical botany and found a botanical garden, after the German naturalist Fuchs declined his offer to leave Tübingen, gives us a precise sense of Ghini's standing in 1543. Fuchs had just published his *De historia stirpium commentarii insignes* (1542), a monumental botanical encyclopedia with more than 500 illustrations which set a new standard for a descriptive, illustrated natural history and celebrated the virtues of *autopsia* in imitation of the ancient medical writer Dioscorides. He was a natural first choice for the grand duke, especially after Vesalius recommended him.[46] With this publication, Fuchs became the most celebrated European botanist of the moment. That Ghini would be a close second is a sure sign of his growing reputation as a superb instructor and researcher on the ground if not in print. He had published absolutely nothing.

[44]Lobel and Pena 1571, 271: "nec dum sibi notas, in chartas ipse conderet et affabre pingendas curaret."

[45]Frati 1907, 24.

[46]Meyer et al. 1999; Kusukawa 2012, 100–136, esp. 134–136; and Garbari 1992, 226.

Disappointed in the uneven support he was receiving in Bologna and enticed by the prospect of having the resources to create a well-stocked public garden, Ghini happily accepted Cosimo I's invitation. He began to set up a garden as early as summer 1543, when he had not yet resigned his position in Bologna (Fig. 7.2). Anguillara assisted Ghini in amassing materials for the garden; after becoming prefect of the Paduan garden, he ultimately traveled much farther and more often than his master but his experience of seeing nature with Ghini remained a touchstone for understanding of how to look at nature with a patient and practiced eye. Ghini proudly described him as *Aloiusius Romanus olim discipulus meus.*[47] While Anguillara did not consider Ghini infallible, he admiringly described "master Luca" as answering to "no greater authority than nature" itself.[48] Years later, he continued to cite their discoveries, and recall their conversations and even their disagreements about what they observed.

In spring 1544 Ghini inaugurated his new position by lecturing on Dioscorides's *De mineralibus*. A nucleus of loyal students – among them Anguillara, Maranta, and Cesalpino – moved with him from Bologna.[49] Another former student, Cesare Odone, replaced Ghini in Bologna, teaching medicinal simples. In Pisa he encountered even more colleagues and students eager to learn from him in this far more promising institutional setting where he could truly demonstrate and perfect his techniques for knowing plants. Maranta especially praised Ghini's "public lectures." The Nuremberg physician Georg Marius fondly remembered hearing him explicate Galen and Dioscorides. Aldrovandi later recalled that Ghini spent an hour or two after each lecture, taking students into the garden to demonstrate what he described.[50] This disciplined approach to pedagogy that envisioned natural history as course for advanced medical students, while also giving a broader public access to its resources, was without precedent. Ghini firmly believed that nature should be studied for the common good.

By July 1545 Ghini described his satisfaction at having "collected many really beautiful plants, which I planted with great diligence in a garden in Pisa." during the previous year. He encouraged Cosimo I's head steward Pier Francesco Riccio to look after them during his latest expedition, lest the summer heat dry them out, "because I want to make, and if no adverse fortune befalls me, I hope that I will make a garden that will please His Excellency and be useful for scholars."[51] Ghini enlisted Riccio's help in securing specimens, including a sprig from "that prickly fruit that is all over the island of Elba" that he recalled seeing in Cosimo I's *scrittoio*

[47]De Toni 1907a: 40.

[48]Anguillara 1561, 127.

[49]Galassi 1992, 197–198; Garbari et al. 1991, 29–33; Engelhardt 2011, 231, 236.

[50]Maranta 1559, 3; Mattioli 1561, 202 (2 September 1558): "me superioribus annis Medicaniam doceret, et subinde nobis Dioscoridem com Galeno commune esse vellet." See also BUB, *Aldrovandi*, ms. 91, c. 49; Chiarugi 1953, 827.

[51]Archivio di Stato, Florence, *Mediceo* 1171, cc. 256–257 (Luca Ghini to Pier Francesco Riccio, Bologna, 4 July 1545), as transcribed in Garbari et al. 1991, 277.

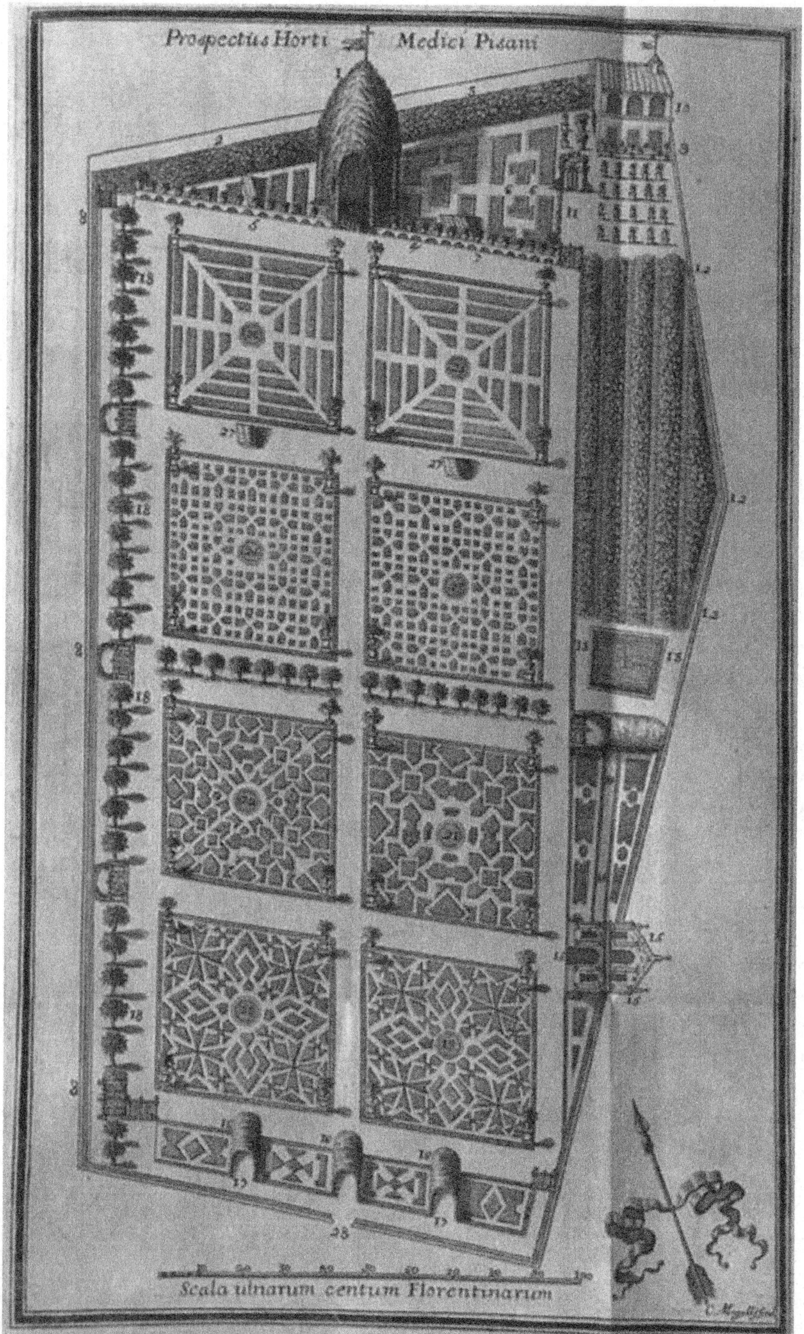

Fig. 7.2 Pisa botanical garden (*Source*: Michel Angelo Tilli, Catalogus plantarum horti Pisani (Florence, 1723). Courtesy of Huntington Library, San Marino, CA)

when he first entered his service, offering "anything that is here in the garden in exchange.[52] Possibly this agreement inspired the detailed paintings of orchids and other plants decorating the grand duke's study done by the Medici court painter Francesco Bachiacca around 1545. They seemed to mirror the very nature captured by Ghini in the botanical garden; the year that Bachiacca began to decorate Cosimo I's *scrittoio*, the grand duke approved the creation of a second garden at the convent of San Marco in the heart of Florence.[53] The Tuscan empire of nature was growing rapidly.

Ghini's annual summer expeditions now enjoyed Medici sponsorship. He traversed the hills and valleys between Tuscany and Bologna, and explored inland mineral deposits and the marine life of the Tuscan coast, including the island of Elba. Collecting Tuscan plants took on special importance, in light of the political function of the garden. An extraordinary garden emerged that contained plants, as Mattioli told readers of his popular commentary on Dioscorides, "that one doesn't find elsewhere in Italy until now."[54] Ghini integrated these new materials into his lectures. Northern European naturalists began to seek him out when they came to Italy, and not only to discuss Italian flora. While traveling in Italy in 1544, Gesner admired Ghini's shell collection.[55]

In order to enhance the new garden Ghini expanded the scholarly, commercial, and informal networks by which he obtained specimens. He conspicuously did not privilege any single source of information since his goal was to add to knowledge of nature's diversity. A careful reader of Theophrastus, Dioscorides, and Pliny, Ghini paid special attention to plants from the Eastern Mediterranean. When in doubt about the relationship between modern plant identifications and ancient Greek terminology, Ghini asked Greek sailors, Tuscan peasants, and even his "Greek maid" to provide him with the current names of various plants while accumulating seeds, leaves, roots, flowers, and of course entire plants.[56] He did not believe than learned physicians and humanists pouring over books and lexicons had all the answers. Folk wisdom and intimate knowledge of the plants in a particular region also mattered. Copying Ghini's lectures, Aldrovandi observed how much he valued the lessons of "daily experience."[57]

Ghini's access to the nascent port of Livorno, as well as the longstanding world of maritime commerce in Pisa, multiplied his range of contacts. A Greek monk who traveled in Syria and Egypt and a "Florentine merchant from the Capponi family" supplied him with samples of true balsam. A Pisan merchant gave him an example

---

[52]ASF, *Mediceo* 1173, fasc. XVIII, c. 395 (Ghini to Riccio, Pisa, 3 November 1547), as quoted in Chiarugi 1953, It is unclear from this description whether Ghini is referring to some sort of thorny vegetation or a kind of non-native fruit like the prickly pear later introduced to the island.

[53]Signorini 1993; Rossi and Signorini 1994.

[54]Mattioli 1555,, 2.

[55]Mazzini 1949, 187.

[56]De Toni 1907a, 18, 41.

[57]BUB, ms. 98, III, c. 11r.

of papyrus with Arabic script.[58] His brother stationed in Crete became a regular long-distance supplier. Ghini also encouraged physicians, merchants, clerics, and other people curious about nature to contribute to the garden. Overwhelmingly, specimens came from Tuscany and more generally northern and central Italy, the Eastern Mediterranean, and the Levant. His description of leaves used by West African slaves to wrap sugar loaves for the Portuguese on the island of São Tomé in the Gulf of Guinea, which led him to reflect on Brazilian plants with similar uses, was a rare instance when the new Atlantic economy materialized in his discussion of plants in the Pisa garden.[59]

On the whole, however, Ghini made no special effort to collect New World flora. His location in the Mediterranean and his contacts with northern Europe primarily defined his vision of nature. He knew of duke Cosimo I's and his wife Eleonora of Toledo's passion for growing maize, and tested an American medicinal syrup to see if it replicated the fabled properties of ancient balsam. He experimented with guaiacum to see if this New World remedy really did cure syphilis. Mattioli recalled that Ghini was the first to show him samples of sarsaparilla, after determining that it also alleviated some of syphilis's most unpleasant effects. He may have also assisted Mattioli in identifying a Peruvian fruit we now call the tomato.[60] In short, Ghini did not ignore what the New World offered. However, there was so much still unknown close to home, so many puzzles to be resolved in reading the works of ancient naturalists whose geographic horizons did not encompass the entire world. Ghini never privileged exotic and curious specimens over humble vegetation; he felt that the most ordinary and useful aspects of nature deserved the Renaissance naturalist's full attention.[61]

As the Pisan garden developed, Ghini fully realized his vision of sustained collaborative research. His efforts to educate the next generation of naturalists were paying off handsomely and he was well-supported by Cosimo I. Ghini's dynamic pedagogy and intellectual generosity forged strong connections with students and colleagues that yielded further dividends. Ghini found himself at the center of a rapidly expanding scientific network that looked especially to him for that ideal combination of reliable information and good advice. He began to collect and commission illustrations. One of his German disciples admiringly recalled how Ghini emphasized the importance of simultaneously "painting and writing about plants" as the best method for advancing botanical knowledge.[62] By the end of his tenure at Pisa, people described the consensus about nature that emerged from his garden as the "Pisan opinion" (*tal opinione che viene da Pisa*).[63] By amassing so

[58]De Toni 1907a, esp. pp. 17–19.

[59]Ibid., 14.

[60]Smith 1994, 13–14; Gentilcore 2010, 1–26, *passim*; and Mattioli 1557, 45. On guaiacum, see De Toni 1907a, 41; and for sarsaparilla, Mattioli, 1573, 156.

[61]Cooper 2007

[62]Mattioli 1561, 202 (2 September 1558)

[63]Anguillara 1561, 224.

many living and dried specimens, field notes, images, and carefully annotated readings of the important works of natural history, Ghini became the supreme glossator of nature.

Ghini also felt that botanical gardens should cooperate rather than compete. Anguillara's appointment as the garden prefect in Padua became a golden opportunity to establish relations between the two botanical gardens. In January 1547 Cosimo I informed the Venetian ambassador that Ghini would be sending letters through the diplomatic pouch, "so that one of his students whom he writes may send certain plants which will be of use to him in this studium." The duke was personally interested in the growth of the garden and also enjoyed discussing the properties of minerals and drug compositions with Ghini during the summer months. At the end of the year, Ghini reported to the duke's head steward Riccio that he was having Anguillara "send me all those plants that are foreign in these countries" in exchange for "those that are in these surrounding regions here."[64] As new personnel joined the Paduan garden, they continued to inquire whether Ghini needed anything from their garden.[65]

By the end of the 1540s Pisa became an international center for natural history. In 1548 Ghini created the first seed index (*Index seminum*) as an important step in cataloging and disseminating his archive of nature.[66] The list began to circulate and became a way for other naturalists to know what the Pisa garden contained so that they might request samples from Ghini. Pier Antonio Michiel (1510–76), a Venetian patrician famed for his garden of exotic plants on the island of San Trovaso, enthusiastically participated in these exchanges, initiating his own herbarium with Ghini's encouragement in 1550. Michiel tellingly described Ghini as "my illustrator of plants" in reference to Ghini's insistence on the importance of seeing for oneself what the ancients described.[67] In order to supplement his specimens, Ghini began to create a visual archive of nature, focusing especially on botanical illustrations.

A growing number of colleagues came to see the garden and discuss natural history with Ghini. In 1549 the papal physician Andrés Laguna (1499–1559) visited and found the garden most impressive. His Spanish commentary on Dioscorides appeared in 1555, the same year that he persuaded Philip II to found the botanical garden at Aranjuez.[68] Towards the end of 1549, Guillaume Rondelet (1507–66) also came to Pisa. Observing Ghini teach natural history inspired his own approach when he returned to Montpellier. Like the Roman physician Ippolito Salviani (1514–72), Rondelet warmly thanked Ghini for sharing images of unusual fish found off the Tuscan coast.[69] Given the high cost of producing quality color images, Renaissance naturalists speculated whether it might be possible "to find a way to

---

[64]Sabbatini 1923, 307, 309; Garbari et al. 1991, 19; Mazzini 1949, 192.

[65]De Toni 1911: 156 (20 September 1554).

[66]Garbari 1980, 527.

[67]Pier Antonio Michiel, *Cinque libri di piante*, book 2, n. 66, as quoted in De Toni 1907a, 30n1.

[68]Olmedilla y Puig 1887.

[69]Rondelet 1554, 487; Salviani, 1554–58, 155r. Reeds 1991, 56, 60.

preserve dried fish in their own shape, like one does with these herbs."[70] This was indeed a moment of rich experimentation in how to study nature. Ghini was tasting, distilling, drying, and drawing everything he could, and seemed willing to share things freely.

The most extensive testimony we have about Ghini's impact on the next generation of naturalists comes from Aldrovandi. Aldrovandi first encountered Ghini around 1551, when he was completing his medical degree in Bologna. They probably met during one of Ghini's regular trips home, possibly introduced by Aldrovandi's first botany instructor Francesco Petrollini, whose herbarium was inspired by Ghini's method of drying and organizing plants.[71] Aldrovandi later recalled conversations with Ghini in his in-law's garden "at the Sarti home" and learned about "many foreign plants" in Paolo Poeti's garden, another Bolognese site of scientific inquiry that Aldrovandi associated with "the good memory of the most excellent Luca Ghini."[72] As their relationship evolved, Ghini increasingly shared the materials of his research and teaching with his young and enthusiastic colleague. Aldrovandi meticulously copied Ghini's lecture notes from his famous botany course in Pisa.[73] He became a virtual student.

Aldrovandi's early contact with Ghini inspired his desire to learn as much as possible from the famous master. In the summer 1553, just before completed his medical degree, he traveled to Pisa. Aldrovandi described his pleasure of wandering "with that Signor Ghini through those mountains where he had the opportunity to see many beautiful things that were the subject of his studies." In Pisa Aldrovandi "picked and described all the rare herbs located in that garden of the Most Excellent Messer Luca Ghini, prefect of that garden of simples."[74] His *Catalogue of Every Plant that was in the Public Garden for Scholars in the Time of Luca Ghini* became a precious archive of what it contained. Many years later Aldrovandi recalled, "The number of plants in his time was 620."[75] Ghini's garden indeed became a valuable repository of carefully chosen specimens, designed to resolve ancient debates about the murkier issues of plant identification while also representing the growing awareness of nature's variety and diversity, both within Italy and beyond.

While a handful of influential physicians studied with Ghini in Bologna, in Pisa he had the opportunity to fully realize his program of teaching and research. Aldrovandi belonged to this next generation. The letters that these two naturalists exchanged manifest the depth of Ghini's admiration and affection for this particular

[70]On preserving dried fish, see Chiovenda 1909, 447 (15 November 1553).

[71]Chiovenda 1909; De Toni 1910–11.

[72]BUB, *Aldrovandi*, ms. 136, c. 29r; BUB, *Aldrovandi*, ms. 91, c. 175r; Tugnoli Pattaro 1981, 53, 58.

[73]BUB, *Aldrovandi*, ms. 98, II, cc. 69r–148r; ms. 98, III, cc. 2r–106r (*Ex lectionibus D. L. Ghini in Academia Pisana legentis collecta*).

[74]Frati 1907, 24.

[75]BUB, *Aldrovandi*, ms. 136, XIV, cc. 17r–26v. This manuscript is dated June 1590. See also De Toni 1907b.

student. "By my actions you will know that I love you as a son," Ghini confessed in October 1553, "and God knows that I want for you the same good things that I wish for my own children."[76] The knowledge that emanated from his time in Pisa became his most precious legacy.

7.3 How to Publish Useful Information

Around 1551, Ghini made a crucial decision not to publish his own illustrated natural history. By then he had amassed a great deal of useful material that deserved to be shared. The Pisa garden was rich in plants, his study overflowed with books, notes, images, and dried specimens, and he wanted to do something with them beyond the lectures he prepared for students. Ghini had neither the time nor the inclination to complete a book.

To be sure, Ghini contemplated the possibility. The previous two decades produced a steady stream of impressive scholarship, and he read it all with great care. Ghini had much to contribute to natural history but was undecided what form it might take. There was no urgent need to gloss Pliny the Elder since that had been done effectively by the previous generation. He could write a commentary on Dioscorides, like so many others, or possibly engage with some of newer work on Theophrastus's philosophical writings about plants that he avidly discussed with Aldrovandi.[77] Perhaps he might write up his own observations, subsuming his opinions of ancient and modern authorities within the empirical project of documenting nature? The only thing we know with certainty is that Ghini's hypothetical book involved images as well as words. He was observing, describing, representing, and preserving nature while trying to see what might grow in his garden. Inspired by Brunfels and Fuchs, he knew that the best natural histories required good images.

Publishing an illustrated book was a costly and laborious enterprise.[78] It demanded not only time but potentially resources that Ghini either did not have or preferred to use for other things. As a result, he made a decision that consigned him to historical obscurity because we have a hard time seeing the value of work published under someone else's name.[79] Instead, Ghini recognized that publication and more generally scientific research was the outcome of distributed knowledge. The natural histories printed in the final decade of his life, and in the period immediately following his death, became a means of presenting aspects of his life's work in print. He published in the works of leading contemporaries, allowing the seeds of his research to germinate in multiple locations.

[76]De Toni 1905, 11 (16 October 1553).

[77]Ibid., pp. 11 (16 October 1553), 13 (14 December 1553).

[78]Kusukawa 2012; Fahy 1993; and Johns 1998.

[79]Biagioli and Galison 2003.

Fig. 7.3 Image of a purple orchid that Luca Ghini shared with Leonardt Fuchs (*Source*: Österriechische Nationalbibliothek, Vienna, Codex 11118, fol. 453. Courtesy of Österriechische Nationalbibliothek, Vienna)

Ghini's reputation as a naturalist who willingly shared what he knew and possessed became the basis of a long-distance relationship with Fuchs. Ghini read Fuchs' *De historia stirpium*, taking note of entries he might correct or add to the next edition of this great illustrated botany. Knowing that irises and orchids were still uncommon in Germany, Ghini posted several images of them to Tübingen (Fig. 7.3). Unfortunately none of their correspondence survives but an exchange of specimens, images, and information ensued. Fuchs seems to have had some part of Ghini's archive, especially a number of his images, at the time of his Italian colleague's death in 1556. Two years later, other naturalists were still wondering what had happened to this material.[80] Fuchs gratefully remembered Ghini as "a

[80]Mattioli 1561, 202.

very learned man who was always kind toward the Germans."[81] They two of them never met in person but forged a warm and respectful friendship at a distance.

Ghini's most important collaboration in print was with the man who was especially anxious to get his hands on anything Fuchs had from Ghini's archive – Mattioli, the ambitious town physician of Gorizia who became an imperial physician as a result of his bestselling book. Originally from Siena, Mattioli also aspired to Medici patronage while having strong ties to the Venetian Republic. In 1544 Mattioli's commentary on Dioscorides first appeared; published in numerous editions, and translated into multiple languages, it became one of the most significant and widely read natural histories of the sixteenth century. Initially, Mattioli's work rested primarily on his critical examination of Discorides's *De materia medica* and prior Latin commentaries, combined with his own herbal knowledge. It was an unillustrated vernacular work written for healers, apothecaries, and a general public interested in the ancient pharmacopeia but unable to read erudite medical tomes. The success of these early editions encouraged Mattioli to expand the scope of his project. In 1550 the Venetian publisher Vincenzo Valgrisi printed third authorized edition of Mattioli's *Discorsi*. The text had begun to expand but was still unillustrated. The appearance of a pirated version containing purloined images from the German herbals increased anxiety about competition; Mattioli and his publisher were in the midst of discussing which changes to make before the next edition appeared.[82] Into this void stepped Luca Ghini.

By 1550 Ghini had gathered materials for an illustrated commentary on Dioscorides, hoping to do for this ancient text what Brunfels and especially Fuchs had done for modern botany. He was reading the earliest editions of Mattioli's *Discorsi*, watching this book grow in popularity, observing the changes Mattioli made, and taking notes. Ghini's eighteenth-century biographer Giovanni Fantuzzi observed that "Mattioli had not previously thought of plant illustrations before he established an intimate friendship with Luca Ghini, which put him in a position to get all the help he needed from him."[83] Rather than publish a competing commentary, Ghini chose to share his research. Did Ghini make this decision, knowing that Mattioli planned to publish a greatly expanded, illustrated Latin edition, which appeared in 1554? Or did the idea for this significantly revised version designed to reach a European-wide audience emerge as the two naturalists worked together? Since their correspondence did not survive, we can only see their collaboration indirectly, as it appeared in the editions of Mattioli's book and in other correspondence of this period.

The international success of Mattioli's Latin commentary earned him the post of imperial physician to Archduke Ferdinand I of Austria, crowned Holy Roman

[81]Meyer et al. 1999, vol. 1, 185 (from the Vienna Codex, II. 120, 401). For the rediscovery of the original color illustrations in Vienna, see Seybold 1986 and 1990.

[82]Andreoli 2006, 195–199. Palmer 1985 offers an excellent overview of the context in which Matitoli developed his commentary.

[83]Fantuzzi 1784, 135n3.

Emperor in 1558. All subsequent editions in any language improved upon the changes introduced in the first Latin edition – bigger and better illustrations, more information, and detailed descriptions that frequently reflected Mattioli's acerbic disagreements with other naturalists.[84] Ghini's specific contributions and his vision of the subject shaped this work. In return, he found an important venue in which to publish some of his most important botanical research.

In October 1551 Ghini completed his annotated reading of Mattioli's commentary, compiling 69 opinions known as the *Placiti*.[85] This small treatise utilized every technique he had advocated for the previous two decades, and its conclusions rested on Ghini's assessment of the materials he accumulated in Pisa. The task of correlating modern plants with Dioscorides's account of the most important natural ingredients in the ancient Greek pharmacopeia, was not an easy one. Ghini jokingly recalled "a certain old man of Bologna, first among the herbalists of his time in this place, who used to say that the etymology of the name Dioscorides in the vernacular beautifully indicated what kind of man Dioscorides was, since Dioscorides sounds like nothing else but the God of Discord (*Deum Discordiae*)." Beyond this jest, however, lay an important truth: "Indeed, Dioscorides bequeathed many descriptions of plants so succinct, brief, mutilated, and imperfect that, from those few notes, it may be impossible to arrive at knowledge of them; hence, so great are the variety of opinions, and so many the opposing judgments of learned botanists (*rei herbariae studiosorum*) about many plants, that it is doubtful that they will ever be able to be discovered."[86] Nonetheless, Ghini's goal was to resolve as many lingering questions as possible, examining nature and the books written about it with a practiced eye. He, perhaps more than any other naturalist of his generation, perfected the science of description.[87]

Ghini devoted the majority of his career to the cultivation of botany as an empirical, descriptive, and visual science. Naturalists respected Ghini because he insisted on the value of experience but was modest in his claims about the results. "I can neither affirm nor deny if it is true," he remarked about Mattioli's description of *dorycnium*, or canary flower.[88] Knowledge for him existed as a full register of information that needed to accumulated and assayed with care. Ghini utilized multiple techniques of description that drew attention to the details, especially in the case of difficult identifications, and insisted on the importance of correlating diverse information. This was the fundamental lesson of his celebrated lectures, which inspired two decades of students and colleagues to approach botany in a different way.

Ghini's *Placiti* contained valuable information not covered in the first three editions of Mattioli's *Discorsi*. Seeds, roots, leaves, and other cuttings accompanied

---

[84]Nutton 2004.

[85]The original can be found in BUB, *Aldrovandi*, ms. 98, II, cc. 33r–53r.

[86]De Toni 1907a, 37.

[87]Ogilvie 2006.

[88]De Toni 1907a, 35.

the text, since Ghini did not believe that anyone could form an opinion without physical evidence. Unsure whether the aromatic herb *polemonia* (now known as *Silene inflata L.* or sclupit) conformed to Dioscorides's description of a similar plant, Ghini explained: "I am sending you the rear branches to submit to your most exquisite judgment." He described how he grew seeds from Crete in the Pisa garden to see if the plant that emerged resembled one of the three different species of ironwort (*Sideritis*) described by Dioscorides: "I am sending this branch to you to see if you judge it to be that second kind of ironwort."[89] Each entry highlighted how carefully Ghini worked and how prudently he offered his opinion.

Take the perplexing plant that both Dioscorides and Pliny the Elder labelled *crocodilium*. Ghini believed it to be a plant "with broad leaves, and spiny edges on all sides, commonly called *Eryngium marinum*." He judged the confusion over its habitat to be an artifact of textual corruption in earlier manuscripts of Dioscorides's *De materia medica*, where a slip of a scribal pen transformed the sandy shoal where eryngium grew into a wooded forest where it did not. The only answer was to send Mattioli "seeds of my new Crocodilium," while apologizing for not yet making a decoction of the roots. This experiment, Ghini affirmed, would offer the most definitive proof of all since true crocodilium should cause a nosebleed, just as the ancients said.[90] He encouraged Mattioli to cultivate and test the properties of this plant.

Ghini also made the plants in the Pisa botanical garden a fixed point of reference for reliable plant identification, a legacy that his student Cesalpino (1519–1603), who inherited his position, would build upon by developing a way of naming plants that prefigured Linnean binomial nomenclature.[91] Ghini constantly reminded Mattioli that he had living examples of virtually everything he described in his garden as well as dried specimens to share, supplying Mattioli with two different samples of the alpine perennial *horminum* or dragonmouth, "dried and affixed with glue to paper."[92] Ghini also included numerous images to encourage Mattioli to see how representing nature was essential to its description. Regarding a specimen he received from their Ferrarese colleague Brasavola, Ghini wrote that he had it depicted on site to preserve its original appearance as much as possible. In this instance, he felt the image rather than a dried sample provided the best information. "Now I am sending you that picture, and I omitted part of the stone so that you will be able to see the plant remnants or at least its vestiges."[93] He wanted to know if Mattioli agreed with him that the plant found in the Tuscan Maremma was *androsace*, described by Dioscorides as native to Syria.

During the 1550s Mattioli's version of Dioscorides became a dynamic repository for all the empirical discoveries of the age as well as a place in which to

[89]Ibid., 32, 34.

[90]Ibid., 25.

[91]Pavord 2005, 226–241; Nepi and Gusmeroli 2008.

[92]De Toni 1907a, 29.

[93]Ibid, 31.

vociferously debate contested identifications of contemporary plants thought to resemble Dioscorides's descriptions in the ancient pharmacopeia.[94] He needed collaborators to keep successive editions up-to-date, and he needed to instill his commentary with the authority of a Renaissance naturalist who commanded the respect of his contemporaries. Ghini's willingness to play this role earned him Mattioli's undying gratitude. The first time Mattioli mentioned Ghini was in the 1554 illustrated Latin commentary. He effusively wrote that, of all the people who had assisted him in bringing this new version of his commentary to conclusion, none deserved greater acknowledgment than "Luca Ghini of Imola, a physician of ingenuity and singular learning, professing botany in Pisa to the great praise of everyone."[95] Mattioli planned to present Ghini with an illuminated copy to thank him personally.

One year later, when a new Italian edition based on the Latin commentary appeared, Mattioli praised Ghini for creating an important scientific institution by persuading Cosimo I to found "a garden for the ease and public ornament of physicians, scholars, and anyone else who enjoys this discipline." Later Italian editions continued to celebrate "the most renowned Messer Luca Ghini of Imola" and his lessons in "this truly glorious faculty of simples."[96] Valgrisi frequently reprinted this new version of Mattioli's commentary in both languages. Mattioli continued to edit and emend the text, ultimately increasing the number, size, and quality of images to almost 1000 in the 1560s; translations into other European languages further broadcast Ghini's reputation.[97] By publishing in Mattioli's book, Ghini assured his own immortality.

Mattioli was notoriously critical of his fellow naturalists. The German physician Melchior Wieland (1520–89), who studied and collaborated closely with Falloppio, succeeding Anguillara as prefect of the Padua botanical garden in 1561, found Mattioli's polemical remarks in one of his books so outrageous that he scribbled in the margins: "you're still an asshole, Mattioli!"[98] Despite his reputation as man who never missed the opportunity to point out the deficiencies of everyone's conclusions but his own, Mattioli consistently acknowledged the quality and significance of Ghini's work. In an uncharacteristic moment of humility, he told Aldrovandi that corresponding regularly with Ghini about botany ensured "that I don't write lies."[99] Their relationship was mutually beneficial. It certainly helped that Ghini never

[94]Ferri 1997; Findlen 1999; and Fausti 2004.

[95]Mattioli 1554, sig.α4v. For Mattioli's discussion of inserting Ghini in his acknowledgments, see Raimondi 1906, 22 (20 May 1554).

[96]Mattioli 1555, 2; and Mattioli 1559, n. p. (dedication to Queen Catherine of Poland, Archduchess of Austria).

[97]Pesenti 1985. On Mattioli's debts to Ghini, see Tongiorgi Tomasi 1992, 290.

[98]For Wieland's marginalia, see Mattioli 1561, 163 (Bibioteca Marciana, Venezia, 18.D.20): "Tu sei pur un coglione Mattiolo." Many thanks to Hannah Marcus for sharing her wonderful discovery; see Marcus 2016. On Mattioli's strained relations with many Renaissance naturalists, see Findlen 1999; and Delisle 2004.

[99]Raimondi 1906, 21.

Fig. 7.4 Pier Andrea Mattioli's reproduction of Luca Ghini's image of *rhamnus* (*Source*: Pier Andrea Mattioli, *Commentarii in libros sex Pedacii Dioscoridis Anazarbei de medica material. Adirectis quam plurimis plantorum & animalium imaginibus, eodem authore* (Venice, 1554), p. 97

chose to publish a book under his own name, leaving Mattioli the glory of claiming in print that he – and not his friend Ghini who systematized and institutionalized how good botany was done – was the Renaissance naturalist who surpassed Dioscorides's noteworthy achievement.

Starting in 1554, Mattioli cited Ghini's letters, reproduced passages from the *Placiti*, and included woodcuts of illustrations that Ghini sent from Pisa as well as images made from his pressed plants (Fig. 7.4). Even when Mattioli wasn't sure if Ghini's conclusions were correct, he reproduced the words of "the most excellent and most famous botanist Messer Luca Ghini, from whose opinion I could not depart."[100] This including Ghini's view that Dioscorides's *androsace* grew in the Tuscan Maremma, about which Mattioli was rather skeptical, but he published his opinion nonetheless. In numerous editions of his book, Mattioli presented Ghini as his most trusted informant, a second Dioscorides (*Dioscorides redivivus*) whose friendship he cherished and whose judgment he valued. While Ghini lived, he celebrated "the great nobility and liberality of his soul," and after his death he praised Ghini's "integrity, sincerity, humanity, and faith."[101]

Other naturalists, especially Ghini's disciples, frequently praised Ghini's contributions to natural history. Yet none of them celebrated Ghini's learning in print

[100]Mattioli 1557, 434.

[101]Ibid., 353; Mattioli 1561, 205 (12 December 1558).

before his death. They could not offer him the kind of partnership that he developed with Mattioli in the final years of his life, since publishing was indeed a costly, laborious, and uncertain enterprise. After Ghini's death, Mattioli did not forget his friend whom he regarded as a brother. In 1561 he published his posthumous "praises of Luca Ghini" (*Lucae Ghinae laudes*) in the very book in which Wieland scribbled his invective against Mattioli. In this letter, originally written in December 1558 to Ghini's German disciple Marius, Mattioli recalled his surprise and pleasure at receiving a congratulatory letter from Ghini, shortly after his commentary first appeared, followed by an offer to collaborate on subsequent editions. He assured readers that he fully credited Ghini's contributions to his commentary in print, including "very many plants, which I certainly reference as having received from him, when embellishing our Dioscorides with figures of them."[102] Mattioli wanted everyone to know how much he admired Ghini's work and how it important it had been to the success of his commentary.

Thanks to Mattioli even more than Fuchs, a great deal of Ghini's work *is* published. In October 1555 Mattioli's secretary Giovanni Odorico Melchiori sent Ghini a copy of the first Latin edition, not only to thank him but "because I know that he too will have noted many things."[103] They hoped that Ghini would improve the next edition by reading the first Latin edition of 1554 carefully, supplying Mattioli with new information, suggested revisions, and more images. It was around this time that Ghini's student Varchi decided that his master deserved a poem to honor his growing fame as Italy's most important botanist whose knowledge of nature materially improved medicine (Fig. 7.5).[104] Thus, in the last 2 years of his life Ghini enjoyed the kind of public acclaim that recognized the singular role he played in the development of his discipline. He was a man whom others were eager to publish.

7.4 Fighting over Ghini's Papers

In December 1556, Melchiori sent a letter to Aldrovandi from Gorizia. He still hadn't recovered from the terrible news of Ghini's death, writing "truly we have all lost the rarest of friends." He also had a practical question to ask his colleague in Bologna: what had happened to Ghini's papers? He particularly wanted to know if Aldrovandi could help him find "some annotations that I understood that he had made on the Latin Dioscorides of Mattioli."[105]

[102]Mattioli 1561, 205 (12 December 1558). I have modified the translation in Greene 1983, Part 2, 801.

[103]De Toni 1924–25, 615 (10 October 1555).

[104]Varchi 1555, 120.

[105]De Toni 1924–25, 618 (12 December 1556).

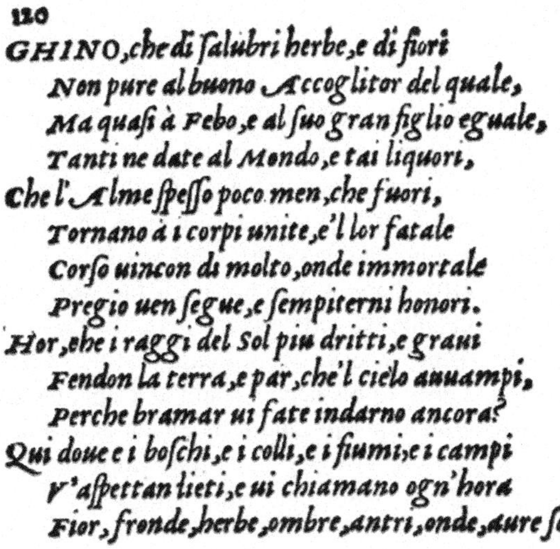
120
GHINO,che di ſalubri herbe,e di fiori
Non pure al buono Accoglitor del quale,
Ma quaſi à Febo,e al ſuo gran figlio eguale,
Tanti ne date al Mondo,e tai liquori,
Che l'Alme ſpeſſo poco men,che fuori,
Tornano à i corpi unite,e'l lor fatale
Corſo uincon di molto,onde immortale
Pregio uen ſegue,e ſempiterni honori.
Hor,che i raggi del Sol piu dritti,e graui
Fendon la terra,e par,che'l cielo auuampi,
Perche bramar ui fate indarno ancora?
Qui doue e i boſchi,e i colli,e i fiumi,e i campi
V'aſpettan lieti,e ui chiamano ogn'hora
Fior,fronde,herbe,ombre,antri,onde,aure ſoaui.

Fig. 7.5 Benedetto Varchi's sonnet in praise of Luca Ghini (*Source*: Benedetto Varchi, *De' sonetti di m. Benedetto Varchi, Parte Prima* (Florence, 1555), p. 120. Courtesy of Special Collections, Stanford University Libraries)

During the final months of his life, Ghini discussed the state of Mattioli's book with Melchiori, as he reviewed the results of their initial collaboration. Mattioli had a tendency to declare victory prematurely; Ghini also knew that he responded badly to criticism. Seeking to avoid a direct confrontation, Ghini told Melchiori that he would feel more comfortable sharing copies of two letters that he had written to Fuchs about the Latin commentary, as an indirect way of conveying his opinion of the 1554 edition. With Ghini dead, Melchiori wondered about the fate of "his annotations" and "his letters." Did Ghini intend to bequeath them to Mattioli? Could Aldrovandi get these notes, or at least the letters from the "heirs of our Ghini of blessed memory"?[106] The response was negative. In his paean to Ghini written in 1558, Mattioli lamented that Ghini's posthumous opinions were unavailable, since Aldrovandi assured them that they did not exist.[107] Instead, Mattioli continued to find new ways to use the materials Ghini originally gave him, publishing portions of the *Placiti* and the illustrations that accompanied them in successive editions of his book. He mobilized Ghini's international network of disciples in an attempt to track down an additional set of illustrations, alleged to be in Fuchs's possession, without any luck.

There was no doubt in anyone's mind that Ghini had left behind a valuable scientific patrimony, while leaving his family virtually destitute. But where was it and who had the right to possess it? Ghini's generosity meant that portions of his archive were dispersed; other parts that remained in his home had yet to be shared with naturalists to whom Ghini promised samples, images, and notes in the final months of his life. From Verona, the apothecary Francesco Calzolari (1522–1609)

[106]Ibid., pp. 618–619.

[107]Mattioli 1561, 205 (12 December 1558).

lamented the loss of a man who had been "like a father."[108] In the early 1550s Calzolari eagerly corresponded with Ghini and guided him up Monte Baldo, braving wind and snow in September 1554 to bring him some of its renowned alpine flora; Ghini thanked him by sending "a book of well-prepared simples" from Pisa.[109] Calzolari was in despair, he told Aldrovandi, because he no longer knew to whom he should write when he had unresolved questions about nature.

Others shared Calzolari's sense of unease and uncertainty. Who would replace the seemingly irreplaceable Ghini? His knowledge commanded their respect and he had so generously shared his archive of nature with others; his moral authority was also unimpeachable. No single person could claim such an inheritance because it seemed impossible to have all of Ghini's virtues. There was great concern about what would happen to his papers, and the years of knowledge and experience they represented. Who ought to inherit his material and immaterial legacy? After May 1556, these were increasingly urgent questions that deeply divided the community of naturalists closely associated with Ghini.

When Mattioli finally heard about Ghini's death in January 1557, he immediately wrote a letter to Aldrovandi in Bologna that stated simply and poignantly: "his death took away half of my heart." Mattioli informed Aldrovandi that his primary consolation lay in the fact that God "substituted you in the place of the most renowned Ghini."[110] The Flemish physician Reiner Solenander, who lived with Ghini for several years when studying with him in Pisa, agreed with this assessment. While mourning the absence of Ghini, because "of the great experience he acquired in a long life which the envious Fates have cut short," he nonetheless declared Aldrovandi to be Ghini's primary heir, declaring, "I don't know whom I can ask for better advice on this subject than you."[111] Falloppio also decided that he would now direct his questions primarily to Aldrovandi, whom he affectionately called his brother.[112] Calzolari agreed with them. The brotherhood of naturalists was in the process of assessing Ghini's academic progeny, in search of a likely heir, and began to make their preferences known.

An intellectual inheritance did not necessarily adhere to the rule of primogeniture. The thirty-something Aldrovandi was Ghini's youngest academic son but socially prominent and academically promising; he had only recently graduated and begun to teach in the medical faculty of the University of Bologna. For several years, Aldrovandi had been a close collaborator who also assisted Ghini's return to Bologna. He was in the midst of indexing Theophrastus's works, which was of great interest to his mentor. When Aldrovandi received his medical degree at the end of

[108]Cermenati 1909, 103 (13 February 1557).

[109]Ibid., 96, 102 (23 September 1554 and 12 July 1555).

[110]Raimondi 1907, 32 (19 January 1557).

[111]BUB, *Aldrovandi*, ms.38^2, vol. 1, c. 201 (5 July 1556). Solenander also described Ghini as "Messer Luca my father"; Tosi 1989, 75 (5 July 1556). See also his paean to Ghini in Solenander 1558, 95; discussed in Engelhart 1995, 241.

[112]Mazzini 1949, 215; De Toni 1913, 35, 37, 40–41; and Di Pietro 1970, 27–28.

1553, Ghini wrote to congratulate him, saying, "I took that solace that I should from this, and for someone whom I love as a son."[113] They read and corrected Mattioli's unpublished Latin commentary together.

Early in 1554, Ghini encouraged Aldrovandi to note "the errors and other things that you don't like in Mattioli" because he felt that he wouldn't have time "to skim his entire book." In April 1554 Ghini spent 4 days completing his own review of Mattioli's manuscript, and compiled a final list of suggested corrections which he sent to Aldrovandi, who was by then mediating Ghini's relationship with the irascible Mattioli, with Melchiori's assistance.[114] There was every reason to believe that Aldrovandi knew the state of Ghini's papers and was in discussions with the family about their future disposition.

In the mid-sixteenth century a great deal of valuable information about nature lay unpublished. This was indeed a material as well as symbolic inheritance, the very stuff from which the next generation made knowledge. In September 1558 Calzolari wrote again to Aldrovandi, hoping that his friend in Bologna might help facilitate discussions with Ghini's brother or widow. Shortly before his death, Ghini promised Calzolari several samples of earths and minerals; he was willing to pay generously to acquire them.[115] Everyone needed Ghini's materials to complete their own projects of *materia medica* and natural history, and everyone was eager to make their publications the mouthpiece for Ghini's still authoritative words about the nature. His legacy was bound up in the fate of his inheritance.

The value of Ghini's inheritance was apparent to his intellectual progeny. Like any good Renaissance family, they fought over it, quite vigorously. While some members of the scholarly community considered Aldrovandi to be the natural heir, in light of the role he began to play just before and after Ghini's death, there was pushback from others who aspired to this position. The Neapolitan physician Maranta studied with Ghini in Bologna long before Aldrovandi, 22 years his junior, ever picked up a plant. Their correspondence was frequent and intense, leading him to boast that Ghini "wrote to me every week, and I him." Maranta was not ready to cede ground to Aldrovandi. While calling himself Aldrovandi's "sincere friend and younger brother," Maranta firmly exercised his prerogatives as the eldest academic son when he negotiated directly with the family about the fate of Ghini's scholarly patrimony. He had urgent reasons to claim this inheritance. Maranta wanted to incorporate Ghini's most recent insights into his forthcoming book on the best method for identifying medicinal simples, *Methodi cognoscendorum simplicium*. It would give his book an authority that it lacked without this material. In January 1557 Maranta informed Aldrovandi that Ghini's brother and widow had decided to give him "the papers ... of Messer Luca."[116]

---

[113]De Toni 1905, 12 (14 December 1553).

[114]Ibid., 13 (n.d. [1554]); 15–18 (10 April 1554).

[115]Ceremenati 1909, 108 (13 September 1558).

[116]De Toni 1911–12, 1525–26, 1520.

By December 1558, some portion of this material was in Padua with the great bibliophile Pinelli, whose family ties to Naples made him Maranta's natural collaborator and ultimately the dedicatee of this book. Aldrovandi was concerned, and Mattioli was enraged; this precious resource had slipped away, just as his book reached the peak of its popularity. Maranta fretted that Pinelli's other commitments would prevent him from completing a copy before sending the originals to Naples. Even worse, he was concerned that whomever Pinelli employed might take advantage of the opportunity to "make a copy of it for himself," dissipating the value of this patrimony. Maranta strongly asserted his right to Ghini's papers. "It is better that they be in my power," he informed Aldrovandi, offering to copy them for his colleague once they arrived in Naples.[117] We will probably never know how much of Ghini's scientific patrimony disappeared as a result of this decision.

Maranta's goal was not necessarily to deny his younger colleague his rightful inheritance but to contest Mattioli's use of Ghini's notes in the latest editions of his commentary on Dioscorides. It galled him to see Mattioli appropriate Ghini's authority. He was one of Ghini's earliest disciples, indeed a loyal and trusted correspondent. Surely Maranta better than anyone knew "Luca's true opinion" (*veram Lucae sententiam*)?[118] Throughout 1558, as he completed his study of simples, Maranta considered how he might create a worthy monument in print to his master. He enlisted Falloppio, opening his book with a letter to his Paduan colleague about "the most bitter and painful death of Luca Ghini." Declaring that he would lament being deprived of Ghini's "authority and prudence," as long as he lived, Maranta rejoiced in his ability to talk with Ghini's other disciples.[119] Falloppio responded by reiterating the great value of Ghini's – not Mattioli's – commentary on Dioscorides.[120] Their celebration of Ghini deliberately sought to undermine Mattioli's privileged relationship with his best informant.

When Maranta's *Methodi cognoscendorum simplicium* appeared in 1559, things finally came to a head. He presented his work as the model application of Ghini's method of distinguishing plants through careful inspection of their seeds, leaves, and roots. Maranta had been a regular beneficiary of Ghini's empirical largesse, receiving numerous specimens, including precious samples of Eastern Mediterranean flora that Ghini's brother collected on Crete. He celebrated his colleagues – mostly notably Falloppio, Aldrovandi, and the Spanish physician Laguna – who shared the knowledge that Ghini imparted, deliberately omitting Mattioli from this community.[121] Ghini also gave Maranta a copy of the *Placiti*, allowing him to look closely at what Mattioli did with this material. The Neapolitan physician declared Mattioli's image of *lonchitis aspera* – a plant the seventeenth-century English botanist John Gerard called "rough spleenewort" and modern botany describes as

[117]Ibid., 1531.

[118]Mattioli 1561, 280 (n. d.).

[119]Maranta 1559, sig.*3r (1 July 1558); p. 4.

[120]Ibid., sig.*4r (11 August 1558).

[121]Ibid., 3–4, 21, 87, 110, 182.

an ancient name for holly fern as well as a variety of plants with spear-shaped seeds – absolutely false.[122] He felt certain that Mattioli had only seen an image but not carefully inspected a specimen of the plant. It became a focal point for their contested inheritance.

Throughout the sixteenth century naturalists quarreled repeatedly about plant misidentification. Mattioli was in the midst of a heated controversy with Gesner about whether he had faked an image of *aconitum primum* to conform to Dioscorides's description of this plant, making him especially sensitive to this kind of accusation.[123] What made the debate between Maranta and Mattioli unique was the use of Ghini's research materials as an authoritative basis for their respective descriptions of *lonchitis*. Maranta triumphantly announced that Mattioli's image lacked the roughness and hairiness of true *lonchitis aspera*. There was of course only one reason for such an egregious mistake. He declared that Mattioli did not possess Ghini's full knowledge of this plant. "I know," Maranta informed Aldrovandi in February 1560, "that until the very end of his life Messer Luca held this plant that I describe as true *lonchitis*, and not that one that Mattioli has drawn." He reminded colleagues that "no one knows the opinion of Messer Luca better than me on this because of our long friendship and continuous correspondence that we had when he was alive."[124]

Mattioli countered by informing correspondents that the illustration he possessed of *lonchitis* was practically a saintly relic, since it came from the "blessed hand of Messer Luca," accompanied by a letter describing it and "writings beneath it in his hand"[125] Each claimed to possess the most authentic version of the kind of knowledge that Ghini so liberally dispersed among his colleagues. Each believed that he truly knew the "mind of Luca Ghini."[126]

How to resolve the debate? While involving most of the Italian community of naturalists, both Maranta and Mattioli hoped that Aldrovandi would arbitrate their disagreement. He also possessed a copy of Ghini's *Placiti*. In the spring of 1560, Maranta attempted to coerce Aldrovandi into signing a sworn testimonial, ideally witnessed by Ghini's nephew Vincenzo and several independent parties, stating his agreement with Maranta's interpretation of Ghini's notes. "Some feel that this is *lonchitis aspera*," Ghini had written in 1551, "but I truly recognize it as a genus of fern rather than true *lonchitis aspera*."[127] Maranta envisioned Aldrovandi's sworn testimonial as the final coup de grâce in his effort to best Mattioli. He was sure that Anguillara and Solenander would back him up.

Fundamentally, the problem was that there was more than one kind of plant with spear-shaped seeds. Aldrovandi enraged Maranta when he decided that Mattioli

---

[122] Ibid., 152; Quattrocchi 2000, vol. 2, 1526.

[123] Delisle 2004; Kusukawa 2012, 164–170.

[124] De Toni 1911–12, 1534.

[125] Raimondi 1906, 46, 49.

[126] Mattioli 1561, 280 (n. d.).

[127] De Toni 1907a, 16.

was describing *lonchitis seconda*.[128] He reminded Maranta that he also possessed Ghini's memory but in different way. Aldrovandi sided with Mattioli based on personal conversations with their deceased mentor, invoking oral testimony to supplement the written record. He also possessed papers in Bologna that were not in Naples. When pressed, Aldrovandi recalled discussing "some errors" in Mattioli's commentary with Ghini, but stated that *lonchitis aspera* was not among them. He referred Maranta to Ghini's personal copy of Mattioli's book, which contained no marginalia "noting this as an error."[129] As it turns out, Aldrovandi did have Ghini's annotated version of the 1554 Latin commentary! Perhaps he found it after his conversations with Melchiori but, more likely, he reserved it for himself, as an important talisman of his own relationship with Ghini. It became the source of his authority.

Fundamentally, Ghini's memory was at stake, as his heirs fought over who had best known their botanical father. Mattioli instead insisted that fraternal friendship (*fraterna … amicitia*) was the most powerful bond of all.[130] Maranta invoked over 400 letters in his possession, unfortunately lost to us today. He felt confident that he knew the mind of their master better than anyone else, and was closest to Ghini's heart because he received a letter written the day before Ghini's death. With Maranta convinced that Ghini's last thought on this earth was for him, the Ghini family's decision to bequeath Ghini's scholarly materials to him offered definitive proof of the special nature of their relationship. "He made me the heir of his writings … and his minerals," he reminded Aldrovandi, who nonetheless reserved some portion of these materials in his possession in Bologna.[131]

Comparing their specimens, Maranta sought to show Aldrovandi who was the superior botanist by demonstrating how Aldrovandi's conclusions rested on defective skills. Aldrovandi's *lonchitis* was yellowish green and smooth; it lacked the rough hairiness, reddish undercoating, and larger leaves of the *lonchitis* described by Maranta as a true specimen of this plant.[132] He simply could not reconcile Aldrovandi's description with the one by Dioscorides. The more Maranta debated the merits of Mattioli's illustration with Aldrovandi, the less he respected the younger disciple's botanical knowledge. Was he a keen observer of nature like their mentor Ghini? Maranta wasn't convinced that he had earned the right to take his place. Mattioli of course was satisfied that his honor had been restored by the most promising member of the next generation. He would eventually revise his illustrations of this plant to include not only two different versions of the plant he identified as *lonchitis*, the smaller one the specimen sent by Ghini, but also Maranta's plant which he labelled *pseudolonchitis* (Fig. 7.6).[133]

---

[128]Perhaps this was a version of the distinction between *Lonchitis hirsuta* and *Lonchitis pubescens* in modern botany? See Encke et al. 1993, 356.

[129]De Toni 1911–12, 1542 (26 May 1560).

[130]Mattioli 1561, 204 (12 December 1558).

[131]De Toni 1911–12, 1545 (26 May 1560).

[132]Ibid., p.1546.

[133]Mattioli 1573, 603.

Fig. 7.6 Pier Andrea's Mattioli's images of the plants he identified as *lonchitis*, and his classification of Maranta's *lonchitis* as an example of misunderstanding Luca Ghini's *Placiti* (*Source*: Pier Andrea Mattioli, *I Discorsi di M. Pietro Andrea Matthioli, sanese, medico cesareo, et del*

7.5 Ghini's Posthumous Authority

Unlike Mattioli and Maranta, Aldrovandi did not rush his own natural histories into print. Instead, he taught and filled his home with all the materials that Ghini valued. His herbarium grew to 17 volumes – and amazed Maranta when he finally visited in 1570.[134] He built his famous museum of natural history, while laying the groundwork for the eventual foundation of Bologna's botanical garden. He created a significant scholarly library filled with his annotated readings of books.[135] This was why Mattioli rushed to befriend Aldrovandi in order to gain access to this new and growing archive of nature, inspired by Ghini's own. At some point after Ghini's death, Aldrovandi sketched a brief history of Ghini's career: "He was the first reader to teach this faculty of the history of plants in Europe," he recalled, forgetting that Giulio da Foligno briefly taught the subject in Rome in 1514 while the Venetian Senate appointed Francesco Buonafede to teach medicinal simples in Padua in 1533, a year before Ghini began to teach of this subject in Bologna.[136] Nonetheless, Ghini was the first professor of medicinal simples worth remembering. There were other aspects of Ghini's legacy to cultivate that avoided the inevitable problems of claiming a complex inheritance.

Aldrovandi was not the only disciple to avoid debating in print who had the strongest claim Ghini's posthumous authority. Asked to reflect on the value of medical travel around 1560, the Ingolstadt physician Lorenz Grill (1480–1560) fondly recalled the 7 years he studied in Italy, including 9 months in Pisa. Undoubtedly aware of Ghini's recent demise, Grill recalled his master "Luca Ghini, who in botany and the universal material of medicinal simples is second to no one in Italy."[137] In his *Voyage to Monte Baldo*, first published in 1561 as a field guide designed to complement Mattioli's commentary, Calzolari created a fascinating portrait of an entire generation of naturalists who accompanied him on trips up this

Fig. 7.6 (continued) Serenissimo Principe Ferdinando Archiduca d'Austria & c. Nelli sei libri Di Pedacio Dioscoride Anazarbeo della materia Medicinale. Hora di nuovo dal suo istesso autore ricorretti, & in più di mille luoghi aumentati. Con le Figure tirate dalle naturali & vive Piante, & animali, & in numero molto maggiore, che le altre per avanti stampate. Con due Tavole copiosissime spettanti l'una à ciò, che in tutta l'opera si contiene: & l'altra alla cura di tutte le infirmità del corpo humano (Venice, 1573), p. 603

[134]Camus 1895, 303. At the time of Maranta's visit, Aldrovandi had compiled fourteen of the seventeen volumes, containing approximately 2000 of the circa 5000 plant species that his herbarium ultimately contained; Antonino 2003.

[135]See bibliography in n.7; and most recently Duroselle-Melish and Lines 2015.

[136]BUB, *Aldrovandi*, ms. 91, c. 429 ("Informatione del giardin public," ca. 1560–80). As transcribed in Bertoloni 1890. On the Roman chair, see Ubrizsy Savoia 2014.

[137]Grill 1566, 5v; see Cunningham 2010, 5, 13.

famous mountain, but he especially singled out Ghini as one of two physicians (the other being a local celebrity, Girolamo Fracastaro) who taught him the most about nature.[138]

That same year Ghini's disciple Anguillara published his vernacular guide to understanding medicinal simples, filled with recollections of his experiences botanizing with Ghini. He concluded that the *lonchitis seconda* he found in 1545, while botanizing with Ghini near Montenero, in the vicinity of Livorno, did not sufficiently resemble samples of this Dioscoridean plant that he subsequently collected from Dalmatia, Greece, and the Ionian islan of Zakynthos. Prudently, Anguillara did his best not to get too entangled in the debate between Maranta and Mattioli.[139] Instead, he left an indelible portrait of his own relationship with Ghini in the field, and his ongoing efforts to expand the kind of empirical knowledge of nature that this represented. Ghini the practitioner of the craft of observing nature mattered far more to Anguillara, as prefect of the Padua garden, than any debate about a single plant.

Throughout the early 1560s, Mattioli and Maranta continued to spar over the question of *lonchitis aspera* and other problems in Renaissance botany, even as they claimed to make the peace. Both continued to cite Ghini copiously in print as the most authoritative source for their opinions – to such a degree that when Maranta published his work on theriac and mithridatum in 1572 he was maliciously accused by one reader of simply plagiarizing it from Ghini's notes in his possession.[140] Aldrovandi continued to mediate their relationship, with increasingly greater circumspection to avoid being drawn further into the dispute. He succeeded in holding onto his part of what turned out to be a divided inheritance. He had the satisfaction of seeing another former Ghini student sign the visitors's book to Aldrovandi's museum as *discipulus Lucae Ghini*.[141]

By commanding so many of the resources with which to study nature, Aldrovandi eventually succeeded Ghini in his position within the community of naturalists. He demonstrated his ability to increase the stock of information from which natural history could be written and knowledge arbitrated. Aldrovandi's own archive of nature vastly exceeded the one Ghini created that inspired him as a young student. He combined specimens, images, and descriptions in the service of an intellectual project designed not only to generate greater knowledge of plants but of the entire natural world. It is telling that Aldrovandi chose not to write about plants, after surviving the vicious debates of the 1550s and 60s. Instead, he turned his attention to birds, initiating a multi-decade project that eventually became the first

[138]Calzolari 1561, 15; see Findlen 1994, 179–184. On his relationship with Ghini, see De Toni and Forti 1906; Cermenati 1909–10.

[139]Anguillara 1561, 241.

[140]Mazzini 1949, 198.

[141]BUB, *Aldrovandi*, ms. 110, as cited in Cermenati 1909, 941. Ghini introduced the French physician Gervasio Mastarelli to Aldrovandi in December 1552, noting how much he had traveled in Spain; De Toni 1905, 10 (19 December 1552).

published volume in his own natural history, the *Ornithologiae* (1599–1603). One of the lessons that Aldrovandi learned from Ghini was that authority did not necessarily lie in the printed word or well-made image, since both could be contested, but in the possession of good information.

In the first posthumous edition of his Italian commentary, Mattioli firmly declared that Ghini "is justly considered to be one of the greatest botanists of our times." He expanded this comment in the 1573 edition, fulsomely describing Ghini as "not only the most singular medical botanist in his time but candid, sincere, truthful, and faithful in everything else."[142] What did his legacy look like by the end of the sixteenth century?

In 1583 Ghini's distinguished disciple Cesalpino finally completed his own philosophical study of plants, having completed his herbarium 20 years earlier, where he first began to improve upon Ghini's methods of naming and classifying plants, eschewing an alphabetical order for groupings based on the fruit and the seed. "All science consists in the collecting of things that are alike, and the distinguishing of the unlike," he proclaimed.[143] Cesalpino's *De plantis libri XVI* was surely the most important intellectual legacy to emerge from Ghini's school. By insisting that the empirical program of natural history needed to benefit from the theoretical insights of natural philosophy to truly become a science, Cesalpino fully realized the program inspired by Aristotle's disciple Theophrastus that Ghini discussed with Aldrovandi. His goal was not only to describe nature well but to use description as a basis for classification and ultimately understanding. And yet, as Anna Pavord sagely observes, Cesalpino's nomenclature endured longer than his system of classifying by fruits and seeds.[144] It preserved a key feature of Ghini's legacy for posterity that remained unpublished in the work of his other disciples, including Aldrovandi.

In the dedication to the Grand Duke of Tuscany Francesco I, Cesalpino described the development of botany in the mid-sixteenth century, generously citing several generations of naturalists to whom he was indebted. Cesalpino explained that his book was "the work of many friends and the bounty of the garden," including his schoolmate Anguillara whose *placiti* he cited; he reserved his most fulsome praise for "my teacher Luca Ghini, the prince in this faculty."[145] Seeking to set the record straight, Cesalpino described how Ghini improved Mattioli's commentary on Dioscorides by sharing his visual archive and knowledge of plants. In 1583 Mattioli had been dead for 6 years. Cesalpino wanted Ghini to emerge from Mattioli's shadow, as the better botanist with a larger vision of his field. He pointedly eschew the history of the debate about *lonchitis* among modern naturalists, briefly describing two varieties with reference to Dioscorides.[146]

---

[142]Mattioli 1557, 537; Mattioli 1573, 156.

[143]Cesalpino, 1583, n.p. (dedication to Grand Duke Francesco I). See also Pavord 2005, 232.

[144]Tongiorgi Tomasi 1992, 278–280; Pavord 2005, 230, 235.

[145]Cesalpino 1583, n.p. (dedication to Grand Duke Francesco I).

[146]Ibid., 412, 596.

Cesalpino also expressed great pride in the visual archive he assembled for this project. He felt that his images achieved a new standard in botanical illustration by portraying “the minutest differences.” Unfortunately, the expense of publishing an illustrated botany prevented him from including these beautiful and accurate images, offering “the most faithful testimony” of nature. Cesalpino hoped that the current grand duke would eventually publish these drawings, while observing that, in his experience, images could not capture nature as well as words: “a picture cannot effect greater certainty: for indeed a picture cannot express all the differences like speech (**oratio**).”[147] This was his memorial to Ghini; he sought to demonstrate how he embodied all the methodological virtues of the master who created the garden in which he worked, reaping the benefits of Ghini’s pragmatic approach to the value of experience by transforming scientific practice into the basis for a new natural philosophy.

Two years after Cesalpino composed Ghini’s final epitaph written by a disciple, Tommaso Garzoni published a different kind of Renaissance bestseller – *La Piazza universale di tutte le professioni del mondo* (1585). In this encyclopedia of professions, he included the medical botanist who trafficked in simples. Garzoni felt obliged to name Mattioli, but not because he necessarily exemplified the best qualities of the Renaissance naturalist. Garzoni informed his readers that Mattioli had learned “an infinity of things from Luca Ghini . . ., in the science of simples indubitably the prince.”[148] Perhaps he read Cesalpino’s preface. It is far more likely, however, that Garzoni absorbed the many different ways in which his generation commemorated the founding father of their discipline when he composed this public epitaph. Aldrovandi must have been pleased to find this reference in a book that he owned. Thus, Ghini’s name survived in Garzoni’s universal marketplace of all the professions of the world, as a talisman of the Italian medical republic of letters that came of age in the mid-sixteenth century. Soon thereafter, the world of books that Vesalius, Fuchs, and Mattioli shaped with their influential publications precipitated the decline of the kind of authority that Ghini embodied at the end of the Renaissance.

Bibliography

Andreoli, Ilario. 2006. *Ex officina erasmiano. Vincenzo Valgrisi e l’illustrazione del libro tra Venezia e Lione alla metà del ’500*. PhD dissertation, Università Ca’ Foscari Venezia and Université Lumière Lyon 2.

Anguillara, Luigi. 1561. *Semplici*. Venice: Valgrisi.

Antonino, Biancastella, ed. 2003. *L’erbario di Ulisse Aldrovandi. Natura, arte e scienza in un tesoro del Rinascimento*. Milan: F. Motta.

[147]Ibid, n.p. (dedication to Grand Duke Francesco I). Also discussed in Tongiorgi Tomasi 1992, 283.

[148]Garzoni 1996, vol. 1, 329; see McClure 2004.

Arber, Agnes. 1986. *Herbals, Their Origins and Evolution: A Chapter in the History of Botany, 1470–1670*, 3rd ed. Cambridge: Cambridge University Press.

Arrizabalaga, John, John Henderson, and Roger French, eds. 1997. *The Great Pox: The French Disease in Renaissance Europe*. New Haven: Yale University Press.

Azzi Visentini, Margherita. 1984. *L'Orto botanico di Padova e il giardino del Rinascimento*. Milan: Edizioni il Polifilo.

Bernabeo, R. 1992. *Il 'De morbo gallico' di Luca Ghini*, ed. Galassi, Cristofolini and Garbari, VIII, 237–243. Imola: Cassa di Risparmio di Imola. Special issue of *Museologia scientifica* 8.

Bertoloni, Antonio. 1890. Notizie storiche sull'origine dello studio dei semplici in Italia. *Nuovo giornale botanico storico* 22: 84–90.

Biagioli, Mario, and Peter Galison, eds. 2003. *Scientific Authorship: Credit and Intellectual Property in Science*. New York: Routledge.

Blair, Ann. 2010. *Too Much to Know: Managing Scholarly Information Before the Modern Age*. New Haven: Yale University Press.

Bronzino, Giovanni, ed. 1962. *Notitia doctorum sive Catalogus Doctorum qui in Collegis Philosophiae et Medicinae Bononiae Laureati fuerunt ab anno 1480 usque ad annum 1800*. Milan: Giuffrè.

Calzolari, Francesco. 1561. *Il viaggio di Monte Baldo*. Venice: Valgrisi.

Camus, Jules. 1895. Historique des premiers herbiers. *Malpighia* 9: 283–314.

Carlino, Andrea. 1999. *Books of the Body: Anatomical Ritual and Renaissance Learning*. Trans. John Tedeschi. Chicago: University of Chicago Press.

Castiglioni, Arturo. 1953. The School of Ferrara and the Controversy on Pliny. In *Science, Medicine, and History*, ed. E. Ashworth Underwood, vol. 1, 269–279. Oxford: Oxford University Press.

Ceremenati, Mario. 1909. Francesco Calzolari e le sue lettere all'Aldrovandi. *Annali di botanica* 7: 83–138.

———. 1909–10. Intorno il Ghini e i suoi rapporti con Francesco Calzolari. *Atti del Real Istituto Veneto di scienze, lettere ed arti* 69(2): 939–961.

Cesalpino, Andrea. 1583. *De plantis libri XVI*. Florence: Georgium Marescottum.

Chiarugi, Alberto. 1953. Le date di fondazione dei primi orti botanici del mondo. *Nuovo giornale botanico italiano* 60 (4): 785–839.

Chiovenda, Emilio. 1909. Francesco Petrollini botanico del secolo XVI. *Annali di botanica* 7: 339–447.

Cooper, Alix. 2007. *Inventing the Indigenous: Local Knowledge and Natural History in Early Modern Europe*. Cambridge: Cambridge University Press.

Cordus, Valerius. 1561. *Annotationes in Pedacii Dioscoridis de Materia medica libros V.*, ed. Conrad Gesner. Strassbourg: Rihel.

Cortesi, Giovanni Battista. 1629. *Pharmacopoeia seu Antidotarium Messinense*. Messina: Pietro Brea.

Cunningham, Andrew. 2010. The Bartholins, the Platters and Laurentius Gryllus. In *Centres of Medical Excellence? Medical Travel and Education in Europe, 1500–1789*, ed. Ole Grell, Andrew Cunningham, and Jon Arrizabalaga, 3–16. Farnham: Ashgate.

Daston, Lorraine, and Elizabeth Lunbeck, eds. 2011. *Histories of Scientific Observation*. Chicago: University of Chicago Press.

De Toni, Giovan Battista. 1905. *Cinque lettere di Luca Ghini ad Ulisse Aldrovandi*. Padua: Tipografia Seminario.

———. 1907a. I Placiti di Luca Ghini (primo lettore dei semplici in Bologna) intorno a piante descritte nei Commentarii al Dioscoride. *Memorie del Reale Istituto Veneto di scienze, lettere ed arti* ser. 8, vol. 27/8: 3–49.

———. 1907b. Spigolature aldrovandiane VI. Le piante dell'antico Orto Botanico ai tempi di Luca Ghini. *Annali di botanica* 5: 421–440.

———. 1910–11. Spigolature aldrovandiane IX. Nuovi documenti intorno a Francesco Petrollini. *Atti del Reale Istituto Veneto di scienze, lettere ed arti* ser. 8, vol. 12/2, tome 69: 815–825.

———. 1911. Spigolature aldrovandiane XI. Intorno alle relazioni del botanico Melchiorre Guilandino con Ulisse Aldrovandi. *Atti dell'Istituto Reale Accademia di scienze, lettere ed arti degli Agiati in Rovereto* ser. 3, vol. 17, fasc. 2: 149–171.

———. 1911–12. Nuovi documenti sulla vita e sul carteggio di Bartolomeo Maranta. *Atti del Reale Istituto Veneto di scienze, lettere ed arti* ser. 8, vol. 14/2, tome 72: 1505–1564.

———. 1913. Spigolature aldrovandiane X. Alcune lettere di Gabriele Falloppia ad Ulisse Aldrovandi. *Atti e memorie della Reale Deputazione di storia patria per le province Modenesi* ser. 5, vol. 7: 3–15.

———. 1924–25. Spigolature aldrovandiane XXI. Un pugillo di lettere di Giovanni Odorico Melchiori Trentino a Ulisse Aldrovandi. *Atti del Reale Istituto Veneto di scienze, lettere ed arti* ser. 8, tome 84: 599–624.

De Toni, Giovan Battista, and A. Forti. 1906. Intorno alle relazioni di Francesco Calzolari con Luca Ghini. *Bollettino della Società botanica italiana* 15: 151–157.

Delisle, Candice. 2004. The Letter: Private Text or Public Place? The Mattioli-Gesner Controversy about the *aconitum primum. Gesnerus* 61: 161–176.

———. 2008. Accessing Nature, Circulating Knowledge: Conrad Gesner's Correspondence Networks and His Medical and Naturalist Practices. *History of Universities* 23: 35–58.

Di Pietro, Pericle. 1970. *Epistolario di Gabriele Falloppia*, Quaderni di storia della scienza e medicina, 10. Ferrara: Università degli Studi di Ferrara.

Duroselle-Melish, Caroline, and David A. Lines. 2015. The Library of Ulisse Aldrovandi (+1605): Acquiring and Organizing Books in Sixteenth-Century Bologna. *The Library* ser. 7, vol. 16: 133–161.

Egmond, Florike. 2007. Correspondence and Natural History in the Sixteenth Century: Cultures of Exchange in the Circle of Carolus Clusius. In *Cultural Exchange in Early Modern Europe. Correspondence and Cultural Exchange in Europe 1400–1700*, ed. Francisco Bethencourt and F. Egmond, vol. 3, 104–142. Cambridge: Cambridge University Press.

Encke, Fritz, Günther Buchheim, and Sigmund Seybold. 1993. *Zander Handwörterbuch der Pflanzennamen*. Stuttgart: Eugen Ulmer.

Engelhardt, Dietrich von. 1995. Luca Ghini (um 1490–1556) und die Botanik des 16. Jahrhunderts. *Medizin-historische Journal* 30: 3–49.

———. 2011. Luca Ghini (1490–1556) il padre fondatore della botanica moderna nel contest dei rapporti scientifici europei del sedicesimo secolo. *Annali del Museo Civico di Rovereto, Sezione: Archeologia, Storica, Scienze naturali* 27: 227–246.

Fahy, Conor. 1993. *Printing a Book in Verona in 1622*. Paris: Fondation Custodia.

Fantuzzi, Giovanni. 1784. Luca Ghino. *Notizie degli scrittori bolognesi* 4: 134–136. Bologna.

Fausti, Daniela, ed. 2004. *La complessa scienza dei semplici*. Siena: Accademia delle Scienze di Siena detta de' Fisiocritici.

Ferrari, Giovanna. 1990. Gli errori di Plinio. Fonti classiche e medicina nel conflitto tra Alessandro Benedetti e Niccolò Leoniceno. In *Sapere e/è potere*, ed. Andrea Cristiani, vol. 2, 173–204. Bologna: Comune di Bologna.

Ferri, Sara, ed. 1997. *Pietro Andrea Mattioli (Siena 1501 – Trento 1578). La vita, le opera*. Perugia: Quattroemme.

Findlen, Paula. 1991. The Economy of Scientific Exchange in Early Modern Italy. In *Patrons and Institutions: Science, Technology and Medicine at the European Courts, 1500–1750*, ed. Bruce Moran, 5–24. Woodbridge: Boydell and Brewer.

———. 1994. *Possessing Nature: Museums, Collecting and Scientific Culture in Early Modern Italy*. Berkeley: University of California Press.

———. 1999. The Formation of a Scientific Community: Natural History in Sixteenth-Century Italy. In *Natural Particulars: Nature and the Disciplines in Renaissance Europe*, ed. Anthony Grafton and Nancy Siraisi, 369–400. Cambridge: MIT Press.

———. 2004. Ereditare un museo. Collezionismo, strategie familiari e pratiche culturali nel Cinquecento. *Quaderni storici* 115: 45–81.

Frati, Lodovico. 1907. La vita d'Ulisse Aldrovandi cominciando dalla sua natività sin'a l'età di 64 anni, vivendo ancora. In *Intorno alla vita e alle opera di Ulisse Aldrovandi*. Bologna: Libreria Treves di L. Beltrami.

Galassi, Nazario. 1992. Luca Ghini, una vita per la scienza. In *Luca Ghini. Cinquecento anni di scienze botaniche*, ed. Nazario Galassi, Giovanni Cristofolini, and Fabio Garbari, 187–205. Imola: Cassa di Risparmio di Imola. Special issue of *Museologia scientifica* 8.

Galassi, Nazario, Giovanni Cristofolini, and Fabio Garbari, eds. 1992. *Luca Ghini. Cinquecento anni di scienze botaniche*, 185–326. Imola: Cassa di Risparmio di Imola. Special issue of *Museologia scientifica* 8.

Garbari, Fabio, 1980. Nasce presso lo studio pisano nel XVI secolo la botanica moderna. In *Livorno e Pisa*. Pisa: Nistri-Licini e Pacini.

———. 1992. Luca Ghini a Pisa, cardine della cultura botanica del XVI secolo. Galassi, Cristofolini, and Garbari, 223–236.

Garbari, Fabio, Lucia Tongiorgi Tomasi, and Alessandro Tosi. 1991. *Giardino dei semplici. L'orto botanico di Pisa dal XVI al XX secolo*. Pisa: Cassa di Risparmio.

Garzoni, Tommaso. 1996. *La Piazza universale di tutte le professioni del mondo*, ed. Paolo Cherchi and Beatrice Collina, 2 vols. Turin: Einaudi.

Gentilcore, David. 2010. *Pomodoro! A History of the Tomato in Italy*. New York: Columbia University Press.

Ghini, Luca. 1589. Morbi Neapolitani curandi ratio perbrevis. In Philip Schopf, *Practica theorica empirica morborum interiorium*, ed. Johan Marquard, 422–464. Spirae.

Greene, Edward. 1983. *Landmarks of Botanical History*, ed. Frank N. Egerton, Robert P. McIntosh and Rogers McVaugh, 2 vols. Stanford: Stanford University Press.

Grill, Lorenz. 1566. *Oratio ... quam in solennis professionis publicae principio recitant Ingolstadii*, in *De sapore dulce et amaro libri duo*. Prague: Georgium Melantrichum.

Johns, Adrian. 1998. *The Nature of the Book: Print and Knowledge in the Making*. Chicago: University of Chicago Press.

Keller, A.G. 1970. Luca Ghini. In *Dictionary of Scientific Biography*, ed. Charles Gillespie, vol. 5, 383–384. New York: Scribner's.

Kusukawa, Sachiko. 2012. *Picturing the Book of Nature*. Chicago: University of Chicago Press.

Levi, Giovanni, 1988. *Inheriting Power: The Story of an Exorcist*. Trans. Lydia G. Cochrane. Chicago: University of Chicago Press.

Lobel, Mattias de, and Pierre Pena. 1571. *Stirpium adversaria nova*. London: Christophe Plantin.

Maclean, Ian. 2008. The Medical Republic of Letters before the Thirty Years War. *Intellectual History Review* 18: 15–30.

Maranta, Bartolomeo. 1559. *Methodi cognoscendorum simplicium libri tres*. Venice: Valgrisi.

Marcus, Hannah. 2016. *Banned Books: Medicine, Readers, and Censors in Early Modern Italy, 1559–1664*. PhD dissertation, Stanford University.

Mattioli, Pier Andrea. 1554. *Commentarii in libros sex Pedacii Dioscoridis Anazarbei de medica materia. Adirectis quam plurimis plantorum & animalium imaginibus, eodem authore*. Venice.

———. 1555. *I discorsi di M. Pietro Andrea Matthioli medico sanese, ne i sei libri della materia medicinale di Pedacio Dioscoride Anazarbeo. Con i veri ritratti delle piante & de gli animali, nuovamente aggiuntivi dal medesimo*. Venice.

———. 1557. *I discorsi di M. Pietro Andrea Matthioli Medico Sanese, ne i sei libri della materia medicinale di Pedacio Dioscoride Anazarbeo. Con i veri ritratti delle piante & de gli animali, nuovamente aggiuntivi dal medesimo*. Venice.

———. 1559. *I discorsi di M. Pietro And. Matthioli sanese, medico del Sereniss. Principe Ferdinando Archiduca d'Austria & c. ne i sei libri di Pedacio Dioscoride Anarzabeo della materia medicinale. I quai discorsi in diversi luoghi dall'auttore medesimo sono stati accresciuti di varie cose, con molte figure di piante, & d'animali nuovamente aggiunte*. Venice.

———. 1561. *Epistolarum medicalium libri quinque*. Prague.

———. 1573. *I discorsi di M. Pietro Andrea Matthioli, sanese, medico cesareo, et del Serenissimo Principe Ferdinando Archiduca d'Austria &c. Nelli sei libri Di Pedacio Dioscoride Anazarbeo della materia Medicinale. Hora di nuovo dal suo istesso autore ricorretti, & in più di mille luoghi aumentati. Con le Figure tirate dalle naturali & vive Piante, & animali, & in numero molto maggiore, che le altre per avanti stampate. Con due Tavole copiosissime spettanti l'una à ciò, che in tutta l'opera si contiene: & l'altra alla cura di tutte le infirmità del corpo humano.* Venice.

Mazzini, Giuseppe. 1949. Luca Ghini medico e naturalista. *Rivista di storia delle scienze mediche e naturali* 40: 176–227.

McClure, George W. 1991. *Sorrow and Consolation in Italian Humanism*. Princeton: Princeton University Press.

———. 2004. *The Culture of Profession in Late Renaissance Italy*. Toronto: University of Toronto Press.

Meschino, Franco Aurelio. 1999. Luca Ghini. In *Dizionario biografico degli italiani*, vol. 53, 767–771. Rome: Enciclopedia italiana.

Meyer, Frederick G., Emily Emmart Trueblood, and John L. Heller, eds. 1999. *The Great Herbal of Leonhart Fuchs*, 2 vols. Stanford: Stanford University Press.

Minelli, Alessandro, ed. 1995. *The Botanical Garden of Padua 1545–1995*. Venice: Marsilio.

Mugnai Carrara, Daniela. 1991. *La biblioteca di Niccolò Leoniceno*. Florence: Olschki.

Nauert, Charles G. 1979. Humanists, Scientists, and Pliny: Changing Approaches to a Classical Author. *American Historical Review* 84: 72–85.

Nepi, Chiara, and Enrico Gusmeroli, eds. 2008. *Gli erbari aretini da Andrea Cesalpino ai giorni nostri*. Florence: Firenze University Press.

Nutton, Vivian. 1997. The Rise of Medical Humanism: Ferrara, 1464–1555. *Renaissance Studies* 11: 2–19.

———. 2004. Mattioli and the Art of the Commentary. In *La complessa scienza dei semplici*, ed. Fausti, 133–145. Siena: Accademia delle Scienze di Siena detta de' Fisiocritici.

Ogilvie, Brian. 2006. *The Science of Describing: Natural History in Renaissance Europe*. Chicago: University of Chicago Press.

Olmedilla y Puig, Joaquin. 1887. *Estudio historíco de la vida y escritos del Sabio español Andrés Laguna*. Madrid: El Correo.

Olmi, Giuseppe. 1991. 'Molti amici in varii luoghi': Studio della natura e rapporti epistolari nel secolo XVI. *Nuncius* 6: 3–31.

———. 1992. *L'inventario del mondo. Catalogazione della natura e luoghi del sapere nella prima età moderna*. Bologna: Il Mulino.

Palmer, Richard. 1985. Medical Botany in Northern Italy in the Renaissance. *Journal of the Royal Society of Medicine* 78: 149–157.

Park, Katharine. 2011. Observation in the Margins, 500–1500. In *Histories of Scientific Observation*, ed. Lorraine Daston and Elizabeth Lunbeck, 15–44. Chicago: University of Chicago Press.

Pavord, Anna. 2005. *The Naming of Names: The Search for Order in the World of Plants*. New York: Bloomsbury Publishing.

Pesenti, Tiziana, 1985. Il 'Dioscoride' di Pier Andrea Mattioli e l'editoria botanica. In *Trattati di prospettiva, architettura militare, idraulica e altre discipline*, 61–103. Venice: Neri Pozza.

Pomata, Gianna, and Nancy G. Siraisi, eds. 2005. *Historia: Empiricism and Erudition in Early Modern Europe*. Cambridge, MA: MIT Press.

Quattrocchi, Umberto, ed. 2000. *CRC World Dictionary of Plant Names*. Boca Raton: CRC Press.

Raimondi, Carlo. 1906. Lettere di P. A. Mattioli ad Ulisse Aldrovandi. *Bullettino senese di storia patria* 13, fasc. 1–2: 3–67.

Reeds, Karen M. 1991. *Botany in Medieval and Renaissance Universities*. PhD dissertation, Harvard University, 1975; New York: Garland.

Riddle, John. 1980. Dioscorides. In *Catalogus Translationum et Commentarium: Medieval and Renaissance Latin Translations and Commentaries*, ed. Paul Oskar Kristeller and F. Edward Cranz, vol. 4, 1–144. Washington, DC: Catholic University of America Press.
Rondelet, Guillaume. 1554. *Libri de piscibus marinis*. Lyon: Matthiam Bonhomme.
Rossi, Walter, and Maria Adele Signorini. 1994. An Old Orchid Painting from the Palazzo Vecchio in Florence. *Curtis's Botanical Magazine* 11 (1): 28–31.
Sabbatini, Luigi. 1922–23. Alcuni documenti su la vita di Luca Ghini. *Atti e memorie della Reale Accademia di scienze, lettere ed arti in Padova* n.s., 39: 243–250.
———. 1923. Il Ghini e l'Anguillara negli orti di Pisa e di Padova. *Rivista di storia delle scienze mediche e naturali* ser. 3, n. 11–12 (1923): 307–309.
———. 1926. La cattedra dei semplici fondata a Bologna da Luca Ghini. *Studi e memorie per la storia dell'Università di Bologna* 9: 14–53.
———. 1927a. Acqua vitae Lucae Ghini. *Archivio di storia della scienza* 8: 477–484.
———. 1927b. La cura del morbo gallico nelle lezioni di Luca Ghini. *Atti del Real Istituto Veneto di scienze, lettere ed arti* ser. 8, vol. #, tome 87: 77–109.
Salviani, Ippolito. 1554–58. *Aquatilium animalium historia*. Rome: Apud eundem Hippolytum Saluianum.
Santing, Catrien. 2010. Pieter van Foreest and the Acquisition and Traveling of Medical Knowledge in the Sixteenth Century. In *Centres of Medical Excellence? Medical Travel and Education in Europe, 1500–1789*, ed. Ole Grell, Andrew Cunningham, and Jon Arrizabalaga, 149–170. Farnham: Ashgate.
Schmitt, Charles. 1976. Science in the Italian Universities in the Sixteenth and Early Seventeenth Centuries. In *The Emergence of Science in Western Europe*, ed. Maurice Crosland, 35–56. New York: Science History Publications.
———. 1980. Theophrastus. In *Catalogus translationum et commentariorum*, ed. Paul O. Kristeller and F. Edward Cranz, vol. 2, 239–322. Washington, DC: Catholic University of America Press.
———, 1984. *The Aristotelian Tradition and Renaissance Science*, vol. XV, 297–336. London: Variorum.
Seybold, Seymour. 1986. *Die Orchideen Leonhart Fuchs*. Baden-Württemberg: Arbeitskreis Heimische Orchideen Baden-Württemberg.
———. 1990. Luca Ghini, Leonhard Rauwulf und Leonhart Fuchs. Über die Herkunft der Aquarelle in Wiener Kräutbuchmanuskript von Fuchs. *Jahrshefte der Gesellschaft für Naturkunde im Württemberg* 145: 241–164.
Signorini, Maria Adele. 1993. Sulle piante dipinte del Bachiacca nello scrittoio di Cosimo I nel Palazzo Vecchio. *Mitteilungen des Kunsthistorischen Institutes in Florenz* 37 (2–3): 396–407.
Siraisi, Nancy G. 1997. *The Clock and the Mirror: Girolamo Cardano and Renaissance Medicine*. Princeton: Princeton University Press.
———. 2013. *Communities of Learned Experience: Epistolary Medicine in the Renaissance*. Baltimore: Johns Hopkins University Press.
Smith, Andrew F. 1994. *The Tomato in America: Early History, Culture, and Cookery*. Columbia: University of South Carolina Press.
Solenander, Reiner. 1558. *De caloris fontium medicatorum causa, eorumque temperatione*. Lyon: Ioannem Franciscum de Gabiano.
Speyer Wecker, and Johann Jacob. 1617. *Antidotarium generale et speciale*. Basel.
Stannard, Jerry. 1966. Dioscorides and Renaissance Materia Medica. In *Materia Medica in the XVIth Century*, ed. M. Florkin, Analecta Medico-Historica 1, 1–21. Oxford: Pergamon.
Tongiorgi Tomasi, Lucia. 1992. Dall'essenza vegetale agglutinata all'immagine a stampa: il percorso dell'illustrazione botanica nei secoli XVI–XVII. In *Luca Ghini. Cinquecento anni di scienze botaniche*, 271–295. Imola: Cassa di Risparmio di Imola. Special issue of *Museologia scientifica* 8.
Tosi, Alessandro, ed. 1989. *Ulisse Aldrovandi e la Toscana*. Florence: Olschki.

Tugnoli Pattaro, Sandra. 1981. *Metodo e sistema delle scienze nel pensiero di Ulisse Aldrovandi*. Bologna: CLUEB.

Turner, William. 1881. *The Names of Herbes* (London, 1548), ed. James Britten. London: English Dialect Society.

Ubrizsy Savoia, Andrea. 2014. 500 anni fa iniziava l'insegnamento della Botanica, s. l. all'Università 'La Sapienza' di Roma. *Annali di storia delle università italiane* 18: 341–354.

Varchi, Benedetto. 1555. *De' sonetti di m. Benedetto Varchi, Parte Prima*. Florence: M. Lorenzo Torrentino.

———. 1827. *Questione sull'alchimia*, ed. D. Moreni. Florence: Stamperia Magheri.

Velschius, H. 1676. *Curationum et observationum medicinalium chilidaes duae*. Ulm.

Chapter 8
The Dedication Strategies of Conrad Gessner

Ann Blair

The 102 dedications composed by the sixteenth-century physician and polymath Conrad Gessner between 1541 and 1565 offer a rich trove of insight into many aspects of his particular career but also into the workings of the Republic of Letters more generally. Although Gessner never benefitted from a major patronage relationship and probably received limited financial support from his dedicatees, he nonetheless managed to publish a number of major works on his initiative, including folio volumes of philology, bibliography, and especially expensive works of illustrated natural history. Crucial to Gessner's success was his accumulation of smaller contributions in kind from a wide range of people who offered him hospitality or sent him information and specimens, manuscripts and images, which Gessner used in his publications. Gessner rewarded contributors not only by private expressions of thanks, but also in print, and especially visibly in his dedications. Gessner was also unusual in calling attention to the role of learned printers for his work, by composing dedications to them and by advertising that various of his publications were initiated by requests from printers or bequests of manuscripts by recently deceased scholars. Gessner thus used the high visibility of the printed dedication to invite further contributions from learned readers, bequests of unfinished manuscripts, and proposals from printers with which to fuel his remarkable productivity.

I am indebted to generous colleagues for their comments and corrections: Anja Goeing, Anthony Grafton, Urs Leu, Paola Molino, Mikhail Sergeev, and the editors of this volume. I am grateful to those who organized and attended my talk at CalTech, especially Mordechai Feingold and David Hall. Warm thanks to Samantha Wesner for help with research and formatting, and to Tom Keeline for his expert reading of the Greek dedications.

A. Blair (✉)
Harvard University, Cambridge, MA, USA
e-mail: amblair@fas.harvard.edu

G. Manning, C. Klestinec (eds.), *Professors, Physicians and Practices in the History of Medicine*, Archimedes 50, DOI 10.1007/978-3-319-56514-9_8

8.1 *Dedicatio*

Before proceeding I must follow Gessner in making three introductory points which Gessner often made in his dedications and which also apply to this study of mine: praise of the libraries that made this work possible, explanation of the limitations of the work and request for corrections and feedback, and praise of the scholar who inspired the project.

Gessner frequently lauded the owners of great libraries of his day for preserving and allowing him access to rare manuscripts of ancient texts that he prepared for publication. Renaissance libraries funded by the wealthy were a new kind of institution, which Gessner hailed in multiple dedications as crucial to humanist scholarship and to the transmission of knowledge. They housed unique manuscripts but also printed books in rapidly increasing numbers. Similarly, in carrying out this research project I found myself relying on a new kind of library—the collective digital repository of early European printed books that has grown remarkably in recent years. The digital collections in the Swiss E-rara and the Bayerische Staatsbibliothek (BSB), which incorporates part of the sixteenth-century Fugger collection that Gessner praised, are especially impressive in their scope and quality, but I also drew on Google Books, Gallica, and Hathitrust (as listed in the Appendix 1).[1] It is more complicated to identify whom to thank for this remarkable collective achievement. Gessner thanked individual owners like Hans Jakob Fugger, Leonard Beck von Beckenstein, and Diego Hurtado de Mendoza, and governing bodies like the City Council of Augsburg for their role in forming and funding great libraries. The formation of our current digital repository has involved countless professionals in selecting and managing the books and in making and curating the scans, and private and public organizations in funding the work. I am grateful to them all.

Just as Gessner issued caveats about the quality of the manuscripts from which he worked and the limits of his corrections or improvements, I too must offer some caveats: this study is limited by the copies I used, and I realize that I may not have identified all Gessner's dedications. Digital copies offer no panacea to the complexities of the early modern printed copies that they reproduce.[2] When the copy used for scanning lacks pages or sections of Gessner's multi-part works or contains manuscript annotations (additions or expurgations), so too does the electronic

[1]On the conception of E-rara in 2008 see Leu 2014a, 53.

[2]Scanning also introduces distortions of its own, including traces of hands, blurred scans, and duplicate pages. The multiple digital copies of *Mithridates* (1555) also omit, without indicating that they do, the fold-out table appended to the work, present in the volume itself. See for example the copy with call number Regensburg 999/Ling1a and its digitization at the Bayerische Staatsbibliothek (hereafter BSB). The table may have escaped the notice of the scanner and certainly did not fit the format of the rest of the octavo book.

copy.[3] As a result it is valuable to view multiple digital copies as well as multiple physical copies, and important to understand the peculiarities of each copy. The copies of Gessner's works from the Zentralbibliothek Zürich are especially interesting in that some of them contain Gessner's own manuscript notes (perhaps made in view of a possible later edition which in most cases did not materialize), but these copies may also represent an early state of production of the book. For example, the printed state of Gessner's Handexemplar of the *Bibliotheca Universalis* digitized on E-rara is not identical to the copy that served as the basis of the 1966 Olms reprint.[4] This is not surprising, given what we know about the variation in states of early modern printed books. But we need to remember that a digital copy is no more "representative" of a print run than the single printed copy it reproduces, even if by virtue of being digitized it is more readily available to modern scholars than copies which have not been digitized.

Since Gessner was tirelessly involved in translating, editing, annotating, compiling, and composing, the publication history of his works is particularly voluminous and complex. I have relied on Hans Wellisch's invaluable bibliography and have defined as my corpus the dedications in the works that Wellisch identified as containing Gessner's work which were published during his lifetime.[5] I searched systematically for dedications only in first editions of these works and included four later editions published in Gessner's lifetime in which I was aware of a new dedication.[6] Given Gessner's propensity to writing many dedications, there may exist other dedications by Gessner that I have not recorded, either in other later editions of his works or in works published by others. Indeed my corpus already brought to my attention one dedication by Gessner of a work composed by someone else: Gessner explained that his friend Benedictus Aretius charged him with finding a dedicatee for the latter's short work on the Bernese Alps which was included in

[3]For example one of the copies of Gessner's 1565 *De omni rerum fossilium genere* on E-rara contains only the first of the ten texts in that collection (see the digitization of call number NG 1910 from the Zentralbibliothek Zurich, hereafter ZB). Or the copy digitized by the BSB of *David Kyber. Lexicon rei herbariae* (1553) contains a thoroughly blacked out sentence fragment on p. 7 concerning the afterlife: "Quanquam enim credendum est, sanctos qui in vera per unicum Christum fiducia obdormiunt, pari inter se omnes amore feruoreque in vita coelesti coniunctos iri." I cite Gessner's works by title alone; further bibliographical details are provided in the appendix.

[4]Compare the E-rara pdf of Gessner's *Bibliotheca universalis*, f. 454v–455r with the edition of 1966.

[5]See Wellisch 1984, section A, 31–100.

[6]The later editions I have included are the fourth edition of the *Dictionarium graecolatinum* (1545) and the second edition of the *Onomaticon* (1549), which were the earliest editions of these works to contain a dedication by Gessner; and the second editions of the *Icones avium* (1560) and *Icones animalium* (1560) which each included a new dedication in addition to reprinting the dedication present in the first editions of these works.

Gessner's edition of Valerius Cordus' annotations on Dioscorides (1561).[7] (Gessner's use of classicized versions of German names—in this case, Aretius for Marti—also complicates the identification of the figures he mentions.[8]) I have only considered dedications, which I define as addresses by Gessner to specific individuals or collectivities; these are usually called *epistola* or *epistola nuncupatoria*, but occasionally also *epistola dedicatoria* or (confusingly, given modern parlance) *praefatio*.[9] I have thus excluded Gessner's addresses "to the reader" (*lectori*, *ad lectorem*) which he wrote in addition to or occasionally in the absence of a dedication.[10] Gessner's paratexts are well worth further study, but in this context I can only conclude, as Gessner did in one of his dedications, that "due to the shortness of time and other occupations it was not possible to add to [these] descriptions."[11] I look forward to learning from the corrections and further work of others.

The scholar who inspired me to investigate this topic has taught us how fruitful it is to study the history of medicine as a branch of general intellectual history rather than a separate sub-discipline. Learned physicians in the Middle Ages and in the Renaissance were trained in philosophy and in the humanistic disciplines (with many variations by context) and brought those skills to their ways of arguing, analyzing, and writing about medical subjects and related fields like the natural history of animals, plants, and minerals. Nancy Siraisi has shown in her most recent books that physicians often wrote on non-medical topics, such as human history, and were active in the Republic of Letters in forming correspondence networks both within but also beyond their professional community. Nancy Siraisi has also brought the history of medicine into the mainstream of intellectual history by making the field accessible to non-specialists through her teaching and her prize-

[7]Gessner chose as dedicatee a mutual friend, Christophorus Piperinus, minister in Sigriswil, Bern. See *Stockhornii et Nessii in Bernatium Helvetiorum ditione montium … brevis descriptio* in *Valerii Cordi Annotationes in … Dioscorides libros V*, 1561, E-rara pdf 486 ff. I refer to image numbers in the pdf listed in the appendix when the printed pagination is difficult to follow, e.g. because of multiple pagination sequences. I am grateful to Urs Leu for bringing to my attention another such example, in which Gessner dedicated to Abel Werdmüller a work by Abel's recently deceased father Otho that Gessner published as part of Johannes Fabricius Montanus 1555.

[8]Aretius was a friend, correspondent and professor of classical languages and theology in Bern; see Gessner 1577, ff. 115v–122v; and Wellisch, 90. On classicizing names, see Bodenmann 2009; and Taylor and Mosher 1951, ch. 2.

[9]For examples of "praefatio" see *Valerii Cordi Annotationes* (1561), *Galeni Opera* (1562). "Epistola dedicatoria" occurs in *Apparatus* (1542), *Sententiae Antonii et Maximi* (1546, Greek volume), *De rerum fossilium figuris* (1565).

[10]Works with a "Gessner to the reader" only: *Actuarius. De medicamentorum compositione* (1540); *Enumeratio medicamentorum* (1543); *Galeni opera* (1549); *Epitome* (1555); *Viaticum novum* (1565).

[11]"Icones quidem propter temporis angustiam aliasque occupationes descriptionibus non licuit." *Valerii Cordi Stiripium descriptionis liber V*, 1563, sig. Aijr. On Gessner's habits of thanking and citing, see Blair 2017a; for a complete list of Gessner's paratexts and some analysis see Blair 2016.

winning textbook—from both of which I benefitted greatly. Like the figures Nancy Siraisi has studied, Conrad Gessner (1516–65) played a crucial role as a learned physician and humanist both in his hometown of Zurich and in the Republic of Letters. Medical learning formed a major part of Gessner's polymathy alongside his other interests in philology, bibliography, and Swiss Reformed piety.[12] Gessner was a master of humanist rhetoric while I am not, so I borrow from one of his dedications for my own: "Receive therefore this little something of a booklet, as a sign of my perpetual admiration for you and a memento of my greatest love."[13]

8.2 The Several Roles of Dedications

The practice of dedicating a work to a powerful and/or wealthy patron can be traced back to ancient Rome, as the humanists knew well. Virgil's dedication of his *Georgics* to Maecenas inspired the coining of that name to designate a generous patron precisely in the early sixteenth century.[14] Dedications are generally interpreted as bids for financial support. In the age of print authors made money from selling their manuscript to a printer, or from selling or gifting (in the expectation of a countergift) the printed copies of their works allotted to them by the printer. The sums involved varied depending on the standing of the author. Optimally a dedication would result in further rewards, in the form of a one-time gift of money or other object of value, or an offer of longer-term employment.[15] The codes of expectations were unwritten and variable by time, place, type of work, and the individual personalities involved, and authors who misevaluated their situation could fail to receive any benefit from their dedication. Gessner's dedications shed light on practices in a particular corner of the Republic of Letters, centered around the Germanic world and focused on learned works of the mid-sixteenth century; especially relevant comparands include the dedications of Erasmus a few decades earlier and those of the Italian natural historian Ulisse Aldrovandi a few decades later.

[12]The 500th anniversary of Gessner's birth has prompted renewed attention to Gessner, including: Leu 2016a; Leu and Ruoss 2016; a special issue of *Gesnerus* (2016); and Blair 2017b. Earlier accounts of Gessner's life include Braun 1990; Wellisch, 1–25; Ley 1929; Hanhart 1824; and Simmler 1566. See also Serrai 1990; Leu 1990; and Leu et al. 2008.

[13]"Accipe igitur, quicquid hoc et quantulumcumque est libelli, meae erga te observantiae pignus perpetuae, et amoris monumentum summi." *Hanno Carthaginensium ducis navigatio* in *Ioannis Leonis Africani de totius Africae descriptionis libri IX*, 1559, E-rara pdf 566 ff.

[14]*OED* records first uses in French in 1526, in English in 1542 in a translation of Erasmus's *Apopthegms*; "maecenas" does not appear in the *Thesaurus Linguae Latinae*.

[15]On the distinction between patron and Maecenas, see Viala 1985. On dedications, see Genette 1997, ch.6; Schottenloher 1953; Moennighoff 2008; Jancke 2002, 141–52; Gilmont and Vanautgaerden, eds., 2003; and Rice 1971.

In this context dedications also served as opportunities to assert authorship. Gessner was often credited in specific ways (e.g. as translator, compiler, editor, author) but sometimes not at all on the title page of the works in his corpus. When Gessner was not named on the title page, his role in the production of the work became visible in the dedication. Conversely, some early modern works were published under the name of a person who either did not compose the work or did not authorize its publication; in these cases the absence of a dedication can serve as supporting (though not in itself sufficient) evidence that the author was not involved in the publication.[16] In Gessner's corpus *De stirpium nominibus aliquot* (1557) contains no dedication since, as the preface explains, the work published an epistolary exchange between Gessner and Melchior Wieland without the knowledge of either writer.[17] Gessner also published two works which he did not acknowledge, at least at first: the *Thesaurus Evonymi … de remediis secretis* (1552) which proved very successful, with multiple editions in Latin, English, French, German and Italian, and a *Historia et interpretatio prodigii* published just once in 1561. Gessner acknowledged both works as his own in his autobibliography, *De libris a se editis* (1562). He explained there that he hid his authorship of the *Thesaurus* because he brought the work to print prematurely in order to please the printer, his relative Andreas Gessner who was just setting up his business, but once he realized that the work was well received, he added his name to it. The *Thesaurus* included a dedication (to Nicolaus ZurKinden, city councilor of Bern), signed pseudonymously by "Evonymus Philiatros"; in 1554 the pseudonym evolved to "Evonymus Gessnerus Philiatros" and by 1558 the title page openly announced Conrad Gessner's authorship.[18] A pseudonymous dedication was unusual and would seem to undermine the effectiveness of honoring the dedicatee. But, as others have shown of anonymity, pseudonymity in print might be used to hide the identity of the author only beyond his local context.[19] Gessner inscribed a copy of the 1552 *Thesaurus* to the leading citizen of Zurich, Heinrich Bullinger[20]; it seems likely that the dedicatee ZurKinden in nearby Bern was also aware of Gessner's authorship. By contrast the 1561 pamphlet describing a wondrous celestial phenomenon contained no dedication; few such pamphlets featured dedications, even when their authorship was explicit. In this case Gessner revealed his authorship only the following year in his autobibliography when he explained that he had

[16]See the case of Jean Bodin in Blair 2013, 138–39.

[17]"Ea quidem lege, ut a fronte libelli significarem Lectori, nec suo [Gessner's], nec Guilandini iussu aut voluntate evulgata haec esse, sed meo unius arbitrio." Nicolaus Philesius, "Ad lectorem," in *De stirpium aliquot epistolae*, 1557, sig. A2r.

[18]ZurKinden/Zerchinta also appears as a correspondent in Gessner's *Epistolarum medicinalium libri III*, 1577, ff. 130v–131r. In his dedication of *Historia animalium II*, 1554, to Johannes Steiger, also a city councilor in Bern, Gessner requested that Steiger send Gessner's greetings to ZurKinden.

[19]See Terrall 2002.

[20]"Et patrono suo colendissimo. Con. Gessnerus d.d." *Evonymi thesaurus*, 1552, ZB call number Md E 377, digitized on E-rara.

withheld his name on the pamphlet lest he be accused of straying "from his last," by writing on a topic outside his usual areas of expertise.[21]

The impact of printing on dedication practices awaits thorough study. Certainly dedications of literary works were common in the Middle Ages and served similar purposes of asserting authorship and seeking patronage.[22] But we can surmise that printing raised the stakes of the dedication by disseminating more widely and unpredictably that public expression of a personal relationship, which lay at the core of the genre. It is clear that dedications and front matter became more common and more voluminous in printed works than they had been in medieval or humanist manuscripts. Niccolò Perotti gives us one indication of this increase when he noted with disapproval in 1470 that editors of classical texts added prose of their own (in dedications and prefaces) which was unworthy of being printed alongside the ancient works. Perotti reported that he erased or cut out these offending passages in the books he owned.[23] Despite such disapproval, the inclusion of paratexts in humanist editions became the norm. The prolific Erasmus set a powerful precedent in composing over 150 dedications in the best humanist Latin during a career which spanned 221 publications (counting first editions only).[24] Early in his career, when he pieced together income from many different sources, Erasmus sought to reach as many potential donors as possible, by composing separate dedications to different people for the various "books" within a single printed work, and by sending inscribed copies of his books to important people in the hope of payment. In exchange for dedications in print and for gifting copies of his books with manuscript inscriptions Erasmus often received money or valuable goods—including silver cups and vases (including one filled with coins, another with rare fruit), and horses. He could either sell the gifts for monetary gain or enjoy them as a sign of his relations with the wealthy and powerful.[25] Nevertheless in surveying the more than 150 dedicatees to whom Erasmus addressed his works, Jean Hoyoux saw no overarching strategy: "students, bishops, kings, and simple humanists intersect [on this list] without order, according to his whim."[26] By 1540 when Conrad Gessner started his publishing career, he could draw for inspiration in his dedication practices on the great prince of humanists who spent the most glorious years of his

---

[21]*De libris a se editis*, 1562, sections 37 ("Hunc librum cum in gratiam Andreae Gesneri patruelis mei, novi tum typographi, immaturum ederem, nomen meum adderem nolui") and 60 ("meum [nomen] enim non ponere placebat, ne quis sutorem ultra crepidam mihi obijceret."). Also Wellisch, 91.

[22]Holzknecht 1966; and, on methods of inscribing authorship into the preliminaries, Brown 1995.

[23]Niccolò Perotti to Francisco Guarneri, as edited and discussed in Monfasani 1988, 26, sentences 20–21. I am grateful to Anthony Grafton for this reference.

[24]On Erasmus's remarkable publishing savvy, see Jardine 1993, and Vanautgaerden 2012. Vanautgaerden enumerates the 221 works in his appendix.

[25]See Hoyoux 1944, 43–49.

[26]See Hoyoux, 41 and 34–41 for the list. Vanautgaerden notes that dedications were important especially before 1516 when Erasmus became counselor to prince Charles; most notably his edition of Lucian was divided into 7 parts, each with a different dedicatee. Vanautgaerden, 82.

career in nearby Basel—Erasmus died there in 1536, when Gessner was 20 years old.

8.3 Gessner, the Abundant Dedicator

Even compared to Erasmus, Gessner took dedicating to new heights. Although he published many fewer books, Gessner reached just as many dedicatees, outfitting all but 11 of his publications with at least one and often more than one dedication. Fifty-seven works of Gessner's contained 102 dedications addressed to a total of 127 individuals and six collective bodies (city councils and university faculties). Almost half of the works with dedications were addressed to more than one person, either through a single dedication addressed to multiple people, or through multiple dedications within one work, or a combination of both.[27] Sometimes Gessner took advantage of the different parts of a single coherent work to compose many dedications, as in the *Pandectae* (1548) with 19 dedications, or *De piscibus* (1556) and *Nomenclator piscium* (1560) with four each.[28] In other cases the publication was an anthology of multiple works, inviting a dedication for each. When Gessner edited the various manuscripts of the recently deceased scholar Valerius Cordus, he added three other texts (two of his own and one by a friend) and composed ten dedications for the multiple parts that resulted. When Gessner was one of multiple authors/editors/translators involved in one publication, each composed a dedication for the section containing his work, so that the dedications by Gessner which I list in the appendix may constitute only a subset of the dedications present in that work.[29] No doubt first dedications were more prestigious; they were more visible to the casual user of the book and were often longer. Gessner's longest dedications were all first dedications; they operated as mini-treatises extolling the significance of the topic. One of these, in one of Gessner's earliest works, *De lacte* (1541), has become famous as a separate text for its celebration of the beauty and utility of the Alps at a time when mountains

[27]I count seventeen works containing more than one dedication and eight cases of dedications addressed to more than one individual.

[28]The record for most dedications in one publication that I have come across is 23 dedications in Paolo Zacchia, *Quaestiones medico-legales*, 1620, as discussed in Pinon 2009, 63, n. 14.

[29]Dedications by others (not Gessner) in this corpus include: *Ephesii Scholia*, 1541 (Magus Leonicus, pdf 109); *Porphyrii institutiones*, 1542 (Joachim Perionius, pdf 6); *Antonij Thylesii opuscula aliquot* 1545 (Joannes Ludovicus Brassicanus, pdf 122); *Sententiae Antonii et Maximi* 1546 (Vincentius Obsopaeus, pdf 278; Conrad Clauser, pdf 317); *Ermolao Barbaro. Compendium* 1548 (Hieronymus Wildenbergius, pdf 128); *Chirurgia* 1555 (too many to list!); *Valentini de anima* 1563 (Vitus Amerbacius, pdf 469; Philip Melanchthon, pdf 711); *De omni rerum fossilium genere* 1565 (Kentmann, pdf 10 and 240; Georg Fabricius, pdf 288; Severinus Goebelius, pdf 358).

were generally viewed only in a negative light.[30] Conversely, some of Gessner's "internal dedications" positioned later in the work were remarkably short.[31]

Gessner's editions of Greek texts with Latin translations offered opportunities for separate dedications for the text in two languages. In three of these cases Gessner composed dedications in Greek to accompany the Greek text he edited—this was a rare feat which even Erasmus did not attempt. Gessner published his Greek dedications late in his career in his editions of 1559 (*editio princeps* of Marcus Aurelius, *De vita sua*), 1562 (Cassius, *Naturales et medicinales quaestiones*) and 1565 (Dioscorides, *De curationibus*).[32] He addressed the first to the nobleman Anton von Werther of Beichlingen, referring to the latter's recent visit to Zurich, and the other two to people he knew well: his former student Antonius Schneeberger, who had become a physician in Cracow, and the imperial physician and frequent correspondent Johannes Crato von Krafftheim, respectively. Gessner's Greek dedications followed the mold of his Latin ones, including praise of the dedicatee and reminiscences of his contact with him, and in the edition of Marcus Aurelius (as in the longer Latin dedications) a mini-treatise relevant to the text being published—to wit, a defense of the merits of studying a pagan work when it is morally edifying. These Greek dedications called attention to the exceptional skills of both their writer and their recipients, who were presumed able to read them, and they would dazzle those more ordinary readers who could not.

As Hoyoux remarked of Erasmus' dedicatees, Gessner's also spanned a wide range of people. No doubt contingent circumstances we cannot recover played a role in many choices of dedicatee. Nonetheless we can recognize the broad arc of Gessner's career through these choices. During the 1540s Gessner first addressed teachers and fellow students from his early years, then turned to colleagues and friends, and finally to city officials responsible for the first stipends and appointments he received. After a few years of teaching Gessner also selected students of his own as dedicatees, starting with the 1549 edition of his *Onomasticon* or dictionary of proper names dedicated to 11 young Zurich patricians, whom he had presumably taught, reminding them that "philosophy is a divine good" and that they should be an ornament to their city. He offered the dedication (dated from

[30]This text is available in Weber 2003, 16–21, translated from the French version in Coolidge 1904, iii–xvii. Also mentioned in Schottenloher, 45.

[31]The appendix lists the lengths of dedications. For other long dedications see: *Apparatus et delectus* (1542), *Stobaeus* (1543), *Martial* (1544), *Antonij Thylesii opuscula aliquot* (1545), *Bibliotheca universalis* (1545), *Historia animalium I* (1551), *Historia animalium IV* (1558), *Marcus Aurelius* (1559), *Ars magirica* (1563), *Dioscorides. De curationibus* (1565), *De omni rerum fossilium genere* (1565). The shortest are to Gregorius Laetus in *Sententiae Antonii et Maximi* (1546), Beck von Beckenstein in *Mithridates*, 1555, and Lemnius in *Nomenclator piscium*, 1560.

[32]One of these was reprinted in the nineteenth century in Botfield 1861, 518–22 (Gessner's Greek dedication to Anton Werther von Beichlingen in his edition of *Marcus Aurelius*, 1559). Botfield also reproduces Gessner's Latin dedication to Johann Jakob Fugger in his edition of *Aelianus*, 1556, 482–85.

the first of January) as a new year's gift, observing that "you'll enjoy its dedication to your names."[33] At midcareer, after his first major works (*Stobaeus* in 1543, the *Bibliotheca* and *Pandectae* of 1545 and 1548) had broadened his reputation beyond the greater Zurich area, Gessner became more ambitious, addressing some of the most powerful figures of his time, especially in his expensive illustrated works of 1551–58. Gessner also diversified his bids for patronage by addressing collective bodies, both local (Zurich and Basel) and less so (Wittenberg and Augsburg). At the same time throughout his career Gessner regularly wrote dedications, for his lesser works or in "internal dedications" to separate parts within a work, to people like him engaged in medical and learned scholarship—25 doctors, 5 apothecaries, 5 churchmen, 4 lawyers, and 8 professors, accounting for more than one third of his dedicatees. Typically Gessner had corresponded with and/or met them, and Gessner thanked them for contributing to his projects by providing access to manuscripts, books, people, or information. In other cases Gessner used the dedication to pressure them publicly to send him materials, notably from their far-flung locations, such as images of birds from England or of fish from the Baltic.[34] In the last 6 years of his life Gessner no longer sought patronage from the great and powerful and instead used his dedications to cultivate relations with scholars and in one case writing a dedication on behalf of the orphans of a deceased author.

8.4 Gessner's Bids for Major Patronage

Gessner followed what was likely a common pattern at the time of dedicating his major works (i.e. the expensive folio volumes with or without illustrations) to the most socially prominent among his dedicatees.[35] These included city officials addressed individually or collectively. Gessner dedicated to two senators from Bern whom he thanked for his stipend to study in Lausanne, his first folio volume, Stobaeus's *Sententiae* of 1543. In his *Bibliotheca* of 1545 Gessner called that work the most useful thing he had published to date; although Gessner is far better known today for his *Bibliotheca*, it was never reprinted, whereas Gessner's *editio princeps*

[33]"Est enim vere philosophia divinum bonum. … [N]on iniucundam eius in nomina vestra inscriptionem vobis futuram existimavi." *Onomasticon*, 1549, sig. ++3r–v.

[34]"Quod si authoritate tua effeceris, ut viri aliqui docti apud vos, illarum avium quae in Anglia reperiuntur ultra eas quas hic exhibui, effigies mihi communicent, librum hunc alias ijs ipsis iconibus, et alijs forte (si quas aliunde interim nanciscar) Domino Opt. Max. vitam largiente, augendum curabo." *Icones avium*, 1560, 128. "Thomas Erastus Helvetius, civis meus … icones aliquot animantium raras, Balthici praesertim maris aquatilium, te possidere adiecit, quas mihi communicaturum te sis pollicitus, si qua commoditas tanto locorum intervallo mittendi daretur." *De piscibus*, 1556, 94.

[35]Hoyoux notes (41) that Erasmus was exceptional in not following this pattern.

of Stobaeus was the basis of a further 15 editions down to 1838.[36] Gessner addressed the first two volumes of his *Historia animalium* to the City Council of Zurich (1551) and to senators from Fribourg and Bern (1554) respectively, and reached higher for volumes three and four, dedicated to Johann Jakob Fugger (in 1555) and the emperor himself (in 1558).

Gessner approached only two grandees more than once. Gessner first addressed Leonhard Beck von Beckenstein, counselor to the emperor Charles V, with a dedication of February 1545 to his edition of manuscripts of Telesio.[37] This fulsome 6-page dedication was devoted almost entirely to praising the dedicatee, his virtues which made him many friends, his frugal life and love of learning, and his great library. In return Gessner promised him the immortality of literature: "from the inscription, Leonhard, you will have like an inseparable link from your name to posterity. In part since these books are small and unworthy of your amplitude; in part because they will become more illustrious from the splendor of your virtue."[38] The response Gessner received was presumably encouraging because just 5 months later Gessner also dedicated to Beck von Beckenstein his most important work to date, the *Bibliotheca universalis* of 1545, an alphabetically arranged bibliography which aimed to include all known works in Latin, Greek, and Hebrew. Gessner devoted much of this long dedication to a "mini-treatise" on the significance of libraries and his own motives and labors in composing the *Bibliotheca*. His most explicit plea for funding (with a pointed criticism of the tight-fistedness of the wealthy of his era) appeared buried in the text, in the entry for himself where he listed his own works published and in progress. There he explained his plan:

> to write a complete history of animals, with their names in various languages, and images. ... If some generous Maecenas would manifest himself and support the project, this whole history will be made more perfect. If not (since now many rich people are avaricious and live for themselves and not for good studies) I will not cease to pursue this laudable topic as my strength allows.[39]

Gessner had already praised von Beckenstein as one who knew how to spend his wealth well in his first dedication to him: "It's most difficult in fact and achieved by

[36]See Wellisch, 40–43. Gessner calls his edition of Martial in 1544 "the most useful of everything I have published after the translation of Stobaeus." "Hunc certe laborem post translationem Stobaei utilissimum omnium quae publicavi obijsse mihi videor." *Bibliotheca universalis*, 1545, f. 182v.

[37]Not much is known about Beck von Beckenstein (whose name Gessner spelled in a variety of ways); on his ownership of Greek manuscripts, see Serrai, 90.

[38]"Haec ex inscriptione mea tibi habeas, clarissime Leonharte, veluti vinculum inseparabile nomini tuo ad posteros propagando futurum: partim quoniam libelli per se sunt eximij, nec indigni amplitudine tua: partim quod virtutis tuae splendore, etiam ipsi proculdubio fient illustriores." *Antonij Thylesii opuscula aliquot*, 1545, sig. a4v.

[39]"Quod si Mecoenas etiam aliquis benignus contingat, cuius auspicijs res peragatur, perfectior tota historia efficietur: sin minus (ut nunc sunt divites plaerique avari ac sibi tantum, non bonis studijs vivunt) non desinam tamen pro virili mea tam plausibile argumentum excolere." *Bibliotheca Universalis*, 1545, f. 182v.

very few [to spend wealth to benefit others]. . . . Your glory comes in no small part from . . . your frugal life, embrace of good letters . . . and promotion of the studies of wisdom."[40] In the *Bibliotheca* Gessner also emphasized von Beckenstein's devotion to building a great library and followed the dedication with a lavish full-page engraving of the latter's coat of arms (which was hand colored in the copy at the ZB Zurich).[41] This was a clear bid for patronage, complete with allusions to the great patrons from antiquity. But Gessner must have been disappointed. He praised von Beckenstein as his "maecenas and patron" in a preface in 1548 but then addressed him just once more, seven years after that, dashing off a 9-line dedication on a foldout table of the Lord's prayer in different languages appended to his *Mithridates* (1555). There Gessner treated him as a potential expert contributor, inviting him to send corrections or examples in old German to his discussion of the diversity of languages.[42]

For his next work after the *Bibliotheca*, in February 1546, Gessner turned instead to the wealthiest man of his day, Johann Jakob Fugger, renowned for his great library and patronage of arts and letters. In summer 1544 Gessner had travelled to Augsburg at Fugger's invitation where he was offered a well-paid post with access to the library and duties tutoring Fugger's children.[43] Gessner declined the position, which would have required moving to Catholic Augsburg and entering into a classic client-patron relationship. Gessner never articulated his reasoning but historians have emphasized Gessner's commitment to Zurich and its religion. Nevertheless Gessner continued to cultivate the relationship. Gessner's Greek and Latin editions of the *Sententiae* of the 11th and 7th-century monks Antonius and Maximus published in two separate volumes in 1546 sported two major dedications and one less visible one inside the Latin volume, all of them building his connections with Augsburg. The Greek edition was dedicated to "Johann Welser and Jakob Herbrot and the whole most prudent Senate of Augsburg" who were praised for the recently formed library in Augsburg which "had no match in all Germany":

[40]"O ego si forem dives, quam recte pecunijs uti, quam beneficus alijs esse vellem? difficillimum tamen est factu, et a paucissimis praestatur, ideoque laudandum constantius, quanto contingit rarius. . . . Accedit gloriae tuae non exigua pars, quod luxu et ocio in tanta occasione post habitis, vitam frugalem degas, et bonas amplectare literas, et literatis summopere faveas, denique pro virili tua disciplinas et sapientiae studia promoveas." *Antonij Thylesii opuscula aliquot*, 1545, sig. a3v–4r.

[41]"Te vero unum ut virtutis et nobilitas excellentia florere, sic etiam doctrina et animi promptitudine ad Bibliothecam instruendam plaerosque omnes tui ordinis homines a tergo relinquere animadverti." *Bibliotheca universalis*, 1545, sig. *5v.

[42]"His ego rationibus defendi poteram, et illustri viro Caesareae maiestatis a consilijs D. Leonardo Beckh a Beckhenstain, Mecoenati meo et patrono optime merito, ... ita hunc etiam iure consecrare." Pandectae (1548), sig. *2r. "Hanc autem qualemcumque tabulam, nobilissime et sapientissime Leonarde, tuo potissimum nomini nuncupare volui, ut et quanti perpetuo te facerem merito tuo inde conijceres: et si quid ad augendum emendandumve Mithridatem nostrum, cum alibi tum circa Germanicae linguae antiquitatem et vetustissima carmina afferre posses, (potes autem pro doctrina et diligentia tua his in rebus plus quam alius quisquam) id promptius citiusque ut faceres, et me tui studiossimum amare pergeres." *Mithridates*, 1555, foldout table.

[43]Leu 2016a, 167–174; Braun, 89–90.

"whereas in most collections . . . most books are vulgar and useless, in the Augsburg library all the books are rare." Gessner had in mind especially the remarkable collection of very old Greek manuscripts there. Gessner also stated very clearly what he hoped for in return: "I dedicate it . . . to you, so that I will enjoy the occasion to conveniently request some books from your library"—a wish that was granted probably more than once.[44] The Latin volume Gessner dedicated to Johann Jakob Fugger with the requisite praise and also a very explicit bid for financial support: "if you like this work you could add encouragement so that I can proceed faster with the Historia animalium with pictures."[45] It is doubtful that the plea to Fugger was effective, since Gessner did not dedicate a volume of the *Historia animalium* to Fugger until 9 years later. Gessner noted then, as if apologizing for the delay, that he had "already many years ago formed the plan of dedicating to you." Gessner praised Fugger's generosity and learning in rather general terms, invoking the testimony of his librarian Hieronymus Wolf and especially his LIBRARY (mentioned in capital letters to highlight it), which rivalled that of many kings and from which Gessner no doubt hoped to gain access to more manuscripts.[46]

In 1555 Gessner also approached Johann Jakob's younger brother Ulrich Fugger with his *Icones avium*, perhaps hoping to leverage sibling rivalry to entice one or both Fuggers to be generous at the same time. It is not clear if they were. In 1556 Gessner again chose Johann Jakob as dedicatee, whom he addressed as "his lord and maecenas," for his Greek and Latin edition of the works of Aelian. But this dedication was clearly couched as a thank you not for funding but for the loan of books—the "oldest Greek manuscript" of Aelian from the library of the city of Augsburg and a further manuscript from Fugger's own collection, which formed the basis for Gessner's edition.[47] The favor was not inconsiderable, since the manuscripts were presumably sent to Zurich for Gessner to work from. In return Gessner

[44]"[Q]uale profecto hactenus, in Germania quidem, nullum videre mihi contigit. Quamvis enim copiosas librorum moles inspexerim alibi, magna tamen illorum pars, aut inutilibus omnino, aut plane vulgaribus libris constabat. Vestra [bibliotheca] vero, ut non adeo multos, sic certe praestantissimos, rarissimos, plaerosque nondum evulgatos, et Graecos, id est antiquos, habet codices. . . . [C]larissimo nomini vestro dedicare volui: ut commode petendi aliquos ex Bibliotheca vestra codices, occasione ista fruerer." *Sententiae Antonii et Maximi*, 1546, Greek volume, sig. +2r–v.

[45]"Quod si hunc laborem tibi non ingratum fuisse declaraveris, animum mihi addideris ut in animalium historia cum picturis singulorum, exquisito et magno opere, alacrius subinde pergam." *Sententiae Antonii et Maximi*, 1546, Latin volume, sig. *2r.

[46]"Te certe unum omnium hoc munere dignissimum iudicavi: et multis hactenus annis, quibus in eo perficiendo elaboravi, animo meo id tibi nec temere nec immerito destinavi. . . . Testatur amplissima illa quam paras omne genus librorum BIBLIOTHECA, tanto quidem studio, ut plerosque etiam reges in eo iam aut aeques aut superes." *Historiae animalium liber* III, 1555, sig. a4r–v.

[47]"D. Ioanni Iacobo Fuggero. . . Domino et Mecoenati suo. . . [Q]uoniam tu Graecum codicem manuscriptum vetustissimum ex publica amplissimae civitatis vestrae Bibliotheca mihi impetrasti, et alterum ex proprio librorum tuorum thesauro addidisti." *Aeliani opera quae extant* (1556), sig. Alpha 2r.

hoped that Fugger would add his new edition of Aelian, "complete, corrected, bilingual" and printed, to the famous library in Augsburg.[48] But Gessner also broached his hope for a gift of money by mentioning that "within our memory François king of France received Aelian's books on animals given in Latin by Petrus Gillius with a grateful and generous mind."[49] Comparing Fugger the merchant to the king of France was a kind of flattery, which invited Fugger to imitate royal largesse. We can surmise that Gessner's bid was not successful, though, since Gessner did not make any dedications to the Fuggers among the 19 further books he published before his death in 1565.

Once Gessner declined Fugger's invitation to join the latter's "court" and become a full-fledged client, it is perhaps not surprising that Fugger never became a steady patron to Gessner. Gessner shifted to a strategy of cultivating a variety of learned men in Augsburg, which after Zurich is the single best represented place among the locations of his dedicatees. Starting in 1555 Gessner selected a number of Augsburgers as dedicatees—all of them physicians: Geryon Seiler (1555 *Chirurgia*) who had drawn Fugger's attention to Gessner 12 years earlier, Laurentius Gryll (1555 *De raris herbis*) whose education was funded by Johann Jakob Fugger, Achille Gasser (1555 *Enchiridion rei medicae*) and Ioannes Crato von Krafftheim (1565 edition of Dioscorides), both longstanding friends and correspondents, and Adolphus Occo (1565 *De omni rerum fossilium genere*) whom Gessner asked to check a point of scholarship in a manuscript of the Bibliotheca of Photius in the Augsburg library.[50] In the 1546 *Sententiae Antonii et Maximi* Gessner issued the first of a few characteristic "group greetings." Deep in the volume Gessner added a translation of an oration by Tatianus, which was the occasion for a short dedication to Georgius Laetus, rector of the Gymnasium of Augsburg. Gessner thanked him for his hospitality (presumably during his trip of 1544) and asked Laetus to convey greetings from multiple figures in Zurich to multiple figures in Augsburg: "Other learned men here greet you, especially Heinrich Bullinger, Rodolph Gwalther, our church bishops and our Ioannes Frisius. And from me please say hello to Wolfgang Musculus, Michael Cellarius, Bernardinus Ochinus Senensis [the father of 1565 dedicatee Adolphus Occo],

[48]Ibid.: "Redeunti igitur ad te Aeliano tuo, iam integro, emendato, bilingui, publico, facilem ac benignum te praebe, et in bibliothecam tuam veluti hospitij iure admitte."

[49]Ibid.: "Huius de animalibus libros a Petro Gillio Latinitate donatos, nostra memoria potentissimum Gallorum regem Franciscum grato et liberali animo accepisse constat."

[50]See Braun, 88 (Seiler); and Häberlein, 167 (Gryll). Crato and Gasser featured prominently in Gessner's *Epistolarum medicinalium libri III*, ff. 1–22v and 22v–44r respectively. On Crato see Louthan 1994. "Peto praeterea ut per occasionem in Photij patriarchae Catalogo authorum quos legit (quem Augustae apud vos in generosi viri D. Io. Iacobi Fuggeri Mecoenatis mei benemeriti nobilissima Bibliotheca Graece manuscriptum vidi) locum, si modo est aliquis, de Epiphanio eiusque scriptis inquiras...." *Sancti Patris Epiphanii de XII gemmis* in *De omni rerum fossilium genere*, 1565, sig. (a)3r. Discussed in Schottenloher, 97, 203. On Gessner's exchanges with Occo, see Delisle 2006, 36–39.

Ioannes Heinrich Helt and Xystus Betulaieus, men whom I love dearly."[51] This greeting appeared in a P.S., positioned after the usual dated sign-off, as it might have appeared in a manuscript letter. While early modern dedicatory epistles may often seem to be letters in name only, Gessner used several dedications as a way of conveying news from one place to another, as one would in a manuscript letter, though print would disseminate them more widely and impersonally. Gessner sent greetings in this way, also on behalf of others in Zurich including Heinrich Bullinger, to English friends who had recently returned to England from their Marian exile in Zurich (in *Icones avium* 1560), to friends in Rhaetia/Graubünden (*Ars magirica* 1563), and to learned men in Danzig (*De halosantho* in *De omni rerum fossilium genere* 1565).

The advantage for Gessner of not having entered a relationship of clientelism, e.g. with Johann Jakob Fugger, is that he could try his luck in addressing those even further up the social hierarchy. Gessner chose dedicatees most ambitiously for the slim folio volumes that reprinted the expensive illustrations of the *Historia animalium* without the bulky scholarly text. These volumes entitled *Icones* were crucial money-makers by which the printers hoped to recoup the vast expense of making the illustrations.[52] They also seemed appropriate for the highest-ranking dedicatees who likely were given hand-colored copies.[53] Gessner dedicated the first of these volumes, the 1553 *Icones animalium*, to Thomas and John Grey, brothers of Henry Grey Duke of Suffolk; the dedication was ill fated in that Henry and Thomas were executed a few months later on suspicion of treason under Queen Mary. In 1555 Gessner dedicated the second such volume to Ulrich Fugger.

Gessner and Froschauer, who printed the *Historia animalium* and associated other volumes, gave themselves a further opportunity for income from these images by publishing a second edition of each of these *Icones* in 1560. In June of that year Gessner addressed the second *Icones animalium* to Elizabeth I who had recently acceded to the throne of England, and demoted the original dedication to the Grey brothers as an internal dedication to the newly added second part of the book. Gessner's dedication to Elizabeth is the one clear case of a failure. Although Gessner composed a fulsome three-and-a half-page dedication topped with a 28-line ode in Greek to Elizabeth (which was perhaps his first published composition in Greek), Elizabeth did not appreciate his explanation in the opening line that he addressed her since the dedicatees of the first edition had been killed in the interim, as if she were some kind of second choice. Apprised of the Queen's displeasure, Gessner wrote a letter to William Cecil, Elizabeth's chief adviser at the time, which is recorded in the Calendar of State papers for March 1561. Gessner explained "that he ha[d] never

[51]"Salutant te cum alij apud nos eruditi viri, tum praecipue Heinrychus Bullingerus, Rodolphus Gualtherus, ecclesiarum apud nos antistites, et Ioan. Frisius noster. Tu ex me Wolfgango Musculo, Michaeli Cellario, Bernardino Ochino Senensi, Ioanni Heinrycho Helt, et Xysto Betuleio, viris mihi maxime colendis salutem dicito." *Sententiae Antonii et Maximi*, 1546, Latin volume, 346.

[52]See Kusukawa 2012, 59.

[53]None of these presentation copies has been identified yet.

received a farthing from any one" of the Greys as a result of that dedication, though he heard that the relations of the deceased were not displeased with it. Gessner also pleaded poverty ("having forty needy relatives") and requested that "the price of the books sent (amounting to four crowns) be transmitted to him." He enclosed another recent publication and requested that Cecil reply through his closest English friend, John Caius, a physician to whom Gessner had recently dedicated a part of his 1556 *De piscibus*. A payment of 6 pounds to Gessner was recorded a few months later, presumably for his expenses—a far cry from the reward he must have hoped for.[54] A month after his dedication to Elizabeth and unaware of its unfortunate reception, Gessner directed the second edition of the *Icones* on birds to another English contact, John Parkhurst, who had spent Mary's reign in exile in Zurich and had recently been named bishop of Norwich on his return after Elizabeth's accession. Gessner had praised Parkhurst in his dedication to Elizabeth for encouraging him to approach the queen, so in this dedication to Parkhurst Gessner offered a thank you for serving as a broker (though Gessner learned later it was ineffective), alongside a request to send more images of English birds gathered from learned Englishmen. Printed dedications preserve for readers far away in time and space only the happy words of praise and hope, not the unhappy aftermath which ensued in this particular case in manuscript letters and archives and face-to-face interactions that we can reconstruct only partially.

Gessner was more successful closer to home. In 1558 Gessner dedicated the fourth volume of his *Historia animalium* to the newly appointed emperor Ferdinand I, calling attention to his "labors which were immense, not to mention the expenses." Gessner also pointed out that he "had so far no patron or maecenas," and argued that he was "more worthy of pardon if I do not satisfy expectation" as a result.[55] The emperor responded by inviting Gessner to Augsburg during a Reichstag and awarding him noble status and a coat of arms; these were even authorized for transmission to a lateral relative (as Gessner had no children).[56] That transfer took place after Ferdinand's death under the new emperor, Ferdinand's son Maximilian II. Gessner had also dedicated a work to Maximilian when the latter was king of Bohemia and archduke of Austria—the third and last of the *Icones* volumes, on fish, published in just one edition in 1560. With all these highest-ranking dedicatees Gessner relied on intermediaries to facilitate the dedication, whom he names (just he had named

[54]See Nutton 1985, 96. See the entry of March 7, 1561 in the 1561–62 Calendar of State Papers http://www.british-history.ac.uk/cal-state-papers/foreign/vol4/pp1-16#fnn1 The Cecil papers record a "short table of divers sums disbursed between 26 May and 4 Aug. 1561, by Sir Wm. Cecil, at the Queen's command. ... 6 *l.* to Conrad Gesner, 'in reward for his book De Animalibus.'"Available online at http://www.british-history.ac.uk/cal-cecil-papers/vol1/pp257-263. I am grateful to Samantha Wesner for this find.

[55]"Labores sane mei (ut nihil de sumptibus pro ea. tenuitate dicam) immensi fuerunt." "Alij etiam qui nuper de aquatilibus scripserunt in Gallia Italiaque, Pontifices aut Cardinales locupletes habuerunt Mecoenates. [E]go unus cum patronum et Mecoenatem hac in re hactenus desyderem, eo venia sum dignior, sicubi expectationi non satisfecero. [S]umptibus enim alicubi opus erat, non paulo quam meae facultates sint maioribus." *Historiae animalium liber IV*, 1558, sigs. a6r and a5v.

[56]See Braun, 71, 136, 143–44.

Parkhurst as his contact to Elizabeth). In the case of Ferdinand these were the imperial "learned doctors Iulius Alexandrinus and Stephanus Lauraeus of Amersfoort [Netherlands]" who both received dedications of their own, though only internal dedications.[57] Gessner dedicated to Alexandrinus a short *De anima* of his own composition which appeared in his 1563 edition of Valentinus' *De Anima*, and to Lauraeus an extensive catalog of German gardens appended to his 1561 edition of Valerius Cordus's notes on Dioscorides. In both these dedications Gessner recalled his visit to Augsburg and the presence of the dedicatee when he spoke to the emperor or dined with "the illustrious J. J. Fugger."[58] In his dedication to Maximilian, Gessner explained that he was "emboldened" to approach the king by Paul Scalichius, of very ancient noble family, and the most learned Ioannes Sambucus, though neither of them received a dedication from Gessner.[59]

The role of these brokers was crucial. Through personal access to the dedicatee the broker could ascertain the interest of the ruler in receiving a dedication. And once the book was finished, if the author lived far away, intermediaries again intervened in presenting the book to the ruler. Skill was required to maximize success by making the presentation under optimal circumstances, e.g. when the dedicatee was in a good mood and well disposed to receive it with pleasure and generosity.[60] Gessner cultivated a number of professionals in the ambit of the emperor, not only as intermediaries but also as potential patrons in their own right. He naturally knew the doctors best, and they could also contribute to his projects from their expertise. In addition to Alexandrinus and Lauraeus, Johannes Backofen von Echt (Echtius) was another imperial physician whom Gessner thanked with a dedication, for contributing images of fish (in his *Xenocrates de alimento ex aquatilibus*, 1559).[61] In his early years when he was

[57] *Historiae animalium liber IV*, 1558, sig. [a5]v.

[58] "[Q]uarum utriusque [incomparabilis virtus tua et doctrina illustris] cum ante paucos annos Augustam Vindelicorum, (ubi tum cum Caesare eras,) venissem, gustum ac fructum non mediocrem cepi." *De anima liber* in *Valentini de anima et vita libri tres*, 1563, 720 (pdf 1019). "Primum ut amicitiam mutuam foeliciter superioribus annis Augustae Vindelicorum, cum ad mensam illustris viri Ioan. Iacobi Fuggeri convenissemus, incoeptam, hoc ceu vinculo inter nos obstringerem." *Valerii Cordi annotationes*, 1561, 236 (pdf 494).

[59] "Mihi quidem frequentem, et pleno omnium ore de virtutibus tuis doctrinaque celebrem famam, Paulus Scalichius Comes Hunnorum, (antiquissima generis nobilitate illustris, et eruditione multijuga...) verissimam esse iampridem asseruit: et eandem nuper cum alij quidam non vulgares et fide digni viri, tum praecipue Ioannes Sambucus ... plane confirmarunt." *Nomenclator aquatilium animantium*, 1560, sig. aa2r. Paul Scalich made an entry in Gessner's liber amicorum; see Durling 1965, 145 and 156. On Sambucus whose dedication practices offer a rich case for comparison with Gessner's, see Visser 2005, especially ch. 4.

[60] On Erasmus' reliance on these brokers, see Hoyoux, 52. Also Kusukawa, 84, notes the importance of intercessors especially for the Emperor. Schottenloher, 197–208, gives examples of works dedicated to those who helped secure privileges too.

[61] "Accepi tandem piscium Oceani Germanici quas ad me dedisti Eicones tam diu desideratas ornatissime Echti." *Xenocratis de alimento ex aquatilibus*, in *Iani Dubravii de piscinis*, 1559, sig. A2r. I have found nothing about Echt beyond the identification in Serrai, 328.

casting about for a patron Gessner made his clearest bids to imperial counselors: not only Beck von Beckenstein, but also Caspar von Niedbruck, the dedicatee of his 1555 *Appendix* to the *Bibliotheca* to whom he wrote bluntly: "I very much want to be among your clients."[62]

Gessner approached another major player in the imperial ambit in his 1545 *Lexicon Graeco-latinum*: Diego Hurtado de Mendoza, a Spanish diplomat who served as imperial ambassador to Venice and took advantage of his location to form a great library of rare Greek manuscripts which he purchased or had copied. The first edition of this dictionary in 1537 had made neither mention nor much use of the many additions and corrections which Gessner had given the printer, as Gessner complained in his autobibliographies.[63] But in this fourth edition of 1545, Gessner was clearly positioned as the author on the title page and in the dedication (which, unusually, was also mentioned on the title page). In praising Mendoza and his library Gessner seemed eager to elicit more information on this rather secret scholarly resource when he wrote: "once the learned Arnoldus Arlenius Peraxylus showed me the catalog of your Greek collection. ... The greater part of the learned will soon know that remarkable books are hiding in your library when they read mention of them in our *Bibliotheca*, which I have just completed, since your name is cited every time that I knew you owned rare Greek books."[64] Gessner also listed Mendoza's library catalog among the major sources of information for his *Bibliotheca universalis*.[65] But it is not clear whether the dedicatee appreciated the publicity—I have found no further mention of Mendoza in Gessner's corpus. Mendoza's library became the property of the king of Spain at his death in 1575.[66]

Very little evidence survives of Gessner's finances, so the pattern and language of his publications and dedications—whom he approached when and whether more than once—is often the best evidence we have about their impacts, though it is only

[62]"Quamobrem et propter tot tantaque divina in te beneficia ex animo tibi gratulor, et inter clientes tuos adnumerari vehementer cupio." *Appendix*, 1555, sig. *3r. In 1565 Gessner mentioned that von Niedbruck had sent him a manuscript some years earlier and bemoaned his premature death; *Sancti Patris Epiphanii de XII gemmis* in *De omni rerum fossilium genere*, 1565, sig. (a)3r.

[63]*Bibliotheca universalis*, 1545, f. 180r–v. And "Sed typographus me inscio et praeter omnem expectationem meam, exiguam duntaxat accessionis meae partem adiecit, reservans sibi forte auctarium ad sequentes etiam editiones." *De libris a se editis*, 1562, section 1, sig. A2r.

[64]"Ostendit nobis aliquando doctissimus vir Arnoldus Arlenius Peraxylus Graecae supellectilis tuae catalogum, et prolixas excellentiae tuae in utraque lingua per omnes philosophiae gradus laudes narravit. ... Etenim brevi magna doctorum pars cognoscet quinam egregij libri superstites apud te lateant, cum eorum mentionem in bibliotheca nostra, sive scriptorum omnium Catalogo iam primum a nobis absoluto legerint, expresso toties fere nomine tuo, quot libros Graecos rariores te possidere cognovi." *Lexicon Graecolatinum*, 1545, sig. +2r–v.

[65]Mendoza's library catalog is one of five catalogs that Gessner named, alongside the Vatican and Medici library catalogs. *Bibliotheca universalis*, sig. [*6]v.

[66]See Braun, 40–41.

suggestive. Gessner was clearly much less successful in eliciting income than Erasmus, who lived well from his publications (including payments from printers, countergifts for dedications, and sales of his books). But Erasmus was such a best-selling author that printers and dedicatees sought him out instead of the reverse. Gessner never approached the level of renown of Erasmus and did not enjoy many years at the height of his reputation, whether we date that height to 1555, Gessner's *annus mirabilis* in which he published seven books and a woodcut portrait was first made of him, or to 1558 which brought his ennoblement and (from an unrelated source) greater financial security.[67] Gessner's straitened finances had long been a source of concern: he worked for the city of Zurich as physician and teacher and complained of being poorly paid. Gessner taught small children in his first year of teaching in Zurich then obtained an appointment at the Carolinum teaching adolescents in 1546. To increase his income Gessner took on a great many editing tasks for publishers in Zurich and Basel who paid him for his work and thus encouraged his prolific output.[68] But Gessner probably did not make much money from his dedications; if he had, given that he wrote so many of them, he would have been a wealthy man. Gessner moved from one dedicatee to the next, at first searching for a patron, then in his later years having seemingly abandoned the hope or the desire to identify a major donor. In 1558 Gessner composed a letter to Heinrich Bullinger appealing for greater support from the city officials, citing the toll that his relentless rhythm of work was taking on his health. An appointment as cathedral canon that resulted "eased his financial situation somewhat," but just 7 years before his death.[69]

Gessner used the term "patron" in a handful of cases, as a term synonymous with dedicatee (e.g. "lest this book arrive imperfect without a patron into the hands of men") whose role was to protect and judge the work.[70] Gessner applied the term consistently and in more heartfelt language to one person in particular, Heinrich Bullinger (1504–75), intellectual and ecclesiastical heir to the Zurich reformer Ulrich Zwingli and, as leader of the city council, the person most responsible for Gessner's employment. Gessner dedicated two works to Bullinger, his edition of Martial's epigrams in 1544 and his edition of Greek

[67]Froschauer used the same woodcut of Gessner in *Icones avium* (1555) at the end of the index and in *Historia animalium* (1555) on the verso of the title page. I have also found it present in one copy of the *Icones animalium* of 1553, placed under the dedication, though the woodcut is not present in the digitized copy from E-rara; see Stanford Library copy KB1553.G4.f.

[68]For a few specifics about these payments see below note 84.

[69]See Wellisch, 17. For Gessner's complaints about his early employment, see *Bibliotheca universalis*, f. 180r.

[70]"Ne vero sine patrono in manus hominum ... liber hic imperfectus perveniret, unius defuncti loco plures excellentissimosque patronos [the University of Wittenberg] delegi." Gessner, *Valerii Cordi annotations*, 1561, sig. a iiiv (pdf 11). Gessner also applied these terms to Steiger ("patronum et spectatorem praecipuum" *Historiae animalium liber II*, pdf 126), Gryll ("iudicem simul ac patronum" *De raris herbis*, sig. A2r), Bullinger (twice, see below), Lauraeus ("patronum ... qui his de rebus et recte iudicare posset," *Valerii Cordi annotationes*, 236, pdf 494), and Alexandrinus ("censorem simul graviorem et patronum meliorem," *Valentini de anima*, 720).

theological works in 1557. We also have the copy that Gessner inscribed to Bullinger of his 1552 *Evonymi Thesaurus*: "to his dearest patron, Conrad Gessner."[71] Gessner addressed Bullinger alongside other city officials in 1544, articulating the hope that they would assign Gessner's expurgated edition of Martial in the schools.[72] In 1557, the year before Gessner wrote his letter requesting better financial support, Gessner showed great respect in addressing Bullinger as his "patron and judge (*vindex*)" asking him for corrections as needed on matters of Greek or of theology—Gessner was thus expressing in advance his willingness to conform to any censure of his interpretations by the chief theologian of Zurich.[73] Gessner wrote with special self-consciousness about Bullinger's generosity: "Long since I should have given you some evidence of my gratitude and memory of your benevolence because of many great kindnesses to me (*beneficia*); although your generosity does not require it."[74] Gessner seemed to attempt to forestall the notion that this dedication was born of Gessner's frustration with the level of generosity Bullinger had displayed recently, although such frustration is perhaps exactly what had motivated it.

Dedications to collective bodies were less common than dedications to individuals, but not unique to Gessner.[75] Gessner's six dedications to collective bodies resulted in the one gift he received that has so far been documented (thanks in part to the record-keeping of such bodies). Gessner dedicated the first volume of the *Historia animalium* to the City council of Zurich in 1551, which voted (likely under Bullinger's leadership) to award him an annual allotment of 10 Malters of grain (rye or wheat) and 10 Eimer of wine (presumably for his lifetime).[76] For his 1561 posthumous edition of works of Valerius Cordus Gessner chose "the most excellent of patrons, indeed the whole medical college of the university of Wittenberg."[77] Collective bodies ranked behind but not far behind the greatest individuals, judging from the order of dedication in Gessner's *Icones* on fish of 1560: the lead dedication was

---

[71]Quoted in footnote 20.

[72]"Et aequissimum est profecto huiusmodi opus, quod gymnasio vestro (cui studiorum meorum incunabula simul ac progressus debeo) aliquando usui futurum sperarem, non alijs offerri quam vobis, quorum beneficio atque liberalitate ferme a puero in hunc usque diem bonis literis vaco." *Martialis epigrammata*, 1544, sig. *4v. On this expurgated edition, see Leu 2014b, 197–208.

[73]"Non solum enim patronum te et vindicem cum Athenagorae, tum Gesneri interpretis, sed arbitrum atque iudicem esse cupio." *Athenagoras*, 1557, 80.

[74]Ibid., 79: "Debueram enim iamdudum aetatem, propter perpetuam in me benevolentiam tuam, propter multa et magna beneficia, aliquod animi grati et memoris argumentum tibi deferre: quamvis liberalitas tua id non requirat."

[75]On dedications to cities, see Schottenloher, 188 ff.

[76]"In den züricher Akten steht als die erste eine Widmung Konrad Gessners verzeichnet, welcher dem Rate ('Meinen Gnädigen Herren') im September 1551 sein Buch von den vierfüßigen Tieren 'verehrt' hatte. Er erhielt dafür als Gegengabe 10 Malter Kernen (Roggen oder Weizen) und 10 Eimer Wein jährlich." Kapp 1886–1923, Band 1, ch. 5, 319 (available from de.wikisource.org) I am grateful to Kusukawa, 56, for this reference.

[77]As quoted in note 70.

addressed to King Maxmilian; the next, for the book on fresh water fish, to the city council of Basel ("I thought of you first as patron because your famous city is founded on the Rhine, the greatest river in Europe..."); a third dedication was addressed to the imperial diplomat Sigismund of Herberstein, and a fourth, short dedication to a fellow doctor and author, Levinus Lemnius.[78] Gessner recalled his dedication to the Basel city council when dedicating 2 years later his edition of Galen's *Opera* to another Basel collectivity, the professors of the University, who awarded him 5 Taler.[79] Gessner approached Augsburg a second time in 1565 in an edition of Dioscorides' *De curiationibus*. This was a Latin translation of a Greek manuscript found in the Augsburg library which Johann Moibanus had started but left unfinished at his death, entrusting it to Gessner to see it through publication. Gessner's dedication was entirely pitched as a plea for support for Moibanus' children who had also lost their mother and were orphaned.[80]

8.5 Gessner as Booster of Learned Printers

Printers played a significant role in Gessner's finances and career. They were mentioned remarkably frequently in Gessner's dedications, and even appeared as dedicatees themselves. In 1548 Gessner experimented with the power of the dedication in an unusual way in his *Pandectae*, which offered a thematic index to the learned literature he had inventoried in his *Bibliotheca Univeralis* of 1545. Gessner divided the *Pandectae* into 19 books, each corresponding to a field of study and for each of which he composed a dedication to a different learned printer of his day. He extended the pattern in a separate volume on theology in 1549, but never completed a projected twenty-first volume on medicine. Gessner explained that the order of the dedications was based not on any judgment of relative merit, but on "their argument"—that is, he selected for each book of the *Pandectae* a printer who had excelled in the area treated in that book.[81] Gessner explained that his purpose in

[78]"Mox autem ipsum argumentum suggerebat, honestissimum Ordinem vestrum, hoc qualicunque dono dignissimum, simul etiam patronum mihi meoque huic operi futurum, imprimis mihi deligendum, quoniam inclyta urbs vestra ad maximum nobilissimumque nostrarum regionum et ferme totius Europae flumen Rhenum condita, summum Helvetiae nostra decus et ornamentum existit." *Nomenclator aquatilium amantium. Icones...*1560, 279.

[79]Letter of Simon Sulzer in Basel to Gessner, Staatsarchiv Graubünden in Chur, shelf mark: D V/37 C 36.06.26. Many thanks to Urs Leu for this information and reference.

[80]"Atqui beneficentia vestra multo magis nunc indigent pupilli, quos iam iam moriturus parens fidei vestrae commisit." *Dioscorides. De curationibus*, 1565, sig. [a6]r.

[81]"Quanquam nemo ira iudicare debet, ac si librorum ordo, quos alijs prius, alijs posterius dedicavi, discrimen aliquod inter typographos constituat. Non enim ordinem, sed argumentum ut plurimum respexi, in quo quis vel numero vel dignitate excusorum a se librorum excelleret: et alicubi forte ut primum quis in mentem venit, ita liber aliquis ei dicatus est." *Pandectae*, 1548, sig. [6]r. See Leu 2014a esp. 59–67.

these dedications was "to incite the [printers] to continued diligence in learned publications."[82] Gessner offered each of them praise of the kind usually extended to more socially prominent dedicatees and in many cases he publicized their activity by listing many or all of their publications following the dedication. While many scholars criticized printers for their errors, haste, or indifference to learning, Gessner viewed learned printers as a crucial resource for the transmission of knowledge, who should be honored and supported.

In his letter of appeal to Bullinger in 1558 Gessner also revealed how commissions from printers were an important source of income when he explained the decisions he faced just then:

> Frobenius (the bookseller) in Basel demands that I compare for him the Latin translation of Galen's complete works with the Greek original text (a work that demands an immeasurable amount of labour), and I have to give him my decision within a month. Froschauer (the famous printer in Zurich) asks me to make for him an excerpt from my three large volumes of the natural history of animals. Exhausted from so many exertions, emaciated, enfeebled, half blind and sometimes not quite conscious of myself (how could it be otherwise, since I am forced to write so many, variegated, and wide-ranging works), I shall now again take this yoke upon me, shall again take on a work that would not let me breathe freely for two or three full years. Would you, my friend, advise me to do this?[83]

Gessner suggested that he would not take on such onerous tasks if he did not need the income they generated. Indeed it appears that Gessner did not engage in the hard work of collating Galen once his letter had resulted in an improvement of his income from the city. In 1559–60 an account book from the Froben-Episcopius printing house recorded a payment of somewhat more than 15 pounds to one Elias Philippinus for doing that work. In fact Gessner became involved 2 years later with this same edition of Galen but in the more glamorous role of writing a preface for which he was paid 15 pounds in 1562 by Froben and another 15 pounds for unspecified purposes by Episcopius (most likely his fee of 30 pounds was split between the two printers).[84] This was a tidy sum, given that early in his career Gessner earned roughly half that amount (30 florins) in a whole year and we know that Heinrich Bullinger, who outranked Gessner in the city hierarchy, was paid

[82]"Etenim in studiosorum gratiam ut Opus hoc universum composui, ita singulorum eius librorum inscriptiones feci in typographorum illorum nomina, qui hoc tempore maxime florerent, ac optime de bonis studijs meriti essent, ut simul hac occasione quinam ex officinis eorum libri prodivissent, quotquot meminisse poteram, commemorarem: simul ut gnaviter pergerent pro virili mea illos animarem. Excitatur enim fere ad maiora, qui anteactos labores suos quamplurimis et utiles et gratos fuisse, seque publicis illorum gratia laudibus, tanquam suae praemio virtutis, affici celebrarique animadverterit." *Pandectae*, 1548, sig. [6]r.

[83]See Wellisch, 14; see the German translation in Ley, 21–22.

[84]See Wackernagel 1881, 20 re Elias Philippinus (paid 15 *l* 2 s 2d "pro Galeni recognitione"); 39 ("Gessnero pro praefatione," 15 *l* from Froben) and 41 ("Gessnero d.d." 15 *l* from Episcopius). I am most grateful to Urs Leu for the information concerning Bullinger and the reference to his "Haushaltungsbuch oder Rechnungsbuch," Zentralbibliothek Zürich, Ms. K 40. For more discussion of Gessner's finances see Blair 2017b and Leu 2016a, 348.

1 pound per 6 folio pages he wrote.[85] Gessner's preface to the Galen edition was no simple affair, but a 73-page bio-bibliography of Galen, which also traced the fortuna of his works down to Gessner's time.[86] In the dedication of that work to Basil Amerbach and the faculty of the University of Basel Gessner explained: "since the printers, from whom I am conscious of many great benefits toward me, decided that I provide prefaces, despite my weakness I could not dare refuse them."[87] The appearance of the modesty trope may in fact result from Gessner's awareness of having been paid an unusually good fee in this case for his writerly services.

Of course Gessner's relentless editorial work was not motivated exclusively by financial need. Throughout his life Gessner worked to promote the transmission and growth of knowledge through high quality editions of ancient and modern texts. But the specific ways in which Gessner did so were often the result of printers asking him for something to publish, or other scholars entrusting him with texts to publish. These themes surface especially in the dedications of his later years (e.g. after 1559). For example when his relative Andreas Gessner was preparing an edition of Leo Africanus' Description of Africa in 1559 and asked him for "something new to add to it, since my leisure was not suited to greater things, I thought of translating" a short text by Hanno of Carthage (fifth century) describing an expedition on the Libyan coast. Gessner characterized the work—5 pages of Hanno's text and 12 pages of his notes—as the "labor of a short day."[88] Similarly requests from Jakob Gessner prompted the bilingual edition of Cassius' *Naturales et medicinales quaestiones* (1562) and a 230-page *De anima* which Gessner added to his edition of Johannes Lodovicus Vives' work on the topic.[89] In 1563 Gessner took the initiative in proposing to Jakob a text by the deceased Jodocus Willich; a relative of Willich's had sent the manuscript (along with a dedication he had written) to Oporinus of Basel, but Oporinus had neglected to print it and was happy for Jakob Gessner to do so. Gessner wrote a new dedication (to Iohannes Pontisella, rector of the Latin school of Chur in the Graubünden) in which he reflected on the proceess: "I would rather have abstained from this not very

---

[85] See Wellisch, 13. On the relations between these different units of currency see Guthrie 2003, 34.

[86] See Wellisch, 91–92. I did not find this edition of Galen digitized and relied on Wellisch's collection of microfilms and a physical copy at the Huntington Library.

[87] "[Q]uoniam typographis placuit me in eos praefari, quibus ego propter multa praeclara eorum in me beneficia, quamvis probe tenuitatis meae conscius, meam operam negare non sum ausus." *Galeni omnia quae extant*, 1562, sig. a + 3v.

[88] "Cum his diebus patruelis meus Andreas Gessnerus, Ioannis Leonis Africam suis typis cuderet: eique auctarium aliquod novum a me addi contenderet, nec ocium ad maiora mihi suppeteret, subito Hannonis Navigationem, qua is maximam Libycae orae partem lustravit, dieculae fere opera, Latinam feci et simul Scholia quaedam, nimis quidem festinanter conscripta, adieci." *Hanno Carthaginensium ducis navigation*, 1559, sig. alpha 2r (pdf 568).

[89] "Patruelis meus typographus hoc ipso tempore sub praelo haberet ... meque rogaret ut novi aliquid adderem de meo..." *Valentini de anima*, 1563, 721 (pdf 1020). "Nunc tandem, postulante typographo patrueli meo, ut praelo ocioso aliquid suppeditarem, ex schedarum mearum acervo hunc libellum deprompsi." *Cassii Naturales et medicales quaestiones*, 1562, sig. A2v.

glorious work, but I am in the habit of being all too ready to gratify the printer and promote the works of the deceased."[90] Gessner's depiction of himself as easily swayed by the requests of printers seems worth taking seriously, even though it also served as a modesty trope by emphasizing that Gessner published these works in haste and at the behest of others.

Gessner cultivated close relations with multiple printers. He worked most often with Christopher Froschauer who published 24 of Gessner's books from his earliest years (*Compendium ... de urinarum differentiis*, 1541; *Catalogus plantarum* 1542) down to his latest ones (*De libris a se editis*, 1562; *Viaticum novum*, 1565). Froschauer also printed all of his largest and most expensive books, in the *Bibliotheca* and *Historia animalium* series. Gessner reserved pride of place in the *Pandectae* to Froschauer, addressing the first of the dedications to him, though he denied that this was a mark of higher esteem: "I have assigned first place to you Froschauer, not deliberately or from a special plan but by chance, as I shall shortly say," but because Froschauer had published so many works relevant to grammar and philology, the themes of the first book of the *Pandectae*. In protesting that he did not, Gessner suggests that he did in fact have in mind the extra honor of giving Froschauer the first and most visible dedication of the volume.[91] Gessner complained about Froschauer in print occasionally, observing in the *Bibliotheca universalis* that in the publication of Stobaeus "not a few [sententiae] were left out that should have been added to the heel of the volume while I was absent in Italy but were omitted by the inattention of the printer."[92] Gessner also disagreed sometimes with Froschauer's commercial decisions. He explained of the Greek and Latin *Sententiae* of Antonius and Maximus printed in 1546 that he "wanted to publish [together] Latin and Greek versions of these loci but the printer Froschauer thought it was better to sell them separately," so they appeared in two separate volumes.[93] Similarly Gessner complained in a letter about Froschauer's decision to have copies of the illustrated volumes colored as rapidly and cheaply as possible.[94] But Gessner

[90]"Maluissem equidem, praesertim in proprijs lucubrationibus, si quisquam alius, occupatissimus, abstinere hoc labore minime glorioso: sed nimium fere ad gratificandum typographis facilis esse soleo, et ad promovendas defunctorum lucubrationes." Willich, *Ars magirica*, 1563, sig. *3r–v.

[91]"Principem vero tibi locum Froschovere inter claros aetatis huius typographos assignavi, non de industria quidem, neque primario instituto, sed casu, ut modo dicam." *Pandectae*, 1548, sig. [6]r.

[92]"Supersunt tamen adhuc non pauca, quae calci voluminis adijcienda me absente in Italia typographorum incuria omisit: sed cum illa, tum veterum exemplarium collationes brevi fortassis aedentur separatim." *Bibliotheca universalis*, 1545, f. 182r. This passage was marked for deletion in the ZB copy but is present in the Olms reprint.

[93]"Interpretationem hanc Locorum communium a Graecis separari non debere, consilium meum fuerat.... Nunc quoniam aliter visum est. Typographo nostro, magis e re sua futurum persuaso, si uterque seorsim vaeneat." *Sententiae Antonii et Maximi*, 1546, Latin volume, sig. *2r.

[94]See Gessner, *Epistolarum medicinalium libri III*, 1577, 22r (to Crato von Krafftheim, 26 March 1564), complaining that the colors were applied "negligenter" and "defunctorie" because of the printer's avarice; as discussed in Kusukawa, 76.

also appreciated Froschauer's need to make ends meet in a difficult business and maintained a strong working relationship with him.

The printers that Gessner mentioned most frequently in his dedications were his relatives Andreas and Jakob Gessner with one or both of whom Gessner published 14 works starting in 1552, when Gessner rushed his *Thesaurus Evonymi* into print under a pseudonym, down to the fat octavo of *De omni rerum fossilium genere* (1565).[95] In 1548 Gessner had dedicated his edition of Ermolao Barbaro's compendium of natural philosophy to young Jakob Gessner when he was still a student, but since he was not yet a printer he did not receive a dedication in the *Pandectae*. Another printer was the object of two dedications from Gessner: Thomas Iunta of Venice, to whom Gessner dedicated both the last book of his *Pandectae* in 1548 and his 20-folio-page discussion of Swiss and German thermal baths within a longer book on spas published by Iunta. This dedication, in which Gessner encouraged Iunta to come visit spas in his area, constitutes an unusual case of an author dedicating a piece of writing to the printer responsible for printing it. Gessner also cultivated relations with a number of other printers with whom he published more than one work: in Basel, Winter in the early years (3 works), then Curio (2), Oporin (4), and Froben (2); in Strasbourg, Rihel (4) and in Geneva, Henri Estienne (2). In addition Gessner published single works with a further seven printers in Lyon, Basel, and Venice.[96] Reflecting in 1562 on the sequels to his *Bibliotheca* which omitted things without his consent, Gessner noted that "printers look to profit rather than to the desires of the authors or the utility of the readers."[97] Gessner's clear-headed understanding of the tensions inherent in the relations between learned authors and printers was no doubt helpful in his ability to work effectively with printers to produce supplemental income for himself and for them through a steady stream of learned publications which also benefitted the common good of the Republic of Letters.

---

[95]This figure includes the reprint of 1557 *Athenagoras* in 1559 as *Theologorum Graecorum libri*. Most of these imprints were octavos, with one quarto (*De raris et admirandis herbis*, 1555), two folios (*De piscibus*, 1556 and *Theologorum Graecorum libri*, 1559), and one 16mo (*Hanno Carthaginensis*, 1559). Gessner refers to Andreas and Jacob as his "patrueles" which means that they descended from his father's brother, making them cousins or cousins-once-removed since they were likely younger than Gessner. For some details, see Gessner's letter to Crato von Krafftheim of 26 March 1564: "patrui mei nomen est Andreas Gesnerus senior (nam et filium eiusdem nominis habuit) annos nunc octoginta natus, tribunitiae dignitatis, senator, liberorum, nepotum et pronepotum utriusque sexus pater, avus et proavus numero centum et triginta quatuor hoc tempore, quorum maior pars vivit." *Epistolarum medicinalium*, 1577, f. 21v. See Serrai, 50.

[96]This count omits the Paduan publication of 1557 in which Gessner was not involved (see note 17).

[97]"Sed suum plerunque lucrum spectant typographi, potius quam vel authorum voluntatem, vel utilitatem Lectorum solidam. "*De libris a se editis*, 1562, section 25, sig. A6r–v. Gessner also showed his awareness of printers' concerns when he invoked the need to fill the blank pages in a quire, as I discuss in Blair 2017a.

8.6 Conclusion

Gessner's strategy of publishing many different kinds of works, in multiple fields of learning and varying sizes (from 16mo to folio), dedicated to a wide range of people and in close collaboration with the interests of printers, can be contrasted with the strategy of another naturalist of the same generation as Gessner but longer-lived. The dedication strategy of Ulisse Aldrovandi (1522–1605), studied by Laurent Pinon, was quite different from Gessner's.[98] With the explicit aim of outdoing Gessner, Aldrovandi sought to publish large multi-volume works of illustrated natural history and sought funding for the expenses of gathering specimens and making images from life especially from the wealthy cardinals of his day. Cardinal Montalto, to whom Aldrovandi dedicated three books of his *Ornithology* (1599, 1600, 1603) funded the second of these volumes but did not become the reliable patron that Aldrovandi sought, despite Aldrovandi's lavish engraved title page and a fulsome dedication. The dedication of Aldrovandi's book on insects (1602) to Francesco Maria Duke of Urbino also failed to yield a major patron. In his search for a patron Aldrovandi gifted 82 copies of his *Ornithology* (worth an impressive sum), including 17 copies he gave to various cardinals, seeking funds from which to publish the many other works of natural history he had prepared. The lack of response to his campaign doomed the bulk of Aldrovandi's vast works to remain unpublished at his death. Only the proceeds of the sales of Aldrovandi's valuable collection of naturalia which he had bequeathed to the City of Bologna made possible the posthumous publication of a further 13 volumes of his natural histories down to 1648. Hundreds of volumes of manuscripts remained unpublished in the University Library of Bologna. Gessner too died leaving manuscripts behind, including a *Historia plantarum*, the publication of which was spectacularly delayed until 1753 and 1771.[99] Nonetheless Gessner's method of accumulating support from many quarters seems to have been remarkably effective by contrast with Aldrovandi's single-minded pursuit of a major donor.

Gessner's case highlights the complex roles of dedications in the Renaissance which were not always limited to bids for patronage. Gessner used his dedications as opportunities to bring to public attention his achievements and aspirations and to navigate both local hierarchies and international networks seeking access to books, images, information, and support across confessional difference and geographic distance. Even when we lack full details about their actual impact, early modern dedications like Gessner's reward careful study with the insight they offer into an author's relationships with printers, peers, and patrons which shaped both the process and the results of research and writing.

[98]On this aspect of Aldrovandi, see the work of Pinon.

[99]Wellisch, 23–25, provides a brief account. A facsimile of Gessner's manuscripts on plants is available: *Conradi Gesneri historia plantarum* 1987, 2 vols.

Appendix 1: Gessner's Publications and Dedications (Ordered by Date of Dedication)

Wellisch number	Date of pub'n	Short title format (fol. 4to, 8vo); privilege if mentioned (imperial, royal [Austria], French, unspecified); pdf source	Place of pub; publisher	Dedication month (+ year if diff from pub yr)	Dedicatee(s); length of dedication	Dedicatee place/origin[a]
Written in Lausanne						
A1	1537	*Lexicon Graecolatinum*. fol. 3-year imp priv. (pdf e-rara)	Basel J Walder		none—no mention of Gessner's contribution; see ed of 1545	
A2	1540	*Actuarius. De medicamentorum compositione*. 8vo. (pdf Google Books)	Basel R. Winter		None	
A7	1542	*Apparatus et delectus simplicium*. 8vo. (pdf archive.org; some expurgations)	Lyon J, F Frellon	Jan 1541	Christophorus Clauserus, archiatros (chief physician of the city). 8p	Zurich
A4	1541	*Actuarius. De differentiis urinarum*. 8vo. (pdf Google Books)	Zurich Froschauer	n.d.	1. Petrus Jacobus Et Stephanus, Hispanus Vincentinus (met in Montpelier). 3p	Spain
		Universalis doctrina Galeni		June	2. (pdf p. 96) Albertus Belfort, medicus (met in Lyon). 4½p	Graubünden
		Sylva experimentorum Galeni		Aug	3. (pdf p. 341) Claudius Milletus, medicus (met in Lyon). 1½p	Lyon
A3	1541	*Historia plantarum*. 8vo. (pdf e-rara)	Basel R. Winter	Aug	Heinrich Billing, stepson of bürgermeister Jakob Meyer. 5p	Zurich
A8	1542	*Catalogus plantarum*. 4to. (pdf e-rara)	Zurich Froschauer	n.d.	Johannes Jacobus Ammianus, bonarum literarum professor—former teacher and lodger. 5p	Zurich

(continued)

Wellisch number	Date of pub'n	Short title format (fol. 4to, 8vo); privilege if mentioned (imperial, royal [Austria], French, unspecified); pdf source	Place of pub; publisher	Dedication month (+ year if diff from pub yr)	Dedicatee(s); length of dedication	Dedicatee place/origin[a]
Written in Zurich						
A6	1541	*De Lacte et operibus lactariis*. 8vo. (pdf e-rara)	Zurich Froschauer	June	Jacobus Avienus [Vogel]. 12p. on the Alps	Glarus (cf De libris a se . . .)
A5	1541	*Michael Ephesius. Scholia in Aristotelis Libros*. 8vo. 7-year priv unspec. (pdf BnF)	Basel: B. Westhemer	Aug	1. Sebastian Singler [Sinkler], medicus et praeceptor. 3½p	Basel
		scholia de longitudine et brevitate vitae		Aug	2. (pdf p. 43) Aegidio Scudo [Tschudi], offered hospitality. 2½p	Glarus
		scholia de divinatione per somnum		Aug	3. (pdf p. 80) Melchior Wirtz, friend since boyhood. 3½p	Zurich
A10	1542	*Moralis Interpretatio errorum Ulyssis Homerici*. 8vo. (pdf BSB)	Zurich Froschauer	Mar	1. Pierre Viret, theologus. 2p	Lausanne
		commentatio de nympharum antro		Mar	2. (pdf p. 35) Beatus Comes [Comte], theologus et medicus. 3p	Lausanne
		lucubratio Procli Lycii		Mar	3. (pdf p. 67) Jean Ribit, prof. of Greek, friend (BU). 3½p	Lausanne
A9	1542	*Porphyrii institutiones*. 8vo. 7-year priv unspec (pdf BnF) with *Compendium de syllogismis authoris incerti*	Basel [R Winter]	Aug	(pdf p. 347) Otho Werdmüller, prof of philosophy. 1½p	Zurich
A11	1543	*Stobaeus. Sententiae ex Thesaurus Graecorum*. fol. (pdf e-rara)	Zurich Froschauer	June	Johannes Jacobus Watenwil [Wattenwyl] and Joannes Franciscus Negelin [Hans Franz Naegeli], noble city consuls of Bern. 8½p	Bern

A15	1543	*Enumeratio medicamentorum*. in Brasavola, *Examen omnium catapotiorum sive pilularum* 8vo. imp priv, duration unspec. (pdf BSB)	Zurich Froschauer		None; starts on pdf p 150 with Gessner's "to the reader"	
A12	1544	*Heraclides Ponticus. Allegoriae in Homeri Fabulas*. 8vo. (pdf BSB)	Basel Oporinus	Jan	Hieronymus Frikker [Fricker], praefectus of Mendrisio, Ticino; helped religious refugees from Italy. 4p	Bern, Ticino
A13	1544	*Martialis Epigrammata*. 8vo. (pdf e-rara, Gessner's own copy, with annotations[b])	Zurich Froschauer	Mar	Felix Fry [Frey], Heinrich Bullinger, Gaspar Megander, Erasmus Fabricius [Schmid], and Rolph [Rodolphus] Gualtherus, ministers, profs of theology, "respected patrons and preceptors." 6½p	Zurich
A14	1544	*Onomasticon*. fol. [used Wellisch mfilm]	Basel [H. Curio]		None; see 1549 edition	
A17	1545	*Antonii Thylesii. Opuscula aliquot*. 8vo. (pdf BSB)	Basel Oporinus	Feb	Leonhard Beck von Beckenstein, consiliarius to emperor. 7p	Augsburg
A16.1a	1545	*Bibliotheca Universalis*. fol. (pdf e-rara, Gessner's own copy with annotations)	Zurich Froschauer	July	Leonhard Beck von Beckenstein. 9½p	Augsburg
A1.4	1545	*Lexicon Graecolatinum*, 4th ed (first to list Gessner on title page). fol. royal priv unspec. (pdf BSB)	Basel [H. Curio]	Aug	Diego Hurtado a Mendozza, imperial ambassador to Venice. 3p	Venice
A18.1	1546	*Sententiae Antonii et Maximi. (Greek)*. fol. (pdf e-rara, Gessner's own copy with annotations)	Zurich Froschauer	Feb	Johann Welser, Jakob Herbrot, and the Senate of Augsburg. 2p	Augsburg

(continued)

Wellisch number	Date of pub'n	Short title format (fol. 4to, 8vo); privilege if mentioned (imperial, royal [Austria], French, unspecified); pdf source	Place of pub; publisher	Dedication month (+ year if diff from pub yr)	Dedicatee(s); length of dedication	Dedicatee place/origin[a]
A18.2	1546	*Sententiae Antonii et Maximi (Latin)*. fol. (pdf e-rara)	Zurich Froschauer	Feb	1. Johann Jakob Fugger, wealthy merchant. "domino in Vuyssenhorn and domino suo." 1p	Augsburg
					2. (pdf p. 360) Gregorius Laetus [Froehlich], archigrammateus. ¼p	Augsburg
A19	1548	*Ermolao Barbaro. Naturalis scientiae totius compendium*. 8vo. (pdf BSB)	Basel Oporinus	Aug	Jakob Gessner, relative. 2½p	Zurich
A16.1b	1548	*Pandectae*. fol. (pdf e-rara) Book I, grammar and philology	Zurich Froschauer	Sept	1. (pdf p. 11) Christophorus Froschauer, Gessner's main printer to date. 1p/ 3p with list of books	Zurich
		Book II, dialectic		n.d.	2. (pdf p. 98) Joan. Bebelius and Michael Isengrin, printers. 1p	Venice
		Book III, rhetoric		Jan	3. (pdf p. 110) Johannes Oporinus, printer. 1½p	Basel
		Book IV, poetics		n.d.	4. (pdf p. 130) Nicolaus Brulinger [Brylinger], printer. ½p	Basel
		Book V, arithmetic		Feb	5. (pdf p. 158) Robert Estienne, printer. ½p	Paris
		Book VI, geometry		Feb	6. (pdf p. 166) Johannes Petreius, printer. 1p	Nurnberg
		Book VII, music		Feb	7. (pdf p. 174) Henricpetri, printer. 1p	Basel
		Book VIII, astronomy		Feb	8. (pdf p. 186) Hieronymus Curio, printer. 1p	Basel
		Book IX, astrology		Feb	9. (pdf p. 202) Joan. Montanus [Berg] and Ulrich Neuber, printers. 1p	Nurnberg

		Book X, divination and magic licit and illicit		Feb	10. (pdf p. 210) Wendelin Rihel, printer. 1p	Strasbourg
		Book XI, geography		Feb	11. (pdf p. 226) Paolus Manutius, printer. 1p; 4½p with list of publications	Venice
		Book XII, history		Mar	12. (pdf p. 246) Sebastian Gryphius. ½p; 6p with list of publications	Lyon
		Book XIII, mechanical and useful arts		Apr	13. (pdf p. 342) Christian Wechel, printer. ½p, 4p with list	Paris
		Book XIV, natural philosophy		May	14. (pdf p. 374) Johannes Herwagius, printer. ½, 1½ with list	Basel
		Book XV, metaphysics		May	15. (pdf p. 486) Johannes Gymnicus [Gymnich], printer. ½p, 2½p with list	Cologne
		Book XVI, moral philosophy		June	16. (pdf p. 534) Johannes Frellon, printer. ½p, 2p with list	Lyon
		Book XVII, economic philosophy		July	17. (pdf p. 618) Vincentius Valgrisius, printer. ½p, 1p with list	Venice
		Book XVIII, politics		July	18. (pdf p. 634) Hieronymus Scotus [Girolamo Scoto], printer. ½p	Venice
		Book XIX, civil law		Aug	19. (pdf p. 670) Thomas Junta, printer. ½p	Venice
A14.3	1549	*Onomasticon* (published with Calepino, *Dictionarium*). fol. royal priv unspec. (pdf BSB)	Basel H. Curio	Jan	(pdf p. 1124) Jacobus and Marcus Roestius [Roeist], Jacobus and Marcus Stapfer, Jacob Habius [Haab], Georgius Grebel, Georgius Escher, Felix Engelhart, Heinrich and Georg Rublin [Rubli], Wilhelm and Geroldus Meier [Meyer von Knonau?], patricians. 5p	Zurich

(continued)

Wellisch number	Date of pub'n	Short title format (fol. 4to, 8vo); privilege if mentioned (imperial, royal [Austria], French, unspecified); pdf source	Place of pub; publisher	Dedication month (+ year if diff from pub yr)	Dedicatee(s); length of dedication	Dedicatee place/origin[a]
A16.1c	1549	*Partitiones theologicae*. fol. (pdf e-rara)	Zurich Froschauer	Feb	Hieronymus Froben and Nicolaus Episcopius, printers. 1p, 2½p with list of books	Basel
A20	1549	*Galenus. Opera*. fol. 5-yr imp priv. (pdf BSB)	Basel Froben		None; but "Gesnerus medicinae candidatis"	
A21	1550	*Aristotelis. Opera omnia*. fol. imp priv unspec. (pdf e-rara)	Basel: Bebel & Isingrin		None; Gessner not mentioned on title page, thanked in the printer's "to the reader"	
A22	1550	*Galeni brevis denotatio dogmatum Hippocratis*. in *Galeni aliquot opuscula*, pp. 103-6. 8vo. (pdf BnF)	Lyon: G. Rovillius		None	
A23	1551	*Historia animalium I: De quadrupedibus viviparis*. fol. (pdf e-rara)	Zurich Froschauer	Aug	Consuls and Senators of Zurich. 8½p	Zurich
A16.2a	1551	*Elenchus* (unauthorized publication, as noted in 1551 *Historia animalium*). (pdf Google Books) 4to	Basel Oporinus		None; Gessner not involved	
A33	1552	*Hieronymus Tragus. De stirpium ... facultatibus*, tr. Kyber. fol. 7-year imp priv. (pdf e-rara)	Strasbourg Wendelin Rihel		None; Gesnerus ad rei medicae studiosos	
A32	1552	*Thesaurus Euonymi Philiatri*. 8vo. (pdf e-rara)	Zurich: A. Gessner + Wyssenbch	n.d.	Nicolaus Zurkinden, consiliarius. 2p	Bern
A34	1553	*De Germaniae et Helvetiae Thermis*. in *De Balneis*. fol. priv from pope and senate of Venice (pdf BSB)	Venice Junta	Mar	(pdf p. 610) Thomas Junta, printer. ½p	Venice

A35	1553	*Kyber*. *Lexicon rei herbariae*. fol. 7-year imp priv. 8vo. (pdf BSB, some expurgations)	Strasbourg W. Rihel	Apr	1. Lucius Kyber, pastor and father of deceased author. 8p	Strasbourg
		Gessner, Tabulae collectionum in genere			2. (pdf p. 624) Nicolaus Speicher, pharmacopola and friend of deceased. 1p	Strasbourg
A29	1553	*Icones animalium*. fol. 8-year imp priv + 10-year French priv] (pdf e-rara) [one issue with woodcut at end of dedication at p. 7 of pdf]	Zurich Froschauer	Aug	Thomas and Johannes Grey, brothers of Henry Duke of Suffolk. 1p	England
A24	1554	*Historia animalium II: De quadrupedibus oviparis*. fol. 8-year imp priv; 10-year F priv (pdf e-rara)	Zurich Froschauer	Feb	1. Valentinus Gravius, decimarius and senator from Fribourg. 1½p	Fribourg, Switzerland
		Appendix Historiae quadrupedum		Mar	2. (pdf p. 126) Johannes Steiger, senator from Bern and quaestor for Savoy. ½p	Bern
A40	1555	*Enchiridion rei medicae triplicis*. 8vo. (pdf BSB)	Zurich A J Gessner	Aug (1554)	Achille Gasser, medicus. 3p	Augsburg
A25	1555	*Historia animalium III: De avium natura*. fol. 8-year imp priv + 10-year F priv. (pdf e-rara) [woodcut of Gessner pdf p. 4]	Zurich Froschauer	Mar	Johann Jakob Fugger, wealthy merchant. 3½p	Augsburg
A36	1555	*Chirurgia*. fol. imp priv. (pdf BSB)	Zurich A J Gessner	Jan	Geryonis [Gereon] Seiler, medicus of the city of Augsburg. 1½p	Augsburg
A16.1d	1555	*Appendix*. fol. (pdf BSB)	Zurich Froschauer	Mar	Caspar von Niedbruck [Nyderbruck or Nydbruck], noble counselor to Ferdinand and Maximilian. 3p	Vienna
A30	1555	*Icones avium*. fol. (pdf e-rara) [woodcut of Gessner pdf p. 137]	Zurich Froschauer	Mar	Ulrich Fugger, count of Kirchberg and Weissenhorn. 1p	Augsburg

(continued)

Wellisch number	Date of pub'n	Short title format (fol. 4to, 8vo); privilege if mentioned (imperial, royal [Austria], French, unspecified); pdf source	Place of pub; publisher	Dedication month (+ year if diff from pub yr)	Dedicatee(s); length of dedication	Dedicatee place/origin[a]
A37 A38.1	1555	*De raris et admirandis herbis. Pilati Montis descriptio*. 4to. (pdf e-rara)	Zurich A J Gessner	July	1. Laurentius Gryllus, medicus, bearing greeting to Ulrich Fugger. 1p	Landshut
				Sept	2. (pdf p. 51) Johann Chrysostom Huober [Huber], medicus. 1p	Luzern
A39	1555	*Mithridates*. 8vo. (pdf e-rara) [appended foldout table is absent in this copy; other digitizations also omit the table]	Zurich Froschauer	Aug	1. Johannes Bale, Englishman, bishop of Ossory, Ireland (spent time in Zurich). 1½p	Ireland
		appended folded table not present in pdf		Sept	2. Leonhard Beck von Beckenstein. ¼ p	Augsburg
Not in Wellisch[c]	1555	Otho Werdmüller, *Similitudinum ab omni animalium genere*, in Johannes Fabricius Montanus, *Differentiae animalium quadrupedum* (pdf E-rara, pp. 137–376)	Zurich A J Gessner	March	Abel Werdmüller, adolescent son of Otho who died in 1552; a relative of Gessner's	Zurich
A41	1556	*Sanitatis tuendae praecepta*. 8vo. (pdf e-rara)	Zurich A J Gessner	Jan	Johannes Wegmann and Felix Peierus [Peier], senators and tribunes of Zurich. 2p	Zurich
A43	1556	*De piscibus et aquatilibus omnibus libelli*. 8vo. (pdf e-rara)	Zurich A Gessner	n.d.	1. (pdf p. 11) Joan. Perrinus, learned young man, who dedicated the book to Sebastian a Loys. 1p	Lausanne
		Catalogus aquatilium ex Plinio		Mar	2. (pdf p. 23) Johannes Caius, medicus et philosophus. 2½p	England

		Aquatilium . . . nomina Germanica et Anglica		Apr	3. (pdf p. 105) Burchardus Mythobius, medicus et philosophus. 2½p	Northern Germany
		De fictis in Germanica lingua aquatilium nominibus		June	4. (pdf p. 256) Melchior Guilandus [Wieland]. 3p	Prussia
A42	1556	*Aelianus. Opera*. fol. (e-rara, Gessner's own copy with annotations)	Zurich A J Gessner	May	Johannes Jacobus Fugger, dominus of Kirchberg and Weissenhorn, and Maecenas. 2p	Augsburg
A44.2/ A49	1557	*Athenagoras*. fol. (e-rara) reprinted in 1559 *Theologorum* (Zurich: A Gessner). fol. (pdf e-rara)	[Geneva] H. Estienne	Feb	(pdf p. 84, following Greek section which has no dedication) Heinrich Bullinger. 2p. (in 1559 repr. same dedication on pdf p. 246)	Zurich
A45	1557	*De stirpium aliquot epistolae*. 8vo. (BnF, Google Books)	Padua Perchacinus		None—Gessner not involved in the publication (comprises one letter by Guilandinus and one by Gessner)	
A26	1558	*Historia animalium IV: De piscium. . .natura*. fol. 8-year imp priv + 10-year F priv (pdf e-rara)	Zurich Froschauer	Aug	Ferdinand I, Holy Roman Emperor. 7½p	Vienna
A46	1559	*Marcus Aurelius Antoninus. De vita sua*. 8vo. 3 year priv unspec. (pdf e-rara)	Zurich A Gessner	Feb	(pdf p. 292) (following Latin translation) Anton Werther [von Werthern] von Beichlingen. In Greek. 9½p	Beichlingen
A47	1559	*Hannonis Carthaginensium ducis navigatio* (in *Leonis Africani de totius Africae descriptio*) 16mo. (pdf e-rara)	Zurich A Gessner	Feb	(pdf p. 566) Joan. Du Choul, royal senator and prefect of the Allobroges. 2p	Allobroges =Dauphiné
A48.1	1559	*Xenocrates de alimento ex aquatilibus*. 8vo. (in *Iani Dubravii De piscinis)* (pdf e-rara)	Zurich unspec	Oct	Joan. Echtius [Backofen von Echt]. 1½p	Cologne

(continued)

Wellisch number	Date of pub'n	Short title format (fol. 4to, 8vo); privilege if mentioned (imperial, royal [Austria], French, unspecified); pdf source	Place of pub; publisher	Dedication month (+ year if diff from pub yr)	Dedicatee(s); length of dedication	Dedicatee place/origin[a]
A29.2	1560	*Icones Animalium (2nd ed.)* fol. imp and French priv unspec. (pdf e-rara) [no woodcut]	Zurich Froschauer	June	1. Elizabeth I, recently crowned queen of England. 4p + 28-line Greek ode	England
					2. (pdf p. 144) Thomas and Johannes Grey (repeated from 1553)	
A30.2	1560	*Icones Avium (2nd ed.)* fol. 8-year imp priv and 10-year F priv (pdf e-rara Gessner's own copy with annotations)	Zurich Froschauer	Mar 1555	1. Ulrich Fugger (repeated from 1555). 1p	Augsburg
		accessio (additions of 2nd ed)		July	2. (pdf p. 130) Johannes Parkhurst, bishop of Norwich; former Marian exile in Zurich. 1p	Norwich, England
A31	1560	*Nomenclator . . . Icones Animalium aquatilium*. fol. 8-year imp priv and 10-year F priv (pdf e-rara, Gessner's own copy with annotations)	Zurich Froschauer	June	1. Maximilian II, King of Bohemia and of Austria. 3½p	Vienna
		Book II		May	2. (pdf p. 309) Senate of Basel. 1½p	Basel
		Ordo II, tome II, on fresh water fish		June	3. (pdf p. 368) Sigismund, Freiherr von Herberstein, Nyperg and Guttenhag [imperial diplomat]. 1p	Vienna etc
		accessio de Germanicis . . . nominibus		June	4. (pdf p. 400) Levinus Lemnius, medicus. ½p	Zierikzee NL
A50	1561	*Josua Maler*. *Dictionarium Germanicolatinum*. 8vo. (partial pdf e-rara)	Zurich Froschauer		None; Gessner ad lectorem praefatio	

A51	1561	*Valerius Cordus. Annotationes in Dioscorides, etc*. fol. 8-year imp priv (pdf e-rara)	Strasbourg Iosias Rihel	June	1. Faculty of Medicine, Wittenberg. 2½p	Wittenberg
		Annotationes in Dioscoridis libros		Aug	2. (pdf p. 11) Son of Joannis Ralla, recently deceased pharmacopola, who transmitted ms to Placotomus. 1½p	Leipzig
		Valerii Cordi historia plantarum		Dec 1559	3. (pdf p. 190) Hieronymus Herold, medicus, proposed dedication #1, supplied legible copy of ms. 2p	Nurnberg from Leipzig
		[extra plant pictures—no title] 8pp		Jan 1560	4. (pdf p. 448) Caspar Collinus [Ambühl], pharmacopaeus. ¼ p	Sion, Switz.
		Sylva observationum variarum (18pp)		Dec 1559	5. (pdf p. 456) Joannes Placotomus, medicus, sent ms to Gessner. 1p	Danzig
		De artificiosis extractionibus (9pp)		Dec 1560	6. (pdf p. 472) Phillippus Bechius [Bächi], medicus, sent ms to Gessner. 1p	Basel
		Compositiones medicinales (5pp)		Dec 1560	7. (pdf p. 481) Sebald Hawenreuter [Hauenreuter], medicus. ½p	Strasbourg
		Stocc-hornii . . . montium descriptio (8pp) by Benedictus Aretius, a friend of Gessner		Jan 1560	8. (pdf p. 486) dedication by Gessner to Christophorus Piperinus [Pfäfferlin], minister. ½p	Sigriswil, Bern
		Horti Germaniae		Jan 1560	9. (pdf p. 494) Stephanus Lauraeus, imperial medicus. 1p	Augsburg
		Appendix . . . de hortis Germaniae		Jun 1561	10. (pdf p. 598) Franciscus Calceolarius, pharmacopola. 2p	Verona
A52	1561	*Historia et interpretatio prodigii*. 8vo. (pdf e-rara)	[Zurich]		None; published under a pseudonym (Conrad Bolovesus)	
A54	1562	*Cassius Iatrosophista. Naturales et medicinales quaestiones*. 8vo. (pdf e-rara)	[Zurich] J Gessner	Jan	1. Johannes Kentmann, medicus. 4p	Torgau in Saxony

(continued)

Wellisch number	Date of pub'n	Short title format (fol. 4to, 8vo); privilege if mentioned (imperial, royal [Austria], French, unspecified); pdf source	Place of pub; publisher	Dedication month (+ year if diff from pub yr)	Dedicatee(s); length of dedication	Dedicatee place/origin[a]
		Greek text		Jan/Feb	2. (pdf p. 70) in Greek: Antonios Niphoreios [Schneeberger] 3½p	Cracow from Zurich
A53	1562	*Claudius Galenus*. fol. *Opera omnia* (mfilm Wellisch)	Basel Froben	Feb	Basil Amerbach and the other professors of the Academy of Basel. 4p	Basel
A55	1562	*Sante Arduino*. *De venenis*. fol. imp priv unspec (pdf BSB)	Basel: Henricpetri & Perna		None; no mention of Gessner	
A59	1563	*Valerius Cordus*. *Stirpium descriptionis liber V*. fol (only 28p). 8-year imp priv (pdf BSB)	Strasbourg J Rihel	Aug 1562	Wolfgang Meurer, medicus et philosophus. 1p	Leipzig
A56	1562	*De libris a se editis epistola ad Guilielmum Turnerum*. 8vo. (pdf e-rara)	Zurich Froschauer	Sept	William Turner. 1p	England
A57	1563	*De anima liber*. in *Ioh. Lod. Vives, De anima et vita*. 8vo. (pdf BSB)	Zurich J Gessner	Feb	(pdf p. 1018) Julius Alexandrinus Tridentinus, medicus to the emperor. 3p	Vienna
A58	1563	*Jodocus Willich*. *Ars magirica*. 8vo. (pdf e-rara)	Zurich J Gessner	Aug	Johannes Pontisella, "moderator Ludi," rector of the Latin school. 11½p	Chur, Graubünden
A60	1564	*Henri Estienne*. *Dictionarium medicum*. 8vo. (pdf BSB)	Geneva H. Estienne		none by Gessner; dedication (inserted as a cancel) by H. Estienne to Philibertus Saracenus, medicus, 2p	
A61	1565	*Dioscorides*. *De curationibus morborum*. ed. Joh. Moibanus and Gessner. fol. 8-year imp priv (pdf Hathitrust)	Strasbourg J Rihel	June 1564	1. Council and Senate of Augsburg. 9½p 2. (pdf p. 38) in Greek. Joannis Crato [von Krafftheim]. 6½p	1. Augsburg 2. Breslau

A63	1565	*De omni rerum fossilium genere*. 8vo. (pdf BSB complete) comprises 10 works by 5 authors. only Gessner's listed here	Zurich J Gessner			
		De bitumine		July	1. (pdf p. 425) Valerando Dourez, pharmacopola. 2p	Lyon
		Valerii Cordi de Halosantho			2. (pdf p. 436) Andreae Pellizero, medicus. 2½p	Kärnten, Austria
		Sancti Patris Epiphanii . . . de XII gemmis		July	3. (pdf p. 514) Adolphus Occo, medicus. 3½p	Augsburg
		De rerum fossilium, lapidum et gemmarum		Aug	4. (pdf p.778) Andreas Schadcovius [Szadkowski], notarius salinarum. 7p	Cracow
A62	1565	*Jacques Houllier*. *Viaticum novum*. 8vo. (pdf e-rara)	Zurich Froschauer		Gessner to the reader; dedication by Caspar Wolf	

[a]Locations are usually explicit in the dedication; when they were not I have relied on Leu 2016b

[b]To identify Gessner's own copies I have relied on Leu, Keller and Weidmann

[c]I am grateful to Urs Leu for calling my attention to this dedication by Gessner in a work authored by another

Bibliography

Blair, Ann. 2013. Authorial Strategies in Jean Bodin. In *The Reception of Bodin*, ed. Howell A. Lloyd, 137–156. Leiden: Brill.

———. 2016. Conrad Gessner's Paratexts. *Gesnerus* 73 (1): 73–123.

———. 2017a. Conrad Gessner et la publicité. Un humaniste au carrefour des voies de circulation du savoir. In *L'Annonce faite au lecteur*, ed. Annie Charon, Sabine Juratic, and Isabelle Pantin, 21–55. Louvain: Presses universitaires de Louvain.

———. 2017b. Printing and Humanism in the Work of Conrad Gessner. *Renaissance Quarterly* 70 (1): 1–43.

Bodenmann, Reinhardt. 2009. L'auteur et son nom de plume. Autopsie d'un choix. Le cas des pays francophones et germanophones du XVIe siècle. In *L'Auteur à la Renaissance. L'altro que è in noi*, ed. Rosanna Gorris Camos and Alexandre Vanautgaerden. *Nugae humanisticae* 9: 19–63.

Botfield, Beriah. 1861. *Prefaces to the First Editions of the Greek and Roman Classics*. London: George Bohn.

Braun, Lucien. 1990. *Conrad Gessner*. Geneva: Slatkine.

Brown, Cynthia Jane. 1995. *Poets, Patrons, and Printers: Crisis of Authority in Late Medieval France*. New York: Cornell University Press.

Coolidge, W.-A.-B. 1904. *Josias Simler et les origines de l'alpinisme*. Grenoble: Allier Frères.

Delisle, Candice. 2006. Une correspondance scientifique à la Renaissance: les *Lettres Médicinales* de Conrad Gesner. In *Réseaux de correspondance à l'âge classique (XVIe–XVIIIe siècle)*, ed. Pierre-Yves Beaurepaire, Jens Häseler, and Antony McKenna, 33–43. Saint-Etienne: Publications de l'Université de Saint-Etienne.

Durling, Richard. 1965. Conrad Gesner's *Liber amicorum* 1555–65. *Gesnerus* 22: 134–159.

Fabricius Montanus, Johannes. 1555. *Differentiae animalium quadrupedum*. Zurich: Andreas and Jacob Gessner.

Genette, Gerard. 1997. *Paratexts: Thresholds of Interpretation*. Trans. Jane E. Lewin. Cambridge: Cambridge University Press.

Gessner, Conrad. 1577. *Epistolarum medicinalium libri III*. Zurich: Froschauer.

———. 1966. *Bibliotheca Universalis*. Hildesheim: Olms reprint.

———. 1987. *Conradi Gesneri historia plantarum : Gesamtausgabe*, ed. Heinrich Zoller and Martin Steinmann, 2 vols. Dietikon-Zurich: Urs Graf Verlag.

Gilmont, Jean-François, and Alexandre Vanautgaerden, eds. 2003. *Offrir un livre, les dédicaces à l'époque humaniste. Nugae Humanisticae*, vol. 3. Turnhout: Brepols.

Guthrie, William P. 2003. *The Later Thirty Years War: From the Battle of Wittstock to the Treaty of Westphalia*. Westport: Greenwood Press.

Häberlein, Mark. 2012. *The Fuggers of Augsburg*. Charlottesville: University of Virginia Press.

Hanhart, Johannes. 1824. *Conrad Gessner*. Winterthur: Steinerische Buchhandlung.

Holzknecht, Karl Julius. 1966. *Literary Patronage in the Middle Ages*. New York: Octagon Books.

Hoyoux, Jean. 1944. Les moyens d'existence d'Erasme. *Bibliothèque d'Humanisme et Renaissance* 5: 7–59.

Jancke, Gabriele. 2002. *Autobiographie als soziale Praxis. Beziehungskonzepte in Selbstzeugnissen des 15. und 16. Jahrhunderts im deutschsprachigen Raum*, 141–152. Cologne: Böhlau.

Jardine, Lisa. 1993. *Erasmus, Man of Letters: The Construction of Charisma in Print*. Princeton: Princeton University Press.

Kapp, Friedrich. 1886–1923. *Geschichte des Deutschen Buchhandels*. Leipzig: Börsenvereins der deutschen Buchhändler.

Kusukawa, Sachiko. 2012. *Picturing the Book of Nature: Image, Text, and Argument in Sixteenth-Century Human Anatomy and Medical Botany*. Chicago: Chicago University Press.

Leu, Urs. 1990. *Conrad Gesner als Theologe: ein Beitrag zur Zürcher Geistesgeschichte des 16. Jahrhunderts*. Bern: Lang.

———. 2014a. Die Bedeutung Basels als Druckort im 16. Jahrhundert. In *Basel als Zentrum des geistigen Austauschs in der frühen Reformation*, ed. Christine Christ-von Wedel, Sven Grosse, and Berndt Hamm, 53–78. Tübingen: Mohr Siebeck.

———. 2014b. Moral Treatment of Immoral Texts from Classical Antiquity: Conrad Gessner's Martial-Edition of 1544. In *Following Zwingli. Applying the Past in Reformation Zurich*, ed. Luca Baschera, Bruce Gordon, and Christian Moser, 197–208. Surrey/Burlington: Ashgate.

———. 2016a. *Conrad Gessner (1516–1565), Universalgelehrter und Naturforscher der Renaissance*. Zurich: Verlag Neue Zürcher Zeitung.

———. 2016b. Conrad Gessners Netzwerk. In *Conrad Gessner 1516–2016: Facetten eines Universums*, ed. Urs Leu and Mylène Ruoss, 66–74. Zurich: Verlag Neue Zürcher Zeitung.

Leu, Urs B., Raffael Keller, and Sandra Weidmann. 2008. *Conrad Gessner's Private Library*. Leiden: Brill.

Leu, Urs B., and Mylène Ruoss, eds. 2016. *Facetten eines Universums. Conrad Gessner 1516–2016*. Zurich: Verlag Neue Zürcher Zeitung.

Ley, Willy. 1929. *Konrad Gessner, Leben und Werk*. Munich: Verlag der Münchner Drucke.

Louthan, Howard. 1994. *Johannis Crato and the Austrian Habsburgs: Reforming a Counter-Reform Court. Studies in Reformed Theology and History* 2(3). Princeton: Princeton Theological Seminary.

Moennighoff, Burkhard. 2008. Die Kunst des literarischen Schenkens. Über einige Widmungsregeln im barocken Buch. In *Die Pluralisiering des Paratexts. Theorie, Formen, Funktionen*, ed. Frieder von Ammon and Herfried Vögel, 337–352. Berlin: LIT.

Monfasani, John. 1988. The First Call for Press Censorship: Niccolò Perotti, Giovanni Andrea Bussi, Antonio Moreto, and the Editing of Pliny's Natural History. *Renaissance Quarterly* 41 (1): 1–31.

Nutton, Vivian. 1985. Conrad Gessner and the English naturalists. *Medical History* 29: 93–97.

Pinon, Laurent. 2009. Portrait emblématique du parfait mécène: comment Ulisse Aldrovandi remercie le cardinal Montalto. In *Conflicting Duties: Science, Medicine and Religion in Rome, 1550–1750*, ed. Maria Pia Donato and Jill Kraye, 59–79. London: The Warburg Institute.

Rice, Eugene. 1971. The Patrons of French Humanism, 1490–1520. In *Renaissance Studies in Honor of Hans Baron*, ed. Anthony Molho and John Tedeschi, 687–702. Florence: Sansone.

Schottenloher, Karl. 1953. *Die Widmungsvorrede im Buch des 16. Jahrhunderts*. Münster: Aschendorffsche Verlagsbuchhandlung.

Serrai, Alfredo. 1990. *Conrad Gesner*, ed. Maria Conchetti. Rome: Bulzoni.

Simmler, Josias. 1566. *Vita...Conradi Gessneri*. Zurich: Froschauer.

Taylor, Archer, and Fredric J. Mosher. 1951. *The Bibliographical History of Anonyma and Pseudonyma*. Chicago: University of Chicago Press, for the Newberry Library.

Terrall, Mary. 2002. The Uses of Anonymity in the Age of Reason. In *Scientific Authorship: Credit and Intellectual Property in Science*, ed. Mario Biagioli and Peter Galison, 91–112. New York: Routledge.

Vanautgaerden, Alexandre. 2012. *Erasme typographe. Humanisme et imprimerie au début du XVIe siècle*. Geneva/Brussels: Droz and Académie royale de Belgique.

Viala, Alain. 1985. *Naissance de l'écrivain: sociologie de la littérature à l'âge classique*. Paris: Editions de Minuit.

Visser, Arnoud. 2005. *Joannes Sambucus and the Learned Image: The Use of the Emblem in Late-Renaissance Humanism*. Leiden: Brill.

Wackernagel, Rudolf, ed. 1881. *Rechnungsbuch der Froben und Episcopius, Buchdrucker und Buchhändler zu Basel, 1557–1564*. Basel: Benno Schwabe.

Weber, Alan. 2003. *Because It's There, a Celebration of Mountaineering from 200 B.C. to Today*. Lanham: Taylor Trade Publishing.

Wellisch, Hans H. 1984. *Conrad Gessner. A Bio-Bibliography*. Zug: IDC.

Zacchia, Paolo. 1630. *Quaestiones medico-legales*. Lipsiae: sumptibus Eliae Rehefeldii.

Chapter 9
Imagination, Maternal Desire and Embryology in Thomas Fienus

Hiro Hirai

9.1 Introduction

A pregnant woman encounters a wolf in the woods. She is so scared that her strong emotion of fear imprints the wolf's morphological traces on the fetus in her womb. Another pregnant woman craves strawberries or cherries so intensely that she leaves certain marks or impressions of these fruits on the fetus. The belief that the power of maternal emotions such as desire and fear can imprint certain marks, signs or signatures on the fetus was widespread in the early modern period. This belief was adopted by Renaissance philosophers such as Marsilio Ficino (1433–1499) and Pietro Pomponazzi (1462–1525), who intimately connected it to the traditional theory of imagination or *phantasia*.[1] Moreover, folkloric tales about the mysterious and extraordinary psychic powers exerted by special groups of women, who were often labeled as witches and healers, might have reinforced this belief, contributing to the propagation of the idea that their forces could exceed the order of nature.

Recent studies have shed new light on the historical and intellectual background of the issue, although its complete history has yet to be written.[2] In this article I will focus on the intervention of early modern learned medicine in the theorization of

The first version of this article was presented in the seminar organized by Concetta Pennuto at the Centre d'études supérieures de la Renaissance (Tours). My warmest thanks go to Clare Hirai for her assistance in its preparation.

[1]See the excellent article by O'Brien 1993, 3–19. On the notion of imagination, see also Bundy 1927; Fattori and Bianchi 1988; and Cocking 1991.

[2]On this belief, see Angelini 1994, 53–69; Pancino 1996; eadem. 1997, 154–162; and Pennuto 2008, 368–379.

H. Hirai (✉)
Radboud University, Nijmegen, Netherlands
e-mail: hhirai2@gmail.com

G. Manning, C. Klestinec (eds.), *Professors, Physicians and Practices in the History of Medicine*, Archimedes 50, DOI 10.1007/978-3-319-56514-9_9

this belief. To this end I will address the case of Thomas Feyens or Fienus (1567–1631) of Antwerp, who wrote a monograph devoted to the power of imagination in the embryological framework.[3]

Fienus was a professor of medicine at the University of Louvain from the end of the sixteenth century until his death and published a series of embryological works. Among these works there is a treatise, entitled *On the Forces of the Imagination* (Louvain, 1608).[4] This was dedicated to Ernest of Bavaria (1554–1612), Count Palatine, Archbishop of Cologne and Prince-Bishop of Liege, now known as an active prince-practitioner and fervent patron of new sciences and arts, including Paracelsian medicine and alchemy.[5]

Fienus's work on the powers of imagination consists of one book divided into twenty-four questions. Fifty-six conclusions, or theses, are disseminated among these twenty-four questions. His main authorities are Aristotle, Augustine and Thomas Aquinas. He also uses a wide range of ancient, medieval and early modern philosophers and physicians, including Galen, Avicenna, Ficino and Pomponazzi. As a preliminary study of this treatise, I will respect the sequence of its discussion around three major themes: (1) the relationship between soul and body; (2) the notion of imagination; and (3) the working of the mother's imagination in fetal formation.

9.2 The Soul and the Body

Fienus begins his discussion with the question as to whether the soul acts on the body (Q1).[6] According to him, the soul separated from the body is an agent *quod*, that is, the true and whole agent of its action (C1). However, this type of soul is not a subject of his inquiry. Then Fienus defines the soul united with the body as an agent *quo*, a "principle" (*principium*) or "reason-principle of acting" (*ratio agendi*) (C2). It is thanks to the soul that the animated body exerts forces on corporeal substances (C3).

Next Fienus asks which kind of bodies the soul can act on and in what way (Q2).[7] To answer these questions, he divides bodies into two groups: "proper" (*proprium*) and "foreign" (*alienum*). The proper body is a subject of the soul and is informed by the soul. This category includes all essential parts of the body, such as the heart, the brain and the hand, which are animated by the soul and in which the soul extends "formally" (*formaliter*) or in the way of a form. By contrast foreign bodies are not

---

[3]On Fienus, see Sondervorst 1958, 1–7; Rather 1967, 340–367; Papy 1999, 317–337; and Hirai 2011, 160–161.

[4]I have used the following edition: Fienus, Thomas 1608. *De viribus imaginationis tractatus*. Louvain: Gerardus Rivius. Its *quaestiones* and *conclusiones* are hereafter indicated respectively as Q and C with numbering.

[5]On Ernest of Bavaria, see Halleux and Bernès 1995, 3–29; and Xhayet and Halleux 2011.

[6]Fienus, Q1, 1–4.

[7]Ibid., Q2, 4–31.

animated nor informed by the same soul of the proper body. The fetus and the body of another person placed near the proper body fall into this category. Fienus further distinguishes foreign bodies into two kinds: "internal" (*internum*) and "external" (*externum*). Internal foreign bodies are the things residing inside the animated body such as blood, humors, spirits, seeds and the fetus, while external foreign bodies are all corporeal entities placed near the proper body.

On the basis of this classification, Fienus argues that the soul has the power of locomotion in the proper body (C4). The animated body is thus directly moved by the soul as its principle of motion. Fienus then addresses the power of change by stating that the soul cannot directly alter the proper body (C5). For him it is only "accidentally" (*per accidens*) that the soul can produce changes in the proper body.

As for internal foreign bodies, Fienus argues that the soul cannot directly change their form (C6). But he accepts the idea that the soul can change them by producing some effects inside them "vitally" (*vitaliter*) and "efficiently" (*effective*) through the temperament of bodily parts and native heat (C7). To explain the issue further, Fienus adds that the soul can change internal foreign bodies such as humors and spirits by the power of concoction (C8). For him the soul can also move these bodies without "informing" them (C9). Indeed he thinks that the soul can directly move humors and spirits as its corporeal instruments or assistants (C10). To elucidate the relation between the soul and these corporeal instruments, Fienus affirms under the authority of Thomas Aquinas, the Coimbra Jesuits and Francisco Vallés (1524–1592) that the soul can "spiritually" (*spiritaliter*) move humors and spirits in the animated body (C11).[8]

Turning to external foreign bodies, Fienus wonders if the soul can also move them. He finds an impressive series of authors, mostly of a Platonic tendency, such as Avicenna, Algazali, Albertus Magnus, Ficino, Pomponazzi and Paracelsus (1493/94–1541), who all ascribed an extraordinary power to the soul and especially to its faculty called "imagination" (*imaginatio*), able to act even at a long distance. However, relying on Aristotle, Thomas Aquinas, Francisco Suárez (1548–1617), the Coimbra Jesuits and Vallés, Fienus clearly rejects this position. For him the soul cannot directly move external foreign bodies or bodies placed at a distance (C12).[9] Like locomotion he also denies that the soul can alter or produce these bodies at a distance (C13). Note that here the notions of sympathy and occult properties, celebrated by Renaissance natural philosophers and physicians such as Girolamo Fracastoro (1478–1553) and Jean Fernel (1497–1558), are totally out of consideration.[10]

Next Fienus asks through which type of powers the soul causes movements and actions in both the proper body and internal foreign bodies (Q3).[11] He first argues that the soul cannot directly act upon these bodies but only through "superimposed"

[8]On Vallés, see Solana 1941, 297–347; Zanier 1983, 20–38; Piñero et al. 1988; and Martin 2002, 1–30.

[9]On Suárez, see especially Hill and Lagerlund 2012.

[10]On these notions, see Hirai 2011, 50–51, 73–75 and 104.

[11]Fienus, Q3, 32–39.

(*super-additae*) powers as its instruments (C14). Among these instrumental powers (C15) Fienus enumerates two: (1) a power called "animal" (*animalis*) by physicians and "motive" (*motiva*), "voluntary" (*voluntaria*) or "progressive" (*progressiva*) by philosophers; and (2) a power called "natural" (*naturalia*) or "involuntary" (*involuntaria*). According to Fienus, animal power resides in the brain and is communicated to the whole body through nerves. This power follows sense and desire and obeys the will. By contrast natural power, residing everywhere in the body, cannot follow sense and desire like animal power, but only assumes a certain natural inclination. On the basis of this classification, Fienus explains that the soul moves the proper body, firstly on its own, secondly through animal power and lastly through natural power (C16). He adds that the soul can also move internal foreign bodies such as humors and spirits though natural power, while the soul's animal power is responsible for their movements only accidentally (C17). As for external foreign bodies, Fienus clearly rejects, once again, the idea that the soul can locally move them or alter them (C18). There is no room for action at a distance for the soul's animal and natural powers.

9.3 Imagination or *Phantasia*

After animal and natural powers, Fienus turns to the soul's higher cognitive and intellectual powers. Here his discourse revolves around "imaginative" (*imaginativa*) power (Q4).[12] Thus he proceeds to the main subject of his treatise: the question as to whether the imagination or *phantasia* can act upon bodies. Fienus first establishes a long list of authorities, who gave an affirmative answer to the issue. Disagreeing with them, he asserts his own view that the soul cannot act upon the body through imaginative power (C20). For him the soul's cognitive powers, derived from cognitive "species" (*species*), are merely passive and cannot produce the movements or changes of bodies even though these powers vitally bring forth higher functions such as judgment and cognition.[13]

Fienus goes further into his inquiry. For him if the imagination can act upon bodies and produce extraordinary effects as is widely believed, it must execute them through a certain movement, either locomotion, change, growth or generation. If the imagination changes bodies, it must alter them "essentially" (*essentialiter*). To verify these points, he concentrates on the problems of locomotion and alteration.

---

[12]Ibid., 39–49. The most important study on the medieval and Renaissance belief in the imagination's extraordianry powers is Zambelli 1996, 53–75.

[13]On the notion of species in the Aristotelian tradition, see especially Spruit 1994–1995. Cf. Lindberg 1976.

First Fienus wonders if the imagination can locally move bodies (Q5).[14] His answer is that this power cannot locally move the proper body (C21) since it is not a cause of natural motion. Internal foreign bodies such as humors and spirits cannot be locally moved by the imagination itself either (C22). However, Fienus expresses a reservation by conceding that this power can cause the movements of humors and spirits "in a certain way" (*in aliquo modo*). According to him, since humors and spirits are directly moved by the soul's natural and vegetative powers, the imagination cannot move them directly but "accidentally." So he suggests that this power can only "direct" (*dirigere*) the movements of humors and spirits in a certain way although it cannot be the efficient cause of these movements (C23).

After locomotion comes alteration (Q6).[15] Although he observes Aristotle, Avicenna and Thomas Aquinas holding the idea that bodies can be altered by the imagination, Fienus argues that this power cannot change them formally or by way of a form (C24). In his opinion the imagination can cause alteration in bodies only accidentally and indirectly, that is, through intermediate factors such as "emotions" (*passiones animi*) and the movements of humors and spirits.

So Fienus wonders whether the imagination itself has no force of acting at all (Q8).[16] His answer is that the imagination can "only accidentally" cause diverse effects, perturbations and alterations in bodies (C27). Indeed these changes are produced by locomotion or alteration, which are not directly caused by the imagination. This leads Fienus to suggest that this power can only influence the soul's other powers, which are capable of moving and changing (C28). He concludes that the imagination can accidentally produce changes in bodies through the soul's lower animal and natural powers (C30). Fienus's use of the term "accidentally" tends to signify "indirectly." Thus the key to understanding his reasoning is the intermediate factors bridging bodies and the imagination. As a simple cognitive power the imagination can only "indirectly" act upon bodies.

What are these intermediate factors? Fienus suggests that the imagination can act upon bodies through desire or emotions (C31). This is not self-evident. Indeed here he is relying on the authority of Thomas Aquinas, Suárez and the Coimbra Jesuits, who regarded emotions as the alterative causes of the body. This is supplemented by another thesis that the imagination can change bodies by the movement of humors and spirits (C32). By combing these theses, Fienus tries to confirm the significance of intermediate factors. According to him, although the imagination itself cannot change bodies through desire or emotion, it can indirectly alter them as follows: desire moves and stimulates the natural powers of organs, especially those residing in the heart, which in their turn move humors and spirits.

That is not all. Fienus adds a third thesis, which makes his discussion more interesting in terms of embryology. First he just suggests that the imagination can change bodies not only through desire and the motive powers of the heart but also

[14]Fienus, Q5, 49–60.

[15]Ibid., Q6, 60–78.

[16]Ibid., Q8, 83–85.

"in another way" (C33). He avows that by this thesis he is disagreeing with Thomas Aquinas, who acknowledges only desire, emotions and cardiac movements. It is exactly here that Fienus refers to the tales of extraordinary changes provoked in the fetus by the mother's imagination. To explain these phenomena, he introduces his theory of "conformative power" (*conformatrix potentia*). This is an adaptation of Galen's idea of "formative power" (*dunamis diaplastikê*).[17] In its Latin form *virtus formativa* or *informativa*, it had been extensively discussed in the tradition of medieval medicine thanks to Arabic authors such as Avicenna and Averroes and their Latin followers such as Albertus Magnus and Pietro d'Abano (1257–ca. 1315). In the Renaissance this notion experienced a fresh rebirth in the intense debates of leading medical humanists such as Nicolò Leoniceno (1428–1524) of Ferrara and Jacob Schegk (1511–1587) of Tübingen.[18] Fienus introduces his theory so as to establish that the imagination can change bodies by directing and moving conformative power (C34). As a synthesis of this line of development, he concludes that the imagination can change the proper body by means of emotions only with the intervention of conformative power (C35). For him the imagination can "direct" the action of conformative power, but when the latter is extinguished or ceases to operate in the body, the imagination alone can no longer alter the body. As for external foreign bodies (Q12), Fienus denies that the imagination can directly act upon them or change them at a distance (C39).

9.4 Maternal Desire and Imagination in Embryology: Imagination and Signs

As has been shown, in Fienus's view, the imagination cannot directly act upon bodies. How about alterations occurring in internal foreign bodies, especially the fetus residing in the mother's womb? From here on, Fienus's discussion revolves around the imagination's impact on the fetus. He first asks how and why this power can imprint changes, also called "marks" (*notae*) and "signatures" (*signaturae*) or signs, on the fetus (Q14).[19] According to him, this question greatly attracted the attention of many people, and leading physicians and natural philosophers acknowledged the imagination's impact on the fetus. Galen, for example, argued in his *On Theriac to Piso* that species conceived by the imagination are transferred to the fetus by means of blood and spirits.[20] To reinforce this testimony, Fienus

[17]For Galen I have used *Galeni opera omnia*, ed. Karl G. Kühn (Leipzig, 1821–1833; repr. Hildesheim: Olms, 1965). On the notion of "molding faculty," see Galen, *De semine*, 2.2, 2.5 = ed. Kühn, *Opera*, IV: 611, 642; *De temperamentis*, 2.6 = ed. Kühn, *Opera*, I: 635–636; *De naturalibus facultatibus*, 1.6, 2.3, 2.6 = ed. Kühn, *Opera*, II: 15, 86, 101.

[18]I have fully examined these debates in Hirai 2011, especially in Ch. 1, 3 and 6.

[19]Fienus, Q14, 122–149.

[20]Galen, *On Theriac to Piso*, 11= ed. Kühn, *Opera*, XIV: 253–254.

quotes Pliny the elder who reported that species can produce a real entity similar to that conceived in the mind by the imagination.[21] Then he turns to Avicenna who taught in *On Animals* that the imagination imprints species on spirits in the brain, which are mingled with the blood that feeds the fetus.[22] As such the species conceived by the imagination are transferred to the fetus through spirits and blood. A more important witness is Augustine, reported by Vallés in his *De sacra philosophia* (1587). Commenting on a passage from *Genesis*, according to which Jacob transformed the colors of lambs, Augustine argued in *City of God* that these animals spiritually received the species which later appeared corporeally in their fetus.[23] As for Renaissance authors, Fienus quotes, among others, Ficino's *Platonic Theology*, in which the imagination is said to imprint signs on the fetus through the vibration of spirits transmitted through nerves.[24]

Unlike these authorities, Fienus tries to defend the idea that the mother's imagination cannot directly change the fetus (C41). For him it is obvious from the preceding discussions. If the imagination can directly act upon internal foreign bodies, it must act by moving humors and their parts spatially, which has been rejected. The imagination, regarded as the instrument of the soul, cannot act upon bodies in this way. Fienus adds that the imagination is neither active nor can formally change internal foreign bodies. Since it cannot locally move the humors and spirits of the mother, this power cannot move those of the fetus either. Since the humors of the fetus are moved by the soul of the fetus, that of the mother is not responsible for their movement.

According to Fienus, the imagination is a cognitive power and is merely passive. If its naked and pure "actuality" (*actus*) or energy were the efficient cause of changes in the fetus, every strong imagination caused by the mother would always produce these changes and imprint marks on the fetus. But this does not happen and is therefore false. Then Fienus returns to his theory of conformative power. For him this power is the direct efficient cause of the fetus and of all its properties such as colors, figures, quantity, number and position. That is why this power must also be the direct cause of marks imprinted on the fetus. Thus there is no direct role to play for the imagination because more than one cause per action is superfluous.

Fienus provides another set of reasoning. First if the imagination directly forms the fetus, it must reside in the seed or fetus as the agent of this action, whereas the imagination is not intrinsic either to the seed or to the fetus. Thus it is not the principle of fetal formation. Second, if the imagination produced in the brain forms the fetus in the uterus, this power must act at a distance without any intermediary. This is not possible. Third, if the imagination changes color in the fetus according to the color imagined by the mother, it must be the efficient cause of this color. But as

---

[21]Pliny, *Natural History*, 7.12.

[22]Although Fienus cites Avicenna, *On Animals*, 5.1, I have been unable to identify his quotation.

[23]See Vallés 1587, 1.11, 125–133. Cf. *Genesis*, 30.41; Augustine, *The City of God*, 12.26 and 18.5. See also Augustine, *Against Julian*, 5.14.

[24]Ficino 1482, 13.1 = Ficino 2001–2006, IV: 111–113. Cf. O'Brian 1993, 8–9.

the imagination does not act at a distance, it must produce the same color everywhere between the mother's brain and the fetus in her womb. Fourth, if the mother's imagination produces certain effects through the movements of the fetus's humors and spirits, it must also move all the intermediate humors in the mother's body, which stretch from the brain to the fetus. But this does not happen.

After these discussions, Fienus addresses a special kind of species, called "phantastic species" (*species phantastica*), which has traditionally been held as an intermediary for the imagination. According to him, even this special species cannot produce changes in the fetus because it does not cause a real change of quality (C42). Here Fienus recognizes that he is disagreeing with the authorities mentioned above, especially Augustine. But if this species causes a real quality change, which directly affects matter, every species strongly conceived in the mother's mind by her imagination must always change the fetus. Although many mothers during pregnancy intensely imagine various things, they do not always imprint particular signs on their fetus.

Fienus explains that emotions and the movement of humors and spirits, mentioned by Thomas Aquinas, can produce certain changes not in the fetus but around the fetus (C43). As the fetus is tied to the mother's body, its humors and spirits are linked to those of the mother and are derived from her. When the mother's humors and spirits are agitated, those of the fetus must also be affected. But Fienus thinks that these alterations are rather deformities, not the particular signs determined after the similarity of things imagined by the mother. For him the production of particular signs cannot be ascribed directly to emotions and the movements of spirits and humors.

9.4.1 Conformative Power

Following his usual *modus operandi*, Fienus suggests that signs similar to the things imagined by the mother are produced "in another way" (C44). As he knows that this disagrees with the view of Thomas Aquinas, he concedes that emotions "concur" in the production of signs. But experience teaches him that it happens rarely. Thus he argues that emotions concur only as a remote cause, while the direct cause of the production of signs in the fetus must be sought elsewhere. According to Fienus, the immediate effect of emotions is the movement of humors and spirits. Transmitted to blood, this movement produces the irregular movements of blood, which cannot establish particular marks well determined in color, figure, position or size. Thus another direct factor is required.

This reasoning leads Fienus to hold that determined particular marks cannot be established without the determined action of conformative power. Since emotions and desire cannot directly affect this power, he wonders through which intermediary they can influence it. Not through species because, according to him, desire does not have species as it is not a cognitive power. Not through emotions because one cause cannot determine various effects as diverse signs follow a single emotion of terror. Not through the movement of humors and spirits because their movement is not determined but irregular.

Since the production of particular signs cannot be ascribed to emotions alone or to the disordered movements of humors and spirits, Fienus concludes that they are produced by conformative power. How? According to him, this power tunes bodies in a determined way to the "directive" (*dictamen*) and "direction" (*regula*) given by the imagination. The mother's imagination imprints particular marks on the fetus by directing conformative power because the latter as the direct efficient cause is responsible for all kinds of deformities, marks and errors in fetal formation (C45). Indeed it is the cause of everything, either good or bad, that happens in this process. The mother's imagination can only corrupt this power's functions. Fienus argues:

> Indeed the imagination sometimes deviates conformative [power] from a right conformation that it should establish according to the species and nature of the seed in which [this power] resides. It determines [this power] to conform something according to the similarity to itself or to that which is imagined.[25]

Fienus argues that since conformative power produces particular signs in a determined way, it can rightly be taken as their direct efficient cause. That is why the products of this power are considerably different from those produced by emotions.

Although the mother's imagination is widely believed to be able to imprint particular signs on the fetus, Fienus does not accept it as the direct cause of these signs. According to him, when conformative power is modified from its normal state, it produces particular signs. Fienus ascribes the cause of this modification to the imagination because these signs follow it. He adds that the imagination cannot alter conformative power if it does not act as its director. Thus the imagination is defined as the director of conformative power.

Fienus provides another line of reasoning. If particular signs are not produced in the fetus by the action of conformative power but by emotions and the movement of humors and spirits, these signs can also be made in the mother's body. But no trace of such signs is found in her. According to Fienus, although conformative power is actually operating in the fetus, it has already ceased to work in the mother's body. That is why the imagination, which alters and directs this power, produces no sign there. So Fienus argues:

> What else can be a reason why these [signs] are produced in the fetus or in a growing person and not in adults or persons already grown up, if it is not that conformative power?[26]

The idea that the formative agent ceases to act and disappears upon the achievement of its work can be traced back to Aristotle's argument in his *Generation of Animals*,

[25]Fienus, Q14, 142: "Verum phantasia aliquando seducit conformatricem a recta conformatione, quam secundum speciem et naturam seminis, in quo est, facere deberet: et determinat ad aliquid conformandum secundum similitudinem sui, seu eius quod est imaginatum."

[26]Ibid., Q14, 143: "Quae enim posset esse alia causa, cur illae in foetu seu homine fiente fierent, in adultis et iam factis non; si non esset ipsa conformatrix?"

in which the formative agent is compared to a coagulant such as fig juice that curdles milk to produce cheese.[27] This argument was well known to Renaissance embryological authors.

9.4.2 Imagination as the Director of Conformative Power

Fienus maintains that conformative power, which forms the fetus, can be directed and altered by the mother's imagination in the production of particular signs. How, precisely, does it happen? This is the question that he addresses next (Q15).[28] His answer is immediately delivered: "the imagination can direct conformative power through species 'in the way of a model' (*exemplariter*)" (C46). Indeed he builds this answer on the basis of the idea found in Avicenna's *Canon*, which explains the cause of resemblance between parents and offspring as follows:

> Not so distanced from a likely judgment, some of the sages said that among the causes of resemblance there is one that is exemplified from a human form [conceived] in the mind of a woman or a man by a strong exemplification at the moment of conception.[29]

A commentary by the fourteenth-century physician Mondino de' Liuzzi (ca. 1270–1326) is useful to elucidate this obscure passage. According to Mondino, "what is exemplified" (*quod exemplificatur*), that is, "imprinted" (*imprimatur*) or "affixed" (*figitur*), "comes from a human form" (*est illud ex forma humana*) conceived "in the mind of a women or a man at the moment of conception" (*in mente mulieris aut viri cum est dispositio conceptionis*).[30] By interpreting the term "exemplification" (*exemplificatio*) as "impression" (*impressio*), Mondino continues by arguing that when a certain image is produced through the imagination, spirits and heat are moved to imprint the imagined thing on the fetus. Referring to the same passage of Avicenna's *Canon*, his elder contemporary Pietro d'Abano also connected the notion of "exemplification" to the imagination.[31]

According to Fienus, Avicenna hinted by the term "exemplification" that the species of things imagined by the parents causes the resemblance of offspring. He deduces that the imagination directs conformative power by way of a "model"

---

[27]See Aristotle, *Generation of Animals*, 2.3, 737a13–16. Cf. Hirai 2011, 98.

[28]Fienus, Q15, 144–154.

[29]Avicenna, *Canon*, 3.21.1.2 (Venice: Giunta, 1555), 381GH: "Et dixerunt quidam sapientum, et non sunt elongati a iudicio possibilitatis, quod de causis assimilationis est illud quod exemplificatur, cum est dispositio conceptionis in mente mulieris aut viri ex forma humana exemplificatione firma." A French translation from Arabic of this passage, in de Koning 1903, 768, reads as follows: "Quelques-uns des savants disent—et qu'il n'est pas invraisemblable que cela soit possible—qu'une des causes de la ressemblance est une forme humaine que l'homme ou la femme a dans l'imagination d'une manière vive au moment de la conception." See also Weisser 1983, 316.

[30]de' Liuzzi 1993, 135. On Mondino and his embryology, see also Vico 2002, 111–121.

[31]See d'Abano, *Conciliator*, diff. 37 (Venice: Giunta, 1565), f. 56Hb.

(*exemplar*). For him this is the real meaning of Avicenna's theory of exemplification. The image of a signet ring might help illustrate the underlying idea.

In any event, referring to Augustine's *On the Trinity*, Fienus interprets somewhat forcefully that Augustine meant by the term "ruler" (*regula*) not "to make a line" or "to establish the lineament of the fetus" but "to direct" or "rule."[32] So the next problem is as to what this "direction" is and how it works. To resolve these issues, Fienus enumerates three ways of direction: (1) through the imagination's "command" (*imperium*) over conformative power; (2) through a singular "sympathy" (*sympathia*) with conformative power; and (3) through the communication of species. According to him, by the term "command" he means something like a "political command" (*imperium politicum*). Then, by the communication of species, he means that species perceived by the imagination are imprinted on the mother's spirits and blood and, by a certain continuity, on the fetus's spirits and blood. By way of a model these species arrange and direct conformative power to modify the fetus in accordance with the "exemplification" and determination of these species.

After this explanation, Fienus rejects the first option under the authority of Thomas Aquinas, Petrus Fonseca (1528–1599) and the Coimbra Jesuits. In their interpretation of Aristotle's *On the Soul*, although the "will" (*voluntas*) commands the soul's other rational powers, it has no command over the vegetative or natural powers. Note that Fienus clearly regards conformative power as a kind of vegetative power. Next he denies the option of sympathy. There remains only the third option: the communication of species.

Here Fienus says: one might object that only cognitive power can act according to a model, whereas conformative power, regarded as a kind of vegetative power, is not cognitive and cannot act after a model. To this objection, Fienus answers by dividing cognition into two kinds: (1) true cognition; and (2) natural cognition or "impulse" (*instinctus*). According to him, vegetative powers, including conformative power, act in the second way, just as plants find their nourishment in the earth. That which acts by a model need not capture something through "true cognition." Fienus concludes that conformative power recognizes a model by its *instinctus*. However, he does not forget to express a reservation by arguing that the *instinctus* of conformative power must be ascribed to "divinity" (*divinitas*) because this power performs the extraordinary effect that cannot be executed by other vegetative powers: to recognize the representation of phantastic species in a certain way.

Fienus further explains that conformative power need not receive phantastic species within itself. These species must be contained in blood or spirits placed

[32] Augustine, *On the Trinity*, 3.8.15: "[. . .] some kind of color should be brought from the variety of the rods to the cattle that were conceived, came to pass in this way: the soul of the pregnant cattle was affected through its eyes from without, but inwardly it drew within itself, according to its own measure, the rule of formation" ([. . .] ut autem de varietate virgarum pecorum conceptorum color aliquid duceret, fecit hoc anima grauidae pecudis per oculos affecta forinsecus et interius pro suo modulo formandi regulam trahens). Cf. Fienus, Q15, 144.

close to this power so that these species as models direct this power. Models need not reside within the "imitator" (*imitans*) but it is enough to stay close. Moreover, these species need not permanently exist but their presence at the initial stage of fetal formation is sufficient.

Next Fienus asks what type of imagination is more efficient in imprinting signs on the fetus (Q16).[33] His answer is that the imagination "with assent" (*cum assensu*), followed by emotions, can produce signs more easily (C49). For Fienus although a simple imagination or intuition, without assent or "inclination" (*inclinatio*), can imprint signs on the fetus, it is more efficient if it is accompanied by strong emotions or with "reflection" (*reflexio*). That is why, argues Fienus, human beings can produce more signs than animals.

To explain how mental activity is associated with the production of signs, Fienus states that emotions concur in various ways: first by troubling and corrupting conformative power, second by reinforcing phantastic species, and third and mostly by connecting these species to conformative power (C50). For Fienus emotions concur in the production of signs as both the auxiliary cause and the necessary cause. In the former case they reinforce the power of phantastic species. In the latter case they "connect" (*applicare*) these species with conformative power. How does this connection take place? Indeed there is a long distance between phantastic species conceived in the mother's brain and conformative power working in her womb. Here, rejecting Ficino's idea that the vibration of spirits through nerves bridges the mother and the fetus, Fienus says:

> That is why I argue that species are sent to remote parts by the act of desire and emotions linked to the imagination. When a sign is produced because of the imagination conceived at the time of intercourse, species are directly sent through nerves into testicles and are immediately imprinted on the seed. When this [seed] is received in the womb, it brings these species with itself and retains them for a certain period. When conformative power starts to fashion the [fetus], this power conforms it to the similitude of these species.[34]

As this quotation suggests, it is not only the mother's imagination that imprints signs on the fetus but also the father's because conformative power first resides in the seed coming from him. Thus in a later question (Q19) Fienus affirms that not only the mother's but also the father's imagination produce signs (C52).[35] To justify this thesis, he calls upon the authority of Bible in *Genesis*, according to which Jacob transformed the colors of lambs, then Aristotle, Pliny and Avicenna.[36]

---

[33] Ibid., Q16, 154–164.

[34] Ibid., Q16, 161: "Quapropter dico, propter actum appetitus et animi passiones, quae phantasia coniunguntur, species in remotas partes irradiare. Quando signatura fit propter imaginationem habitam in hora coitus, species irradiant recta per neruos in testes, et semini immediate imprimuntur: quod semen cum in uterum recipitur, secum defert species illas, et aliquanto tempore retinet, et cum formatrix incipit illud efformare, format ad similitudinem erarum."

[35] Ibid., Q19, 172–173.

[36] For Bible, Pliny and Avicenna, see respective notes above. See also Aristotle, *Problems*, 10.10, 891b33–37: "Is it because man's mental condition is more varied at the moment of sexual intercourse, and so the offspring varies according to the condition of male and female parents?"

After all this development, Fienus concludes that both the mother's and the father's imagination can imprint signs on the fetus at the conception. As he recognizes that the mother is always mentioned but the father almost never, he tries to provide reasons: (1) the production of signs at the conception is rarer than during pregnancy, while the father can produce signs only at the conception; (2) the mother's imagination is stronger than the father's since women are observed to be more easily scared and more heavily frightened; (3) and lastly the imagination provoked during intercourse can last longer in women than in men. Thus by acknowledging the father's role, though much weaker and less prominent than the mother's, Fienus tries to redress the exclusive right traditionally assigned to women. In this sense his theory might have contributed to the demystification of female psychic power in embryology.

9.5 Conclusion

Thus far I have outlined the main arguments of Fienus in his theorization of a widespread belief, according which pregnant women's strong emotions of desire or fear and imagination could imprint particular morphological traces on the fetus in their womb. Although Fienus stressed the importance of experience, he rather relied on a wide range of sources provided by ancient, medieval and early modern philosophers and physicians. For the aspects related to psychology, he was heavily dependent on the Latin commentary tradition of Aristotle's works, perpetuated by a series of authors from Thomas Aquinas, via Suárez to the Coimbra Jesuits. Fienus modified their theories by adopting the ideas of physicians such as Vallés. It is also noteworthy that he was particularly attracted by the teachings of Avicenna at crucial points in the construction of his system. By contrast Fienus discarded a set of recent authors, whose ideas were of a Platonic orientation, such as Ficino, Fracastoro, Fernel and Paracelsus.

Fienus tried to reinterpret the widespread belief in the power of imagination and maternal desire in the framework of the learned medicine of his time. In this endeavor his appeal to the notion of formative power is particularly noteworthy because his immediate forerunners among physicians and natural philosophers with a medical background such as Leoniceno and Schegk carried out intense debates on this notion throughout the sixteenth century. Fienus was a direct heir to this tradition, which was developed at the heart of medical humanism. His embryological theory of imagination might appear odd to the eyes of modern readers but can be understood in this historical and intellectual context.

Bibliography

Angelini, Massimo. 1994. Il potere plastico dell'immaginazione nelle gestanti tra XVI e XVIII secolo: la fortuna di un'idea. *Intersezioni* 14: 53–69.

Avicenna. 1555. *Canon*. Venice: Giunta.

Bundy, Murray Wright. 1927. *The Theory of Imagination in Classical and Medieval Thought*. Urbana: University of Illinois Press.

Cocking, John Martin. 1991. *Imagination: A Study in the History of Ideas*. London: Routledge.

d'Abano, Pietro. 1565. *Conciliator*. Venice: Giunta.

de' Liuzzi, Mondino, 1993. *Expositio super capitulum de generatione embrionis Canonis Avicennae cum quibusdam quaestionibus*, ed. Romana Martorelli Vico. Rome: Istituto Palazzo Borromini.

de Koning, Pieter. 1903. *Trois Traités d'anatomie arabes*. Leiden: Brill.

Fattori, Marta, and Massimo L. Bianchi, eds. 1988. *Phantasia-imaginatio*. Rome: Ateneo.

Ficino, Marsilio. 2001–2006. *Platonic Theology*, eds. and trans. Michael J.B. Allen and James Hankins, 2 vols. Cambridge, MA: Harvard University Press.

Fienus, Thomas. 1608. *De viribus imaginationis tractatus*. Louvain: Gerardus Rivius.

Galen. 1821–1833. *Galeni Opera Omnia*, ed. Karl G. Kühn, 20 vols. Leipzig; repr. Hildesheim: Olms.

Halleux, Robert, and Anne-Catherine Bernès. 1995. La cour savante d'Ernest de Bavière. *Archives internationals d'histoire des sciences* 45: 3–29.

Hill, Benjamin, and Henrik Lagerlund, eds. 2012. *The Philosophy of Francisco Suárez*. Oxford: Oxford University Press.

Hirai, Hiro. 2011. *Medical Humanism and Natural Philosophy: Renaissance Debates on Matter, Life and the Soul*. Leiden: Brill.

Lindberg, David C. 1976. *Theories of Vision from Al-Kindi to Kepler*. Chicago: University of Chicago Press.

Martin, Craig. 2002. Francisco Vallés and the Renaissance Reinterpretation of Aristotle's *Meteorologica* IV as a Medical Text. *Early Science and Medicine* 7: 1–30.

O'Brien, John. 1993. Reasoning with the Senses: The Humanist Imagination. *South Central Review* 10: 3–19.

Pancino, Claudia. 1996. *Voglie materne: storia di una credenza*. Bologna: CLUEB.

———. 1997. La croyance aux envies maternelles entre culture savante et culture populaire. *Ethnologie française* 27: 154–162.

Papy, Jan. 1999. The Attitude towards Aristotelian Biological Thought in the Louvain Medical Treatises during the Sixteenth and Early Seventeenth Century: The Case of Embryology. In *Aristotle's Animals in the Middle Ages and Renaissance*, ed. Carlos Steel et al., 317–337. Louvain: Leuven University Press.

Pennuto, Concetta. 2008. *Simpatia, fantasia e contagio: il pensiero medico e il pensiero filosofico di Girolamo Fracastoro*. Rome: Storia e letteratura.

Piñero, Lopez, José María, and Francisco Calero. 1988. *Las Controversias (1556) de Francisco Valles y la medicina renacentista*. Madrid: Consejo superior de investigaciones scientificas.

Rather, Lelland J. 1967. Thomas Fienus' (1567–1631) Dialectical Investigation of the Imagination as Cause and Cure of Bodily Disease. *Bulletin of the History of Medicine* 41: 340–367.

Solana, Marcial. 1941. *Historia de la filosofía española: época del Renacimiento (Siglo XVI)*. Madrid: Real Academia de ciencias exactas.

Sondervorst, François-André. 1958. Vie et ouvrages des Feyens d'Anvers. *Ypperman: Bulletin de la société belge d'histoire de la médecine* 5: 1–7.

Spruit, Leen. 1994–1995. *Species Intelligibilis: From Perception to Knowledge*, 2 vols. Leiden: Brill.

Vallés, Francisco. 1587. *De iis, qui scripta sunt physice in libris sacris, sive de sacra philosphia*. Turin: Heirs of Nicolas Beuilque.

Vico, Romana Martorelli. 2002. *Medicina e filosofia: per una storia dell'embriologia medievale nel XIII e XIV secolo*. Milan: Guerini.

Weisser, Ursula. 1983. *Zeugung, Vererbung und pränatale Entwicklung in der Medizin des arabisch-islamischen Mittelalters*. Erlangen: Lüling.

Xhayet, Geneviève, and Robert Halleux, eds. 2011. *Ernest de Bavière (1554–1612) et son temps: l'automne flamboyant de la Renaissance entre Meuse et Rhin*. Turnhout: Brepols.

Zambelli, Paola. 1996. L'immaginazione e il suo potere: desiderio e fantasia psicosomatica o transitiva. In *L'ambigua natura della magia: filosofi, streghe, riti nel Rinascimento*, ed. Paola Zambelli, 53–75. Venice: Marsilio.

Zanier, Giancarlo. 1983. *Medicina e filosofia tra '500 e '600*. Milan: FrancoAngeli.

Chapter 10
Gerardus Blasius and the Illustrated Amsterdam *Observationes* from Nicolaas Tulp to Frederik Ruysch

Domenico Bertoloni Meli

In 1677 Amsterdam anatomist and physician Gerardus Leonardus Blasius (or Blaes) published an illustrated collection of *Observationes medicae rariores*, a genre involving reports on rare individual cases that was becoming increasingly popular in those years. His small octavo book consists of 120 pages of text and nine plates with small and rather undistinguished figures, and an appendix on three monstrous births. The text presents just under 100 cases, often, though by no means always, ending with the patient's death and a postmortem. At first it appears unremarkable, and not surprisingly it has attracted little attention.[1]

Yet, I shall argue that Blasius's *Observationes* belongs to a genre that is of considerable interest with regard to medical practice, clinical teaching, and visual representations at the time in Amsterdam and beyond. In order to appreciate the significance of Blasius's work, besides providing a biographical sketch of its author, I discuss it as part of the genre of illustrated *Observationes*. Those we are going to examine were called *medicae*, *chirurgicae*, or *anatomicae*: the first two qualifications usually involved a diseased state often described in a case history that could end with the patient's death and the ensuing postmortem, or with a successful cure; the third involved a dissection, whether of a diseased or healthy human body or of an animal. Here I privilege surgical and medical cases

I thank Joel Silver at the Lilly Library of Indiana University and Ron Sims at the Galter Library at Northwestern for support and assistance. I am grateful to Ellen Dwyer, Ashley Inglehart, Cindy Klestinec, Evan Ragland, Bret Rothstein, Allen Shotwell, and the IU history of medicine reading group for comments on a previous version of this paper. Special thanks to Eric Jorink, who kindly offered a number of valuable comments on an earlier version of this essay.

[1] Blasius 1677, 36–7, Observatio 2, for example, deals with a cleft palate.

D. Bertoloni Meli (✉)
Indiana University, Bloomington, Bloomington, IN, USA
e-mail: dbmeli@indiana.edu

G. Manning, C. Klestinec (eds.), *Professors, Physicians and Practices in the History of Medicine*, Archimedes 50, DOI 10.1007/978-3-319-56514-9_10

and illustrations, rather than purely anatomical ones, which have been analyzed in histories of anatomy.

I study in particular the illustrated *Observationes* published in Amsterdam in the half century between the celebrated collections by Nicolaas Tulp (1641) and Frederik Ruysch (1691), which mark the endpoints of my narrative. Both authors were prominent Amsterdam physicians who taught anatomy to the surgeons' guild. Tulp (1593–1674), immortalized by Rembrandt while teaching anatomy, was one of the most visible and influential figures in Amsterdam, who repeatedly served as one of the city's burgomasters; he set the scene with *Observationum medicarum libri tres* (Amsterdam, 1641). Although collections of *Observationes* had appeared at many locations in the previous decades, his work inaugurated the illustrated Amsterdam tradition and was widely imitated by later medical men. Fostered by its thriving book and publishing trade and by an emphasis on practical matters, the Dutch commercial hub became the main center for this genre, which also benefited from the emphasis on visualization that was especially strong in the Netherlands as a whole. Ruysch (1638–1731) relied on his refined injection techniques and other methods of preservation to assemble the main anatomical museum of its age, which became the basis for many of his publications, including his *Observationes anatomico-chirurgicae*. After Ruysch, illustrated collections of *Observationes* seem to have become less frequent, possibly because of the increasing competition of medical and philosophical journals and the decreasing emphasis on remarkable and "monstrous" cases.[2]

My investigation of Blasius's work in the context of contemporary illustrated Amsterdam *Observationes* addresses a few questions: What was included in the *Observationes* and what do they teach us about the city's medical world? Why were *Observationes* illustrated and what do the illustrations show? What is the status of Blasius's *Observationes* within this genre?

10.1 Illustrated Medical, Surgical, and Anatomical *Observationes*

The early modern period saw the emergence of a new literary genre, the collection of *Observationes*. In a recent impressive series of articles Gianna Pomata has documented this process and its intellectual significance. Briefly, *observationes* and their cognates, *curationes* and *historiae*, stem from the empirical tradition and focus on descriptions of individual cases, often eschewing extensive causal explanations. Some *observationes*, however, may carry implicit broader theoretical and philosophical implications that would have been obvious to a contemporary audience, even if they were not openly discussed. Pomata's investigations bear

[2]Cook 2007; Pomata 2010, 2011a, b, "A Word of the Empirics" and "Observation Rising"; and Park 2011.

profound consequences on the broad transformations of knowledge in the early modern period and the emergence of a new epistemology privileging natural history over more theory-oriented natural philosophy. Here I focus on a geographically and thematically limited set of texts and issues sharing an interest in illustrations.[3]

The illustrated *Observationes* discussed here are usually rather brief, a few pages at most. Most of those that concerned medical cases and postmortems were collegial affairs including physicians, surgeons, and at times even their students. The point of the *observatio* was to record something unusual, remarkable, or even "monstrous" – a term that was used with a range of meanings. Together with the extensive collections of *Historiae anatomicae* and *Epistolae medicinales* by Danish physician Thomas Bartholin and the new philosophical and medical journals, such as the *Philosophical Transactions*, the *Acta medica et philosophica hafniensia*, and the *Miscellanea curiosa medico-physica*, the Amsterdam collections of *Observationes* were the richest repositories of illustrated case histories and postmortem reports of the time.[4]

While I shall not attempt to discuss all the texts listed below, the following synopsis documents my claim about the role of Tulp's 1641 collection; a list of seventeenth-century Amsterdam *Observationes* can be found in Appendix 2. Tulp's work went through several subsequent editions, notably a Dutch translation where the term *observationes* was rendered as *aenmerkingen* (Amsterdam, 1650), and an expanded Latin edition, *Observationes medicae* (Amsterdam, 1652). His associates, surgeons Hendrik van Roonhuyse (1625–1672) and Job van Meek'ren (1611–66), published similar works in Dutch under the title of "medico-chirurgical" or "medical" "aanmerkkingen" (with variant spellings). Van Roonhuyse published two works, both dedicated to Tulp: the first was on the diseases of women (Amsterdam, 1663), the other was a collection of medical and surgical observations (Amsterdam, 1672). Van Meek'ren work appeared first in Dutch (Amsterdam, 1668) and then in a Latin translation (Amsterdam, 1682) by Abraham Blasius, Gerard's son, with a dedication to Gerard. Both works by van Roonhuyse were translated into German (Nürnberg, 1674) and English (London, 1676), the one by van Meek'ren appeared also in German (Nürnberg, 1675). The interplay between Latin and vernacular editions testifies to their appeal to a mixed audience of learned scholars and practitioners.[5]

---

[3]Pomata 2010, 2011a, b and 2005.

[4]Bartholin 1654, 1657, 1661; and 1663–67. Geneva physician Théophile Bonet assembled a huge number of mostly previously published reports in *Sepulchretum* (1679); a second expanded edition was edited by Jean-Jacques Manget (1700). See Siraisi 2001a, b, "Segni evidenti" and "Signs and Evidence"; and 2013; Donato 2011; and Weber 2006.

[5]Lindeboom 1984, esp. 1294–5, 1668–9, 2004–6. In the following century the illustrated collection Cornelis Trioen, *Observationum medico-chirurgicarum fasciculus* (1743a) was rendered as *Genees- en heelkundige waarneemingen* (1743b). Hendrik van Roonhuyse's work on the diseases of women had a new expanded edition (Amsterdam, 1672). The German was *Historischer Heil-Curen in zwey Theile verfassete Anmerckungen* (Nürnberg, 1674). See also van Roonhuyse 1676; and van Meek'ren 1675. I will not discuss *Observationes anatomicae selectiores* by the *Collegium privatum Amstelodamensis*, which are exclusively devoted to zootomy.

Other illustrated works belong to the same genre: Theodor Kerckring (1639–93), *Specilegium anatomicum, continens observationes anatomicae rariores* (Amsterdam, 1670), includes a mixture of anatomical and pathological material. The *Observationes et historiae* (Amsterdam, 1674) by Amsterdam physician Justus Schrader (1646-after 1674) joins an investigation on generation inspired by William Harvey with 40 medico-anatomical observations including postmortems by a number of physicians including, besides himself, Franciscus dele Boë Sylvius, Johannes van Horne, Jan Swammerdam, and Ruysch; here only the report of a postmortem carried out by Schrader and Swammerdam is illustrated. *Observationes medicae rariores* (Amsterdam, 1677) by Blasius discusses diseased states and minor malformations. *Collectanea medico-physica* (Amsterdam, 1680, 1683, 1686) was the first Dutch scientific journal edited by physician Steven Blankaart, including "aanmerkingen" pertaining to the study of medicine and of nature. Much like the German *Miscellanea curiosa*, it was heavily illustrated; it was also translated into German (Leipzig 1680–90).[6]

Some collections appearing in Amsterdam were due to medical men active elsewhere: The Hague physician Cornelis Stalpart (1620–1702) authored *Hondert seldzame aanmerkingen* (Amsterdam, 1682), followed by additions and translations published elsewhere (The Hague, 1686; and in Latin, Leiden, 1687), in which he cited all the figures we have discussed so far, among several others. The Middelburg surgeon and physician Anton de Heide (1646-ca.1690) added a *Centuria observationum medicarum* (Amsterdam, 1683) to his study of the anatomy of mussels, containing a broad range of materials, including diseases affecting humans and animals, experiments on bone regeneration in frogs, and a study of the reproductive parts of female frogs.[7]

Lastly, Ruysch added to the catalog of his rapidly growing museum his *Observationes anatomico-chirurgicae* (Amsterdam, 1691), which are largely though not exclusively devoted to diseased states. Many of the cases discussed by Ruysch concern pregnancy, a condition he had ample opportunity to investigate since in 1669 he had replaced van Roonhuyse as examiner of the city's midwives and in 1672, following van Roonhuyse's death, he had become city obstetrician.[8]

It is not immediately clear what prompted Tulp to publish his work and to include nearly 20 plates. The notable illustrated collections of *Observationes* and *Curationes* by learned surgeon Guilhelmus Fabricius von Hilden or Hildanus (1560–1634) that appeared between 1598 and 1646 in six "centuriae", the last one posthumously, was a relevant precedent known to and cited by Tulp. Moreover,

---

[6]Lindeboom, 152–4, 156–8, 1030–2, 1773–74. See Schrader 1674, 207–8, Figures VII–VIII. I omit here the *Observationes anatomicae selectiores* (Amsterdam, 1667–73) by the Amsterdam *Collegium privatum* in that they are exclusively anatomical. See Cook 1992, 139.

[7]Lindeboom, 807–8, 1865–66. See de Heide 1684, 51–199, esp. 84–5, 112–6, 123–6, 196–9. On de Heide, see Bertoloni Meli 2013a, 219. See also Stalpart vander Wiel 1682, which translates works originally in Dutch.

[8]Lindeboom, 1700–1704. See also Kooijmans, esp. 78–9, 82, 88–90, 95–7.

it is worth recalling here that overall the cognitive orientation of surgeons favored localism: it is not by accident that so many of my protagonists were either surgeons or, as with Tulp and Ruysch, had close contacts with them. Fabricius, however, seemed more interested in highlighting his extensive contacts with prominent physicians and therapeutic skills, whereas Tulp, from the pinnacle of the medical pyramid, focused more on remarkable and curious cases.[9]

Although illustrations were frequent in anatomical publications since the time of Berengario da Carpi (1460–1530) and Andreas Vesalius (1514–64) in the first half of the 16th century, their presence in other medical works was not a given. I focus here on works representing the human body as opposed to medicinal plants, for example, though those too pertained to medicine.[10] Illustrated genres included broadsides announcing remarkable cases, from peculiar births to extraordinary diseases – such as the celebrated 1496 woodcut attributed to Albrecht Dürer, showing a case of syphilis under a celestial configuration for 25 November 1484 that would have been responsible for its appearance; surgical works displaying instruments, procedures, and occasionally lesions, as *Das Buch der Cirugia* (Strasburg, 1497) and the *Feldtbüch der Wundtartzney* (Strasburg, 1517) by surgeons Hieronymus Brunschwig and Hans von Gersdorff (ca. 1455–1529), or *Ophthalmodouleia, das ist Augendienst* ([Dresden], 1583), by Dresden court surgeon Georg Bartisch (1535–1607); lastly, some anatomical works included illustrations of unusual or anomalous cases, such as *De humani corporis fabrica* (Basel, 1543) by Andreas Vesalius, which shows on a plate different shapes of human skulls, or *Theatrum anatomicum* by Basel physician Caspar Bauhin (1560–1624), which includes illustrations of two kidneys joined together is what became known as a horseshoe kidney. Overall, whereas most anatomical publications provided an image of an idealized – usually male – body, other types of medical publications favored individual portraits of specific specimens and conditions.[11]

While there is an extensive and distinguished tradition of scholarship dealing with anatomical illustrations, other types of medical illustrations have received less attention. These illustrations served different purposes, besides seemingly making publications more appealing: they highlighted philosophical points, such as the connections between the microcosm and the macrocosm; documented and certified the authenticity of unusual cases; showed off the surgeons' fancy instrumentation; advertised the medical man's therapeutic successes; and displayed rare and extraordinary events, as in a cabinet of curiosities whose contents were as valuable as they

[9]Guilhelmus Fabricius from Hilden or Hildanus 1646. On Hildanus, see Jones 1960. Also, Tulp, *Observationes* (1652), contains XVIII numbered plates and one that is not numbered at 259; the reference to Fabricius Hildanus is at 275. See Dudok van Heel 1998; Temkin 1951; Bertoloni Meli 2015; eadem., "The Rise of Pathological Illustrations: Baillie, Bleuland, and their Collections," forthcoming in *Bulletin for the History of Medicine*.

[10]See Ogilvie 2006; Kusukawa 2012.

[11]See von Gersdorff (1517) and Bauhin 1605, 1312–1313. See also Dackerman 2011, esp. 60–3, 80–1; Siraisi 2001d, "Vesalius and Human Diversity," 287–327; and Kusukawa, 218–33.

were rare and, indeed, curious. In some cases, as with bony or stony formations, the specimens were permanent and would have been available for inspection; in other instances, however, the specimens were not permanent and therefore the illustrations provided the only lasting visual documentation.[12]

Thus Tulp's move was not a radical break with the past; nonetheless, his prominent position in Amsterdam, joined with the emphasis on visual culture in contemporary Dutch society, ranging from insects and shells to plants and maps, made his book the likely exemplar for subsequent works in the same genre, one with strong Dutch and especially Amsterdam roots.[13]

Tulp's text and illustrations form a rather heterogeneous collection; overall, he had a predilection for preternatural growths, such as polyps found in the nose and heart and urinary stones. One plate shows a *valvula intestinalis* as an anatomical feature displaying divine providence. What was available in relation to the author's practice obviously played a key role, though there were significant differences in what was deemed relevant. In *Observationes medicae*, for example, Tulp relied on both live patients and cadavers but he also included material more pertinent to natural history than to medicine, such as his illustrations of an "orangutan" and a "unicornum marinum" or narwhal tusk; two of his plates show two-headed "monsters". Other authors were to adopt more restrictive criteria, excluding purely natural historical subjects and monsters from *Observationes medicae*.[14]

Tulp's title page (Illustration 10.1), first included in the 1652 edition, is rather elegant and fairly representative of the contents of the book, in that it combines surgical procedures and anatomical practices. At the top it shows the astounding case of smith Jan de Doot bravely performing a lithotomy on himself; clearly this was an idealized reconstruction, since no medical man was present at the scene, let alone an artist. The relevant plate in the book reproduces the egg-sized bladder stone and the knife de Dood employed. The figure on the left of the title page shows a man affected by dropsy removing excess water, an operation called paracentesis, with a tool devised by a surgeon, a "penicillum ligneum"; the device is clearly depicted in the original plate inside the volume, which shows a more pronounced belly protruding with fluid. The figure on the right shows Cathalina Bonnevall, who had died of a bilateral condition whereby the "cornua uteri" had progressively filled with water over 9 years. Although Tulp describes her cadaver as "miserrimum", in the figure, apart from her specific condition, she appears rather flourishing, even more so in the plate inside the book. Following earlier conventions going back to

[12]See Impey and Macgregor 1985; Findlen 1994; Daston and Park 1998; Jorink 2010, esp. 257–399; Acheson 2010; Ogilvie 2003; Baigrie 1996; Anderson 2009; Soojung-Kim Pang 1997; Dackerman 2011.

[13]See Alpers 1983; Olmi 1992, 119–61, esp. 121; Cook 2007, 154–64. On visual genres, see Silver 2006; Swan 2008a, 239–49; eadem 2008, 199–213; eadem., in Dackerman 2011, 186–90; eadem 2008, "The Uses of Botanical Treatises."

[14]Tulp 1652, 271 and title page, 256–8, plate XIII; 258–60, the plate is not numbered, 221–7. See Daston and Park 1998; Bates 2005; Swan 2008b.

Berengario da Carpi and Adrian van der Spiegel, she is shown standing, in a Venus-like pose, her arm only partially covering her breasts. At the bottom the pensive "orangutan" from Angola contemplates its sad fate.[15]

The majority of the plates in Tulp's treatise show affected body parts. In the case of a polyp found in the heart of "Momma Baxius, homo torpidus", or sluggish, Tulp challenged Galen, who had argued that the interior of the heart is not affected by tumors. The plate shows the sectioned heart with the white polyp at *aa*, *bb*, and *cc* (Illustration 10.2). Tulp compared his case to the one studied and illustrated by Basel physician Caspar Bauhin in *Theatrum anatomicum* (Frankfurt/M, 1621), which allegedly consisted of fatty matter (Illustration 10.3, Figures 1 and 2), whereas the one Tulp described was made of pituita or phlegm: his was shown *in situ* in an open heart, Bauhin's was shown in isolation. Later in the century in *De polypo cordis* Marcello Malpighi did not include illustrations but referred to Tulp's observation and plate and cited the opinion that the polyp consisted of fatty matter or of phlegm. Tulp's and Malpighi's references to the literature document the usage of postmortem reports and images in the study of pathological conditions. As Luigi Belloni emphasized in a remarkable essay over half a century ago, the investigations of heart polyps highlight the early focus on showy and rare findings compared to subtle and more common ones, such as atherosclerosis. One may add that structural changes such as polyps and also aneurisms are easier to represent than textural ones, such as ossification.[16]

One of the celebrated cases in Tulp's treatise illustrates "spina bifida", or a split spine, a condition he named and described. Tulp started from the case history of a newborn with a turnip-shaped tumor on his lower back that he attributed to the mother's unsatisfied craving for turnips during pregnancy. This case highlights the problematic and ambiguous boundary between what we call malformations and diseased states, in this case involving maternal imagination. Whatever the cause, the condition seemed to call for treatment; a surgeon tried to remove the growth by tying it very tightly with a thread, until it died and it could be removed with a lancet, but the child succumbed before the operation could be performed. The postmortem revealed the split spine. Tulp sought to describe the spine as much as to defend the surgeon, whose name is not mentioned. He also says that he had seen a similar tumor, presumably with the same spine condition as well, six times, thus offering valuable information on the frequency of the condition. The plate shows both the child with the back partially dissected and, on the left, the split spine much enlarged. In this case Tulp states that the nerves were both protruding and torn (Illustration 10.4).[17]

[15]Tulp 1652, 341–4, at 342; 364–6, 369–71, 283–91, at 284. See Goldschmid 1925, 42–3; Park 2006, ch. 4 and 266; Sappol 2006, 16, Plate 21.

[16]See Tulp 1652, 55–60; Bauhin 1621, Appendix, Plate IX; Malpighi, "*De polypo cordis*. An Annotated Translation", by John M. Forrester 1995, 480–481; Bertoloni Meli 2001; eadem 2011, 142–9; and Belloni 1956, 28–35.

[17]Tulp 1652, 242–4. See Park 1998, 254–71; eadem 2006; Wilkin 2008.

Illustration 10.1 Tulp, *Observationes*, 1652, title page. (Courtesy of the Lilly Library, Indiana University, Bloomington)

Van Roonhuyse was associated with Tulp; his first work reflects his position of city obstetrician and instructor to midwives, in that it deals with a number of complex births. He is celebrated as the surgeon who pioneered novel procedures, such as the attempt to repair vesico-vaginal fistulae resulting from traumatic births. The repair of cleft lips was another of his specialties. One of the most striking plates in his treatise on the diseases of women shows a case of a ruptured uterus with fetus and placenta in the abdomen. The uterus is shown slightly to the right at C, with the rupture at F; the placenta and umbilical cord are slightly to the left at D and F. Unlike the case shown on Tulp's title page and treatise, here it is not immediately clear whether the mother is shown standing or horizontal Illustration 10.5.[18]

The work by Tulp's associate van Meek'ren is especially rich of illustrations that he drew himself; the book consists of a collection of cases and a posthumous appendix that opens with a short illustrated chapter on monsters. Overall, van Meek'ren had a preference for extraordinary cases, including one of an infant whose intestines had grown outside the venter, "spectaculo sane horrendo." He often relied on anatomical knowledge and performed postmortems when he had the chance. Judging from the cases he reported, historian Daniel de Moulin argued that his therapeutic approach was restrained and his scope limited. The remarkable title page (Illustration 10.6) shows an unhappy congregation of diseased patients around a corpse laying on a table; to add further gloom to an already quite dispiriting party, additional cases and surgical instruments are shown in illustrations hanging on the wall. Although van Meek'ren's title page eschews the classicizing architectural elements present in Tulp's, both include a collection of prominent cases readers would encounter inside the books. The figure at the front left is that of a Spanish young man, George Albes, affected by a disease making his skin very loose and stretchy; he was inspected by such luminaries as Johannes van Horne and Sylvius.[19]

The cadaver in the center shows the corpse of Ingeltje Thoveling (1619–1651), wife of celebrated painter Govert Flinck and daughter of a director of the East India Company in Rotterdam. In line with other contemporary cases, van Meek'ren routinely provides the names of the patients, even when, as in this case, a highly visible and wealthy lady was involved. Her corpse is shown reclining on a hard surface, with the head slightly raised. Van Meek'ren provided a case history and devoted two plates to the report. Thoveling had shown signs of the disease already during her puberty, aged 13 or 14, when her venter started to swell. The condition did not prevent her from getting married and bearing a child. Until a few days before dying, her husband reported she was able to accomplish all her domestic duties, and even to climb a tower with no problem. After her death in 1651 her body was dissected by renowned anatomist Johannes Walaeus, who found the navel cartilaginous; he emptied 125 pints of watery fluid from the peritoneum. The first

[18]van Roonhuyse 1672, 20–21. See Ricci 1949, 138–45; de Moulin 1988, 144–7.

[19]van Meek'ren, 49–51, and for the infant affected by omphalocele, 134–6. See the introduction by de Moulin to the facsimile reprint of Job van Meekeren, *Heel- en geneeskonstike aenmerkingen* 1979; Parapia and Jackson 2008, 32–5; and de Moulin, 142–5, esp. 144.

Illustration 10.2 Tulp, *Observationes*, 1652 (originally 1641), 56, heart polyp. (Courtesy of the Lilly Library, Indiana University, Bloomington)

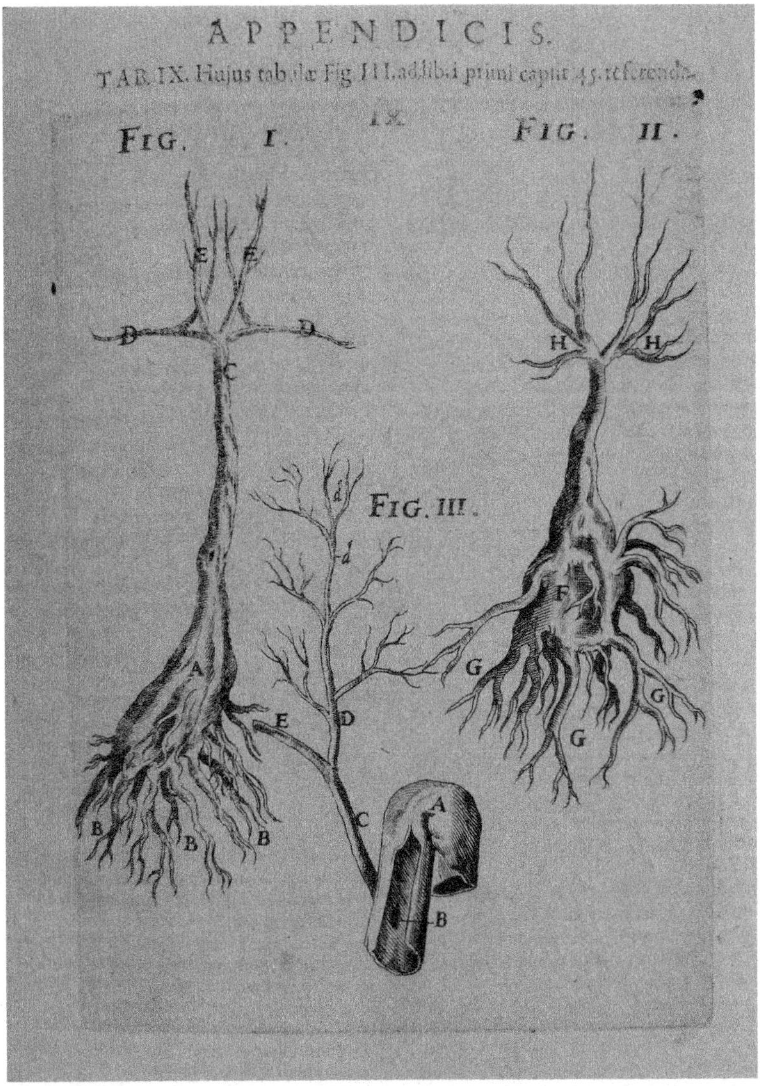

Illustration 10.3 Bauhin, *Theatrum*, 1621, Appendix, Plate IX, heart polyp. (Courtesy of Special Collections, Regenstein Library, University of Chicago)

plate (Illustration 10.7) shows the body whole as on the title page, the second shows the interior after the removal of a huge amount of fluid (Illustration 10.8). The right side presented various malformations, including the lack of the female testicle – or ovary. Medical men judged that the huge amount of watery fluid could not possibly stem from menstrual blood, which would not have produced such a clear liquid. More likely, the watery fluid had descended from the right ureter as a result of the malformation revealed by the postmortem, which would have been the cause of

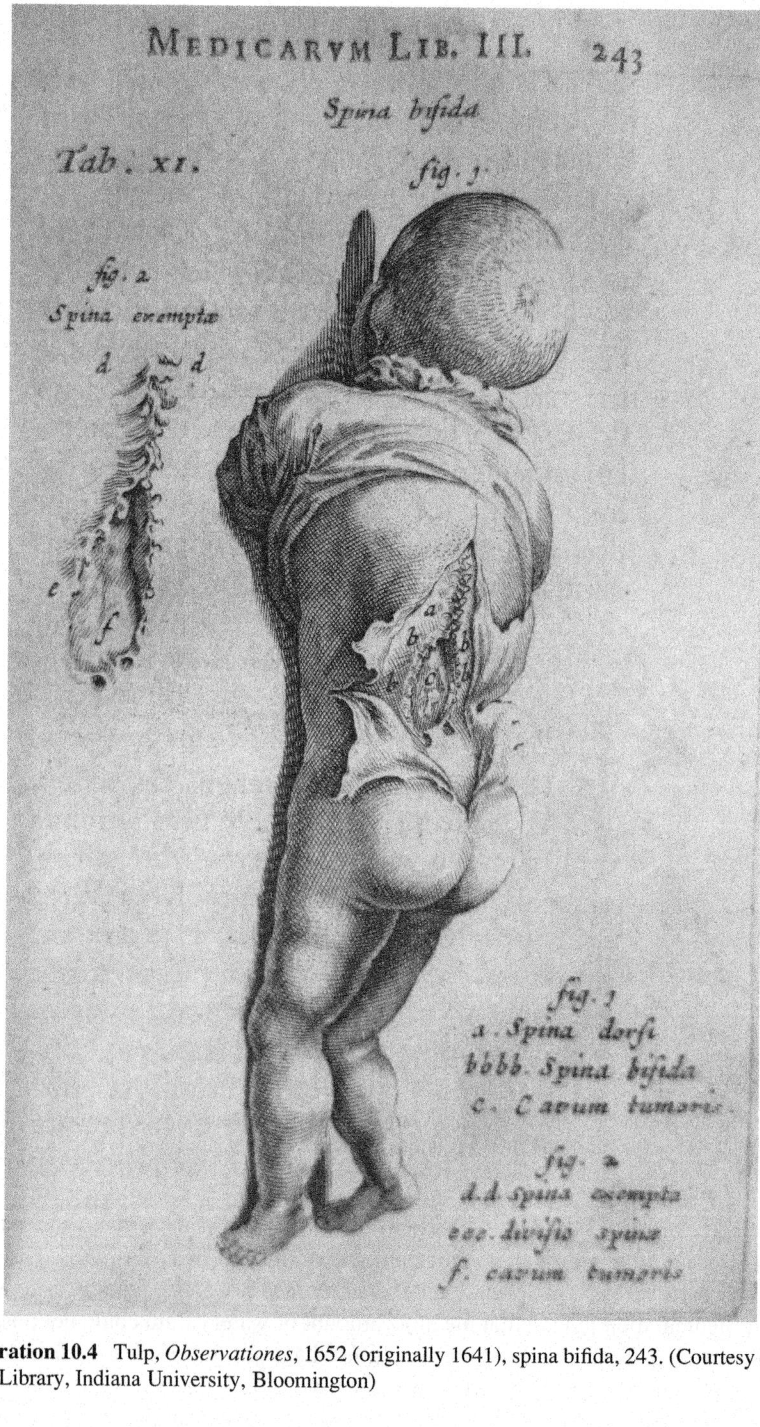

Illustration 10.4 Tulp, *Observationes*, 1652 (originally 1641), spina bifida, 243. (Courtesy of the Lilly Library, Indiana University, Bloomington)

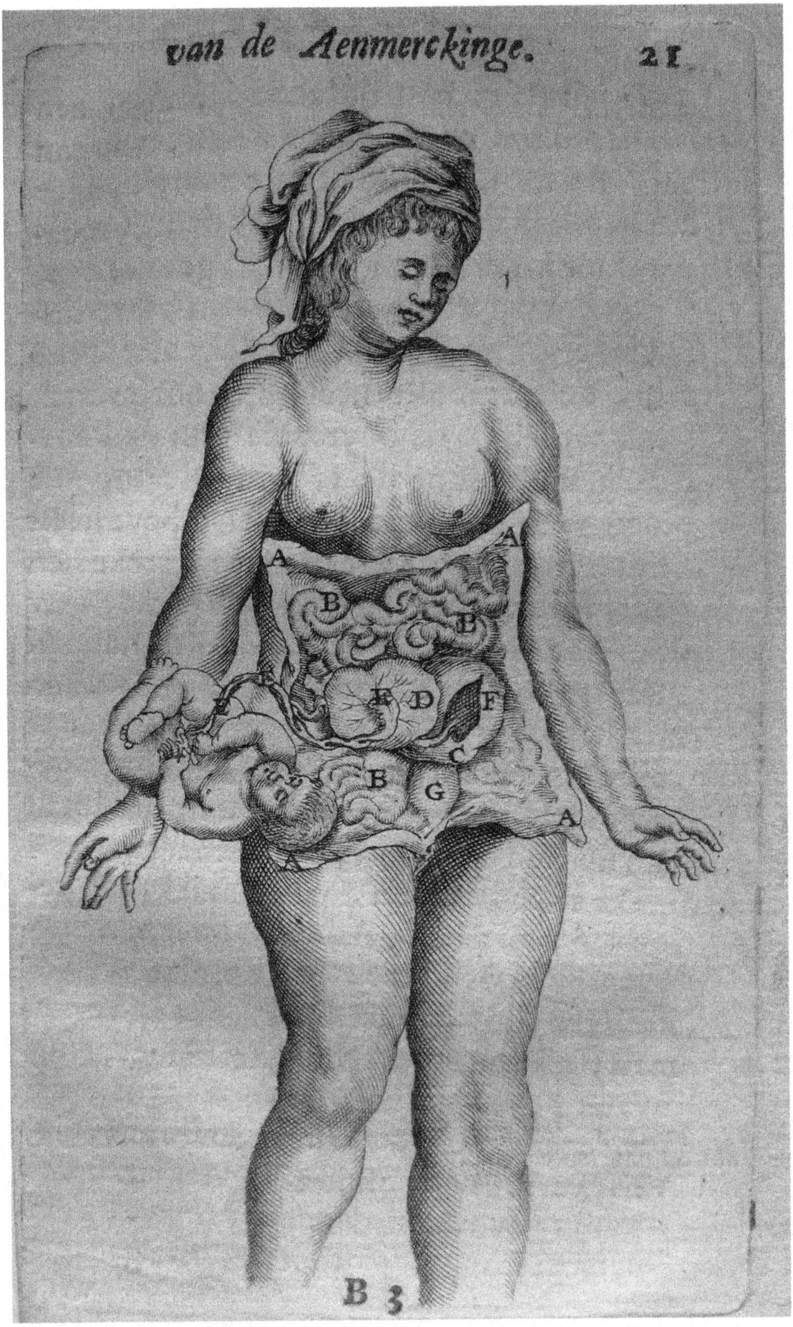

Illustration 10.5 van Roonhuyse, *Observationes*, 1663, extra-uterine pregnancy. (Courtesy of Special Collections, Leiden University Library. NB: the caption should say: van Roonhuyse, *Aanmerkkingen*, 1663, extra-uterine pregnancy)

Thoveling's condition. Following the opinion of eminent physician and anatomist Johannes Walaeus, they also advocated paracentesis or the draining of excessive fluid as the correct therapy in analogous cases. Walaeus was a leading figure in the early reception of Harvey's circulation in the Netherlands.[20]

Not surprisingly, overall the cases discussed by van Meek'ren privilege surgery: most of the live patients on the title page present external growths that would have come under the knife. Not all external growths, however, could be treated that way; the young child sitting on a pillow in the foreground on the title page is a case in point. The corresponding rather classicizing figure, almost a putto not dissimilar from those appearing in Vesalius's initials in the *Fabrica*, shows the child from behind supporting herself on a cube used as a foil to represent her dissected head on one of its sides (Illustration 10.9); this move enables the artist to show simultaneously the live child with the growth from the back of her head and the result of her postmortem with the growth removed. The discussion of the case history defends van Meek'ren's decision not to attempt the removal of the growth, which would have been dangerous; the child reached the age of 2 years showing no physical or mental problem, but then succumbed. The growth was filled with a watery fluid that van Meek'ren identified as pure urine, noticing that he child had not passed any urine for 4 or 5 days before dying.[21]

Van Meek'ren compared the case to that of a 2-month infant described by Fabricius Hildanus, whose work was often referred to in the literature. Hildanus and his correspondents frequently exchanged not only case histories but also images, as in this case, when Basel physician Johannes Screta discussed an image of the 2-month old infant from the Strasbourg hospital that Hildanus reproduced in his work (Illustration 10.10). Hildanus explained the reason why the face of the infant was so wrinkled as to resemble that of an old lady by arguing that all the nutriment went to the growth; indeed the woodcut shows the face of the infant resembling that of a much older person. Moreover, Hildanus surmised that the cause of the fact that the Strasbourg infant did not cry was that the growth pressed the recurrent nerves that control the larynx. The infant died because the surgeons attempted to remove the growth. Van Meek'ren's reference, like those by Tulp and Malpighi we have seen above, provides valuable information on how case histories and images were used by medical men to investigate previous cases, identify similarities and differences, and defend their actions.[22]

Another plate highlights the difficulty in representing the conditions encountered inside the body. The case involves an abscess on the left side of the abdomen affecting Christianus Friderici Holsatus; van Meek'ren also gave details about the medicinals administered, including the recipe for the "decoctum aperiens" commonly used at the Amsterdam municipal hospital where he worked, the Binnengasthuis. At one point he was able to extract a folded membrane that differed from the intestine by being deprived of vessels. Despite the cures administered, the

[20]van Meek'ren, 230–6. See Liedtke 2007, 1: 201; Kooijmans, 64, 106; Schouten 1974.

[21]van Meek'ren, 51–5.

[22]The reference to Fabricius Hildanus is at 52; see also 41, 57, 181–2. See Hildanus, *Opera*, 511–3, Centuria VI, Observationes XVI–XVIII.

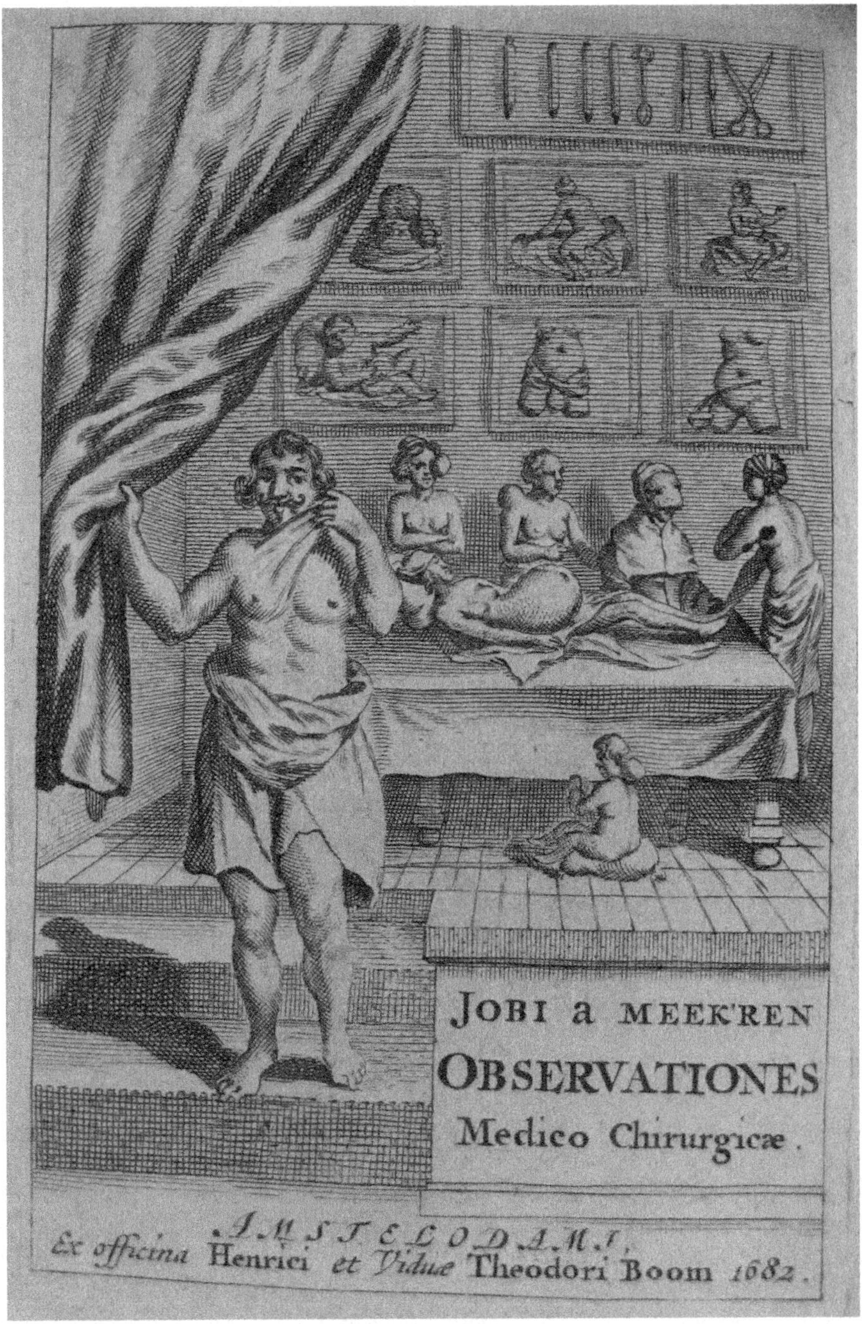

Illustration 10.6 van Meek'ren, *Observationes*, 1682 (originally 1668), title page. (Courtesy of the Lilly Library, Indiana University, Bloomington)

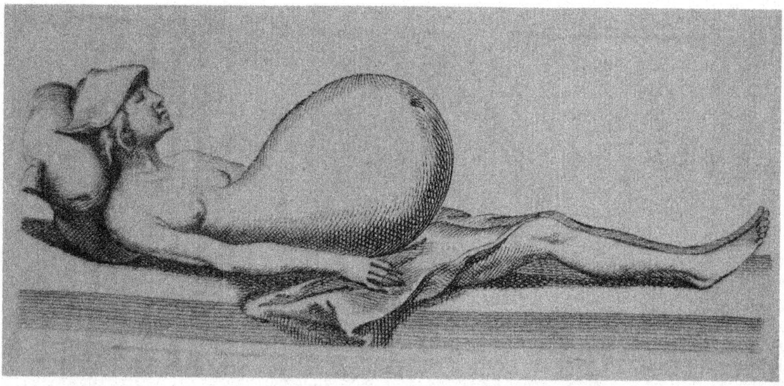

Illustration 10.7 van Meek'ren, *Observationes*, 1682 (originally 1668), water in the peritoneal folding, page 232. [1] (Courtesy of the Lilly Library, Indiana University, Bloomington)

patient died on November 4, just 1 month after van Meek'ren had first seen him. Opening the body revealed a glutinous, yellowish, fetid matter. The spleen was very large, divided in two lobes, and had descended towards the groins; van Meek'ren noticed a vesicle marked G in the plate, filled with a watery liquid and enclosed within a membrane in all respects similar to the one previously extracted. The plate with the open abdomen (Illustration 10.11) framed by the folded peritoneum HH resulting from a cruciform incision shows the enlarged spleen *in situ* on the right. While van Meek'ren's figure succeeds in showing the peculiar shape and location of the spleen, it provides only a very approximate view of the diseased organ at best; the abscess is identified by a faint circle marked *F* on the spleen *EE*.[23]

The works by Tulp, van Roonhuyse, and van Meek'ren we have briefly discussed share common features: they were in a small format, with relatively small figures with a seeming practical intent and no great artistic ambitions. Two years after the Dutch edition of van Meek'ren's *Aanmerkkingen*, Theodor Kerckring published *Spicilegium anatomicum* (1670), including 100 anatomical observations often dealing with medical and surgical cases. Kerckring's work stands out compared to the previous ones for the quality of its printing; it is in quarto, spaciously and elegantly produced and with just over 30 plates, some tiny, many full-size, and even some folding. The elegant frontispiece was signed by distinguished artist Abraham Blooteling, who in the following decade was to work for Govert Bidloo, whose *Anatomia humani corporis* (Amsterdam, 1685) is generally seen as a landmark in anatomical illustrations for its dramatic realism and focus on individual specimens; it is not clear whether he was also responsible for the plates inside Kerckring's volume. *Specilegium* includes several intriguing cases. I wish to discuss one that was written in a dialogue – a critical textual and iconographic one, in this case – with Bauhin and Tulp; no doubt also Malpighi would have been in Kerckring's mind, though his name was not mentioned. Kerckring argued that

[23]van Meek'ren, 364–8.

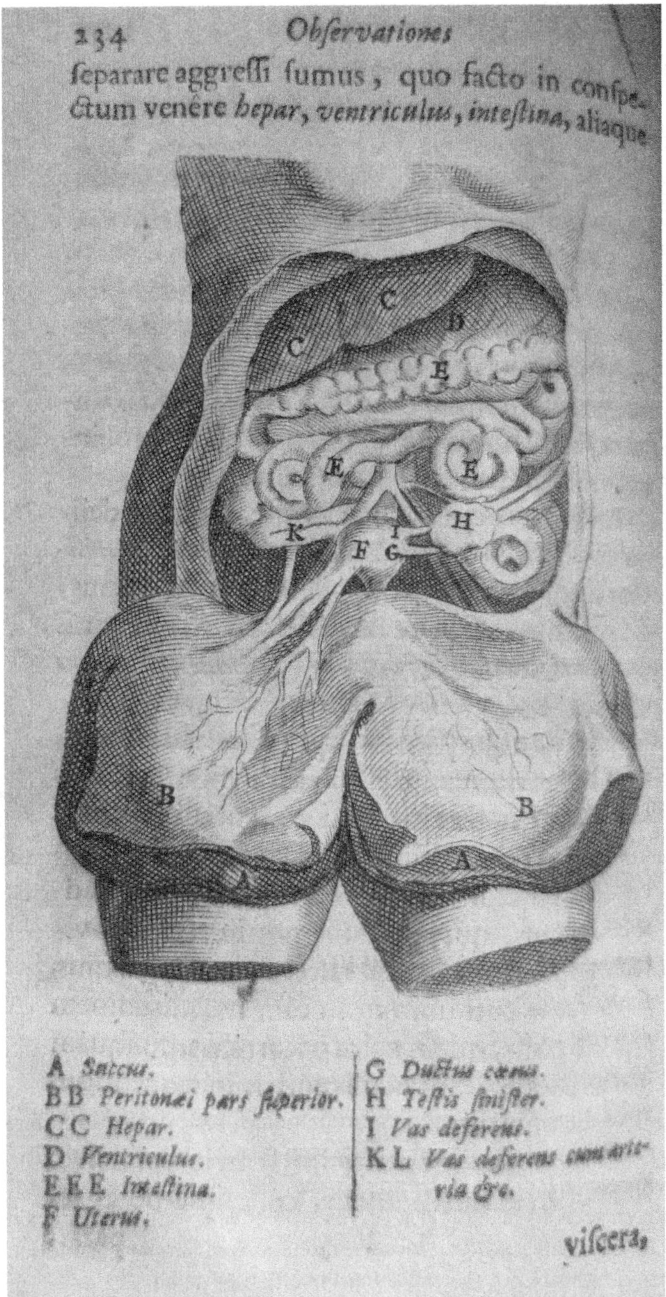

234 *Observationes*

separare aggressi sumus, quo facto in conspe-
ctum venere *hepar, ventriculus, intestina,* aliaque

A *Saccus.*
BB *Peritonæi pars superior.*
CC *Hepar.*
D *Ventriculus.*
EEE *Intestina.*
F *Uterus.*
G *Ductus cæcus.*
H *Testis sinister.*
I *Vas deferens.*
KL *Vas deferens cum arteria &c.*

viscera,

Illustration 10.8 van Meek'ren, *Observationes*, 1682 (originally 1668), water in the peritoneal folding, page 234. [1] (Courtesy of the Lilly Library, Indiana University, Bloomington)

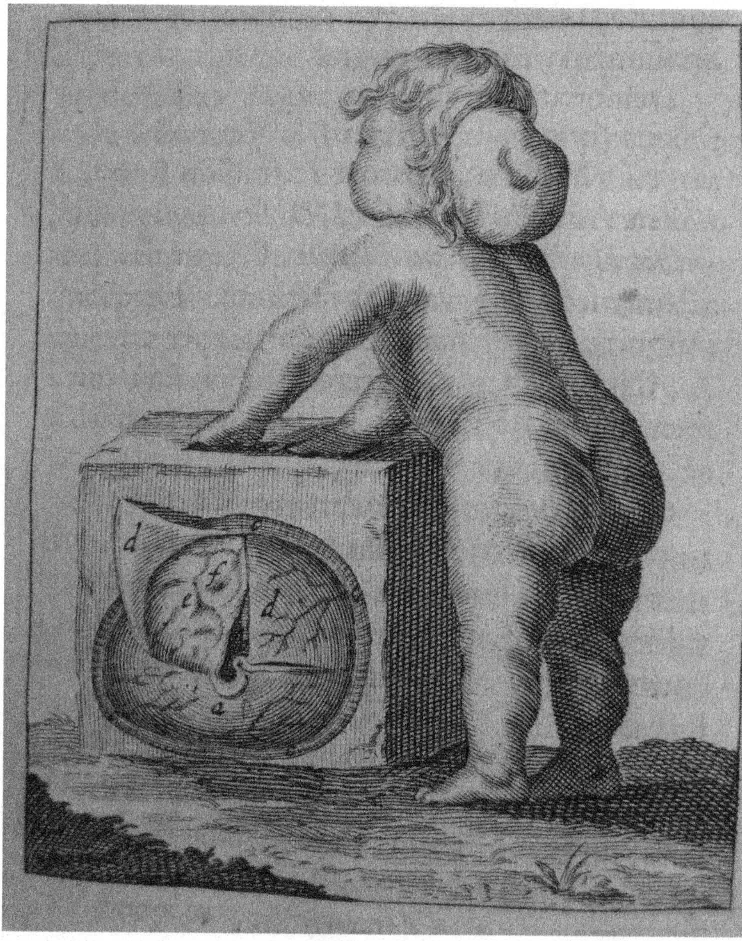

Illustration 10.9 van Meek'ren, *Observationes*, 1682 (originally 1668), head tumor, page 53. (Courtesy of the Lilly Library, Indiana University, Bloomington)

polyps consist of coagulated blood and are formed after death. He defended his claim by arguing that he could form polyps at will, by simply injecting spirit of vitriol into the veins of live dogs, which led to the formation of hard polyps in different vessels (Illustration 10.12), such as the pulmonary vein (*arteria venosa*, I), the liver (II), and the right ventricle (III). Thus Kerckring's plate with its coral-like polyps differs from others we have seen so far in that it serves a polemical purpose; it does not show a diseased state but challenges previous interpretations of similar postmortem findings, which in his view were not the cause but the effect of death.[24]

[24]Kerckring, 145–7. See Margócsy 2011, 185–215, esp. 211; Kemp 2010, 192–208, figures 5, 6, 16; Roberts and Tomlinson 1992, 300–305, 309–19; Bertoloni Meli 2011, 147–8; and Goldschmied, 45.

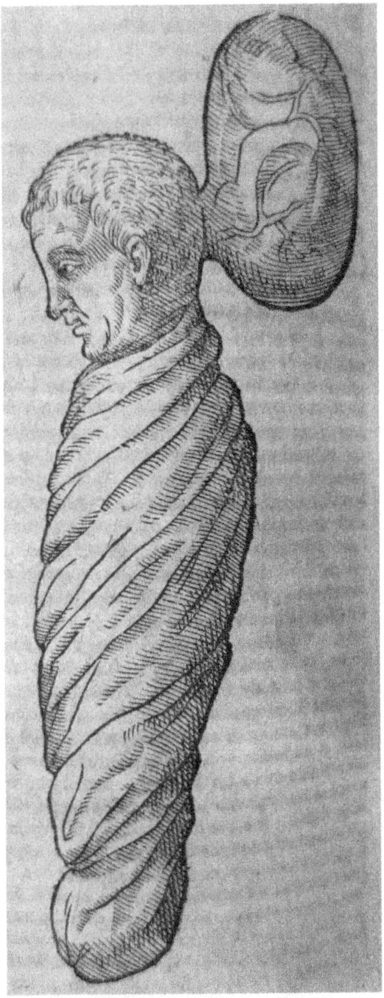

Illustration 10.10
Hildanus, *Opera*, 1646 (originally 1641), head tumor, page 512. (Courtesy of the Lilly Library, Indiana University, Bloomington)

10.2 Gerardus Blasius and the Amsterdam Medical Scene

Despite its size and wealth, seventeenth-century Amsterdam lacked a university. The city had a number of medical officers, from city physician and forensic anatomist to instructor to surgeons and midwives; the medical world was organized in several guilds and supervisory boards. The first institution of higher learning was the *Athenæum illustre*, established in 1632 in the chapel of the former convent of St. Agnes. Unlike a university, the Athenæum could not confer degrees. Since each province in the Netherlands was allowed one university only, and from 1575 Holland already hosted at Leiden the most ancient university of the United

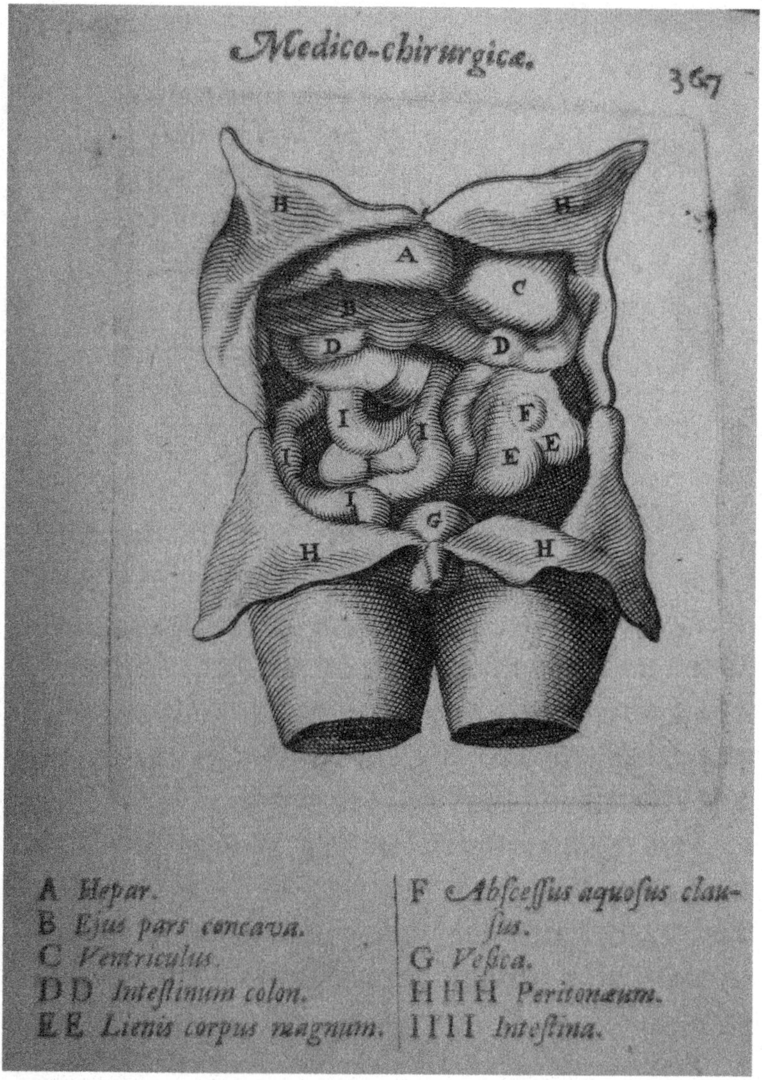

Illustration 10.11 van Meek'ren, *Observationes*, 1682 (originally 1668), spleen abscess, page 367. (Courtesy of the Lilly Library, Indiana University, Bloomington)

Provinces, Amsterdam's ambitions were thwarted. Thus the *Athenæum* was limited to offering courses to students who then went on to take degrees elsewhere. The curriculum did not include medical teaching until 1660, several decades after its establishment. Gerardus Blasius was the first to hold an official position for medical teaching at the *Athenæum* and, setting aside informal teaching and those who lectured to professional bodies of surgeons or midwives, in Amsterdam as a whole.[25]

Blasius (ca.1625–92) was the son of a Dutch architect that at some point worked in Copenhagen, where he studied for a while. In 1645 he matriculated at Leiden University, where he studied with several eminent professors, such as Johannes Walaeus. Blasius graduated MD with a dissertation *De nephritide* (1648), written under the direction of Otto Heurnius. In response to the rise of Utrecht University, Heurnius was an early advocate of clinical teaching in Leiden and he started a *Collegium medico practicum* there in 1636. Blasius was a key figure at the Amsterdam *Collegium privatum*, a group of physicians and medical students devoted to collaborative anatomical investigations.[26]

Two passages in Blasius's 1677 *Observationes* refer to cases from the Leiden (Cecilia) Hospital from 1647, thus from his student days. Both concerned the formation of calculi, in the lip and liver; the former was found in the cancerous portion of a lip extirpated by a surgeon, as Blasius put it, "me aliisque spectantibus", possibly suggesting a clinical setting; the latter was black, spiral shaped, and as hard as flint. Blasius presumably preserved the preparations, or images from them, since he reproduced them in 1677 (Illustration 10.13). On two other occasions he included figures in scale of preparations of a heart polyp and a stone found under the tongue that he preserved "in musæo meo Anatomico" and at his house, "domi meæ," where his museum was presumably located.[27]

In the mid-1650s we find Blasius practicing medicine in Amsterdam, where he spent the rest of his career. In 1660 he was appointed city physician and extraordinary professor at the *Athenæum*, where he single-handedly taught the whole medical curriculum; in 1666 he was appointed ordinary professor.[28]

Early in his career Blasius edited several works by other authors, possibly because of his broad-ranging teaching duties. They include treatises on practical medicine, chemistry, and especially anatomy, such as Johann Vesling, *Syntagma anatomicum* (Amsterdam, 1659), with a much expanded edition including excerpts from many recent anatomical works, such as Nathaniel Highmore, Thomas Bartholin, Thomas Willis, Marcello Malpighi, Lorenzo Bellini, Nicolaus Steno, and Frederik Ruysch (Amsterdam, 1666). Blasius further edited works on the lymphatic vessels by Thomas Bartholin and Lorenzo Bellini on the kidneys (both in 1665). Much later he edited the *Opera omnia* of Thomas Willis (Amsterdam, 1682). In addition, Blasius had an interest in monsters, as evidenced by his edition of Fortunio Liceti, *De monstris* (Amsterdam, 1665), an appendix on "monstrous"

[25]See van Miert 2009; Kooijmans 2010; Mooij 1999, esp. 71–6; Cook 2007, ch. 4.

[26]See Lindeboom, 156–8; van Miert, 92–3, 164; Beukers 1987–8; Cook 2007, 149–50; Huisman 2009, 128–44; Jorink, 283–9; Bertoloni Meli 2011, 33.

[27]Blasius 1677, 82, 84, quotation at 82. See also the *Observata anatomica extraordinaria in homine* in Blasius 1674, 119.

[28]Van Miert, 319–24.

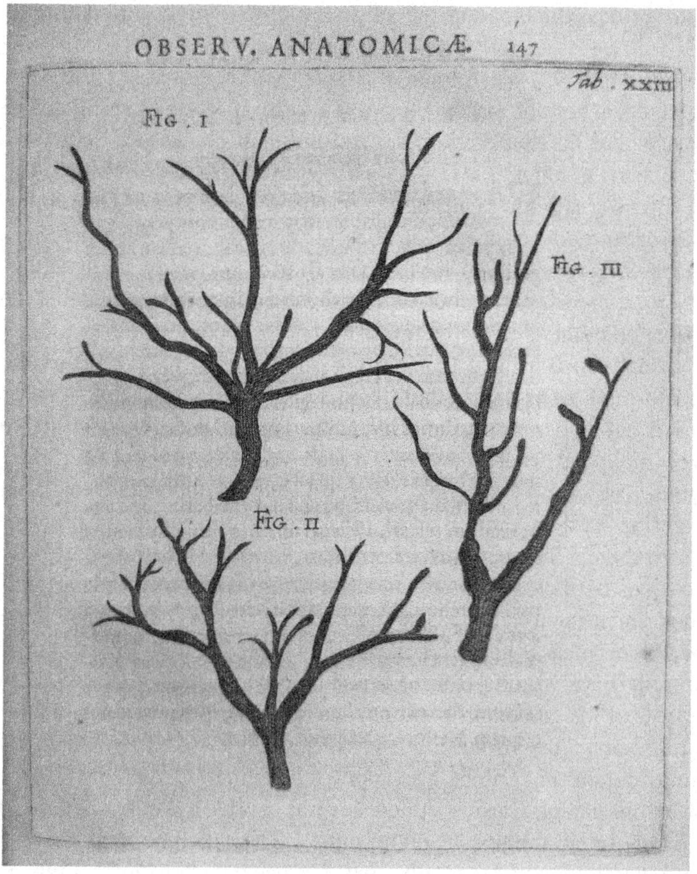

Illustration 10.12 Kerckring, *Observationes*, 1670, pseudo-polyps, plate XXIII. [1] (Courtesy of Special Collections, Regenstein Library, University of Chicago)

kidneys to the edition of Bellini's work, and an appendix reissuing recent publications on three monstrous births added to *Observationes medicae rariores* (Amsterdam, 1677). By "monster" Blasius generally intended a congenital condition, often conjoined twins. In the appendix to Bellini's work the term was used more loosely. A sense of his remarkable productivity can be gained from the partial list of publications in Appendix 1.

Blasius's own publications too covered many areas of medicine: *Medicina generalis* (Amsterdam, 1661) was a textbook whose frontispiece shows Blasius teaching young students. This is the work that led to the controversy over the discovery of the parotid duct with Nicolaus Steno. *Anatome contracta* (Amsterdam, 1666) and *Medicina curatoria* (Amsterdam, 1680) both bear the words "written for the pupils' sake" in the subtitle, highlighting their didactic role. Blasius was also

Illustration 10.13 Blasius, *Observationes*, 1677, liver calculus, Plate IX, Figure XI (Courtesy of Special Collections, Regenstein Library, University of Chicago)

very active in presiding students' disputations, as part of their training rather than the awarding of degrees, and published 50 of them.[29]

Moreover, Blasius had extensive interests in the dissection and study of animal anatomy: *Miscellanea anatomica, hominis, brutorumque variorum* (Amsterdam, 1673) dealt with the topic and, as Francis J. Cole pointed out, was reissued with a different title, *Zootomia, seu anatome variorum animalium* (Amsterdam, 1676); *Observata anatomica* (Leiden, 1674) examined the human body as well as, among the others, the monkey, horse, calf, sheep, tortoise, hedgehog, dormouse, snake, and heron; the more ambitious *Anatome animalium* (Amsterdam, 1681) joins his own researches to an extensive survey of the literature.[30] In addition, the first volume of *Observationes anatomicae selectiores* (Amsterdam, 1667, 1673) by the Amsterdam *Collegium privatum*, included work by Blasius and young Swammerdam on animal dissections; interestingly, the name of the Amsterdam anatomical group echoes that of the Leiden clinical teaching set-up.[31]

Cole complains that Blasius assembled a large number of facts without an organizing principle, "such as the notion of evolution," on the basis of which to order them. Unquestionably this judgment sounds anachronistic, though it does raise the question of what Blasius was trying to do. Anatomists as diverse as Hieronymus Fabricius and Marcello Malpighi, for example, investigated a wide range of animals

---

[29]Ibid., 93, 122, 320–22. See also Steno 1986, *sub indice*; Bertoloni Meli 2011, 111–3.

[30]Cole 1944, 150–5.

[31]Collegium privatum Amstelodamense, *Observationes anatomica selectiores* (1667, 1673). Facsimile edition with introduction by Lindeboom 1975. See also Cole, 330–41.

for different reasons: Fabricius sought to identify the purpose of various organs in relation to other features of the animals, such as whether they lived on land, in water, or they flew. Malpighi sought to exploit differences in the size and structure of their organs in order to investigate their microstructure. Blasius seemingly had a variety of purposes: in *Miscellanea anatomica*, for example, he provided an extensive section on the anatomy of the dog, which was the most common substitute for the human body when human bodies were scarce, thus enabling students to learn about the human body through canine anatomy. In many cases, however, it seems as if Blasius was simply fascinated by the diversity of animal structures.[32]

In an age rich with new major anatomical findings, Blasius has comparatively little to offer once his priority claim against Steno floundered. However, it would be reductive to evaluate his contributions in those terms only. In 1669, for example, Blasius started a limited form of clinical teaching to instruct his students from the *Athenæum* at the Amsterdam town hospital, the Binnengasthuis, and, if the patients died, to perform postmortems.[33] Bedside instruction was available at some point at Utrecht and more extensively at the Cecilia Hospital in Leiden, where Otto Heurnius and especially Sylvius were attracting large numbers of students also for this reason. Even if they could not offer degrees, the Amsterdam authorities presumably wished to retain students by offering something comparable to Leiden.[34]

Blasius, however, did something Sylvius had not done: he published several illustrated collections of medical observations, relying at least in part on his clinical teaching. Most of them appeared in the 1677 *Observationes medicae rariores*, though several were included in his works on zootomy. *Miscellanea anatomica* (1673), for example, ends with a few pages of *Annotata practica in cadaveribus humanis*. Similarly, the full title of Blasius's *Observata anatomica* (1674) ends with the words "Extraordinary things found in human bodies illustrating medical praxis as well as anatomy have been added." Other Dutch scholars had done something similar in the past, though without adding illustrations: Jacobus Bontius had appended some *Observationes aliquot selectae ex dissectione cadaverum* to his *Medicina indorum* (Leiden, 1642); Otto Heurnius had added a section *Historiae et observationes quaedam rariores ex praxi et diario* to his edition of Jean Fernel, *Universa medicina* (Utrecht, 1656).[35] Notice the similarity between Heurnius's and Blasius's 1677 titles, especially the emphasis on case histories, observations, and rarity. Clearly these sections belong to the genre of *Observationes* whether the term

---

[32]Cole, 151, 341. See also Blasius, *Miscellanea anatomica*, address to the reader and 168; Cunningham 1985, 195–222; eadem 2010, 308–40, at 314, 316; Ragland 2008; Bertoloni Meli 2011, 32–3; and eadem 2013.

[33]See Lindeboom, 156–8; van Miert 2009, 92–3, 164, 319–24. Amsterdam clinical teaching declined after Blasius.

[34]See Mooij, 75; Cook, 149–50; Beukers, 139–45; and Huisman, 145–54.

[35]Beukers, 151n25. The full title of the edition of Fernel, *Universa medicina*, by Otto Heurnius, ends with the words "Cui accedunt casus et observationes rariores quas cl. d. d. Otho Heurnius in diario practico annotavit." The Latin full titles of Blasius's works can be found in Appendix 2. See also Bontius 1642, 191–206; and Cook, 191–209, at 194.

occurs explicitly in the title or not. Although Fernel had coined the term "pathologia", the expression "pathological anatomy" was not current in the early modern period; what we find instead are expressions including "medical praxis" and "anatomy," often under the heading of *Observationes.*

Blasius's work in human anatomy, zootomy, and practical medicine through postmortems, joined with his familiarity with the recent literature in these areas, make him an especially rewarding figure to investigate the relations among these anatomy domains and the role of illustrations. In human anatomy the dominant tradition privileged idealized images representative of the average or canonical body.[36] In zootomy, the issue may have depended on the animal in question, dogs being more common than tigers. Further, in human anatomy and zootomy the main emphasis was on structures, whereas in the study of disease textures were often highly significant. Blasius approached zootomy and the study of disease with a similar perspective, with an interest in curios, remarkable, or at least unusual features; further, in both cases he sought to organize the material in similar ways, as we are going to see below.[37]

Together with the smaller collections presented in his zootomy publications, Blasius's *Observationes* are an unusually rich source of information on practices of dissection in Amsterdam around 1670. The work is dedicated to Bonaventura van Dortmond and Gulielmus de Penijn, both physicians at the Amsterdam hospital: the connection with hospital medicine was crucial for the availability of cadavers and patients. In the address to the reader, Blasius presents his work as a companion to his *Zootomia* (1676), that opened with a long chapter on human body in its "natural state" for students; *Observationes* presents to the students examples removed from that state both in living bodies and cadavers, though he specified that he discussed only those cases that occur rarely. Dissections were collegial affairs and Blasius often mentions the names of other medical men involved, both surgeons and physicians: besides van Dortmond and Penijn, he mentioned Ruysch, Bidloo, Hendrik van Roonhuyse, and his son Rogier, among many others. Moreover, in the text Blasius refers to dissections performed in front of students, supporting the view that many of his reports stem from his clinical teaching.[38]

[36]For example, see Kemp 2010; Sappol 2006. Daston and Galison 2007, ch. 2, argue that before the early nineteenth century, images were abstract representations meant to capture the key features of specimens, what they call "truth-to-nature," or "characteristic "images of individual representative specimens. Their analysis seems to draw primarily from botany and anatomy but sits awkwardly with early representations of diseased states and monsters. Kusukawa highlights that the canonical body was not attained exclusively from frequency, 219–21, and discusses anatomical works focusing on individual specimens, 239–40.

[37]Cole includes many reproductions that are analyzed from a strictly anatomical standpoint. See also Guerrini 2010; Goldschmid; Bertoloni Meli 2015. For a rather clumsy attempt at representing textures, see Vesalius 1543, 2, and the discussion by Long 2011, 58–59.

[38]Blasius 1677, address to the reader, 31. See also eadem, *Miscellanea anatomica*, 307. On van Dortmond, see Kooijmans, 96–8; and Lindeboom, 483–4. References to medical men involved in cures and dissections are ubiquitous: see Blasius 1677, 8, 17, 27, 28, 40–1.

Discussing human anatomy in *Miscellanea anatomica*, Blasius followed an approach dating from the old Alexandria school by organizing the material into sections on substance, quantity, number, qualities, figure, place, connection, and purpose. By contrast, he presented his case histories in *Miscellanea anatomica* and *Observata anatomica* in seemingly random order.[39] In *Observationes*, however, he structured the material in six sections, resembling, though not identical to, the scheme he had adopted previously, possibly because he now faced different material. He grouped together diseases (*morbi*) involving magnitude, such as tumors, abscesses, hernias, and dropsy; defects (*vitia*) of figure, such as cleft palate or closed uterus; defects (*vitia*) of the parts contained in a given place, such as prolapsed uterus or the separate entrance of pancreatic and bile ducts in the intestine; diseases (*morbi*) related to number, either lack of body parts, such as the hymen, female testicles, or kidney, or presence of extra-numerary parts, such as a double stomach or gall-bladder; diseases (*morbi*) of the union or cohesion of parts, such as caries or ulcers; and lastly the presence of preternatural (*praeter naturam*) formations, such as polyps, ossifications, and stone formations.[40]

Despite Blasius's usage of distinctive terms such as *morbus* and *vitium*, it is not always clear whether he and his contemporaries would have seen them as substantially different: conditions in the first group involving magnitude, such as tumors, abscesses, hernias, and dropsy were categorized as *morbi*. However, the same term was applied to conditions in the fourth group too, involving the lack of body parts, such as the hymen, or the presence of extra-numerary parts, such as an extra kidney; these conditions may not cause any harm. It is also doubtful whether Blasius would have considered them as all pertaining to pathology; lack of the urinary bladder, for example, led a 35 year-old man to very frequent urination; similarly, in another man of 35 a double stomach connected by a very narrow passage led to frequent vomiting. On the other hand, the presence of an extra testicle allegedly only made a 30 year-old "greatly libidinous."[41] More generally, the presence or lack of the hymen or a peculiar conformation of the kidneys and the urinary vessels, have a dubious status in pathology; as Padua physician Giovanni Battista da Monte had pointed out in the previous century, the presence of an extra finger is not a disease. Similarly, in a case of horseshoe kidney, English anatomist Edward Tyson questioned whether the peculiar conformation was "of much inconvenience to the Patient".[42] Unlike Tulp, however, who had included monstrous births in his *Observationes*, Blasius discussed monsters separately in an Appendix reporting three human and animal congenital cases.[43]

[39]French 2001, 133–4. See Blasius, *Miscellanea anatomica*, 1 and following, which discuss various body parts by dividing the material in sections; and eadem., *Medicina universa*, 20–1.

[40]Blasius 1677: the six sections start at 1 (25 observations), 35 (8 observations), 42 (5 observations), 47 (21 observationes), 61 (7 observations), 73 (25 observations).

[41]Ibid., 52–4, 60–1.

[42]Tyson 1678–9, 1038. See also Siraisi 2001d, 320.

[43]Plate VII in Blasius 1677 shows several malformations of the urinary system.

Many anatomy books in small format exploited folded plates as a way to have larger illustrations. Blasius not only eschewed this practice here, but with one exception also included many figures on each plate; thus they barely attain the size of a postage stamp. The only exception is Plate II, showing the corpse of a 35 year-old woman whose abdomen had been growing for the previous 3 years. Blasius indicates that some form of physical examination had taken place, since the abdomen felt like stone to the touch. This is not the only instance when a physical examination is recorded: on another occasion concerning a man circa 30 years of age, van Dortmond "latus dextrum imi ventris manu premebat", or was pressing with the hand the right side of the lower belly. The inside of the body in our case was found to be dramatically affected; Blasius could find neither fat, nor membranes, nor muscles, since everything had been altered: the outside had turned into a cartilaginous substance, the inside was "cancerous" (Illustration 10.14). This case was also discussed in slightly different language in *Miscellanea anatomica*, where Blasius referred to his Leiden teacher Walaeus as having performed the dissection of a "simile fere exemplum." Despite the similar external appearance, however, the descriptions of the inside of the body in the two cases appear quite different. Blasius's plate closely echoes the one by van Meek'ren showing the corpse of Ingeltje Thovelings, wife of painter Govert Flinck. Both engravings show the reclining body on a hard surface, with the head slightly raised (Illustration 10.7); one wonders whether Blasius just modified the previous plate. Whereas van Meek'ren added an additional plate showing the inside of the body after the removal of a huge amount of fluid (Illustration 10.8), Blasius's shows signs of the dissection, with a dark wedge in the abdomen. Despite this attempt, his plate fails to provide any sense of the body's interior; his task may have been especially challenging, since he could detect no recognizable structure.[44]

This is not the only instance in which Blasius relies on previously published material. His first three *observationes*, for example, correspond to the first three cases reported by Hendrik van Roonhuyse in *Genees en heel-konstige aanmerkingen*. There was considerable overlap between the two accounts, though Blasius's did not merely translate van Roonhuyse's text; they were both present and it would seem that while Blasius relied on the surgeon, he also provided his own account. Blasius's very first figure concerns the case of Pieter Sluysen, a 38 year-old who presented a huge growth on his left ear that he could no longer conceal with his hair (Illustration 10.15). The very same case, dating from 1665, was reported by van Roonhuyse; whereas the surgeon described the growth as "monstrous," Blasius eschewed this term, thus suggesting that he was using it more restrictively. Both included a plate showing the man with the large growth on his left ear; Blasius's was a much smaller version of the surgeon's. It would seem that in this case the figure served different purposes: for van Roonhuyse, who had actually performed the operation, the figure would have served as an advertisement of his skill, since

[44]Ibid., 10–11; quotation at 17. Also, eadem., *Miscellanea anatomica*, 294–5, at 295; and van Meek'ren, 230–6.

the removal of the growth ended successfully, whereas Blasius was seemingly more interested in documenting the case for a students' audience. Neither, however, provides an image exactly showing how the growth was attached.[45]

Other plates adopt different conventions. We have seen that van Meek'ren provided two images showing the outside and the inside of Thoveling's abdomen (Illustrations 10.7 and 10.8); in another case (Illustration 10.11) he showed the diseased spleen of Christianus Friderici Holsatus *in situ*. Blasius faced similar problems in other instances, as in the following figure where he studied the cadaver of a woman of 42 who had died of a liver affliction. In that case he had recourse to three figures showing the affected organ in situ and removed from the body (Illustration 10.16, Figures I–III). Figure I displays the body's interior, with the liver shown *in situ* at *a*; *b* is a steatoma – a fatty tumor – on its left side; *B* the stomach; *cc* and *dd* are portion of a scirrhus or a growth, hard in *cc*, softer and festering in *dd*; *ee* and *ff* the intestine. Figure II is on a slightly larger scale and shows the liver *aa* removed from the body, with the gall bladder at *b*; *c* is the folded umbilical ligament; *ee* and *ff* show the scirrhus, the latter portion being ulcerated. Figure III shows at *bb* yet again the same liver, *a* being the gall bladder, c the cistic duct leading to the intestine *dd*; *e* the pancreas, *ff* the schirrhus, and *g* the spleen. Blasius went to a considerable length to provide a detailed account of the condition, and to tie the visual representation to the textual description, though the diminutive size of the figures and their charming simplicity thwarted somewhat his efforts.[46]

On the same plate (Illustration 10.16, Figure IV, bottom left) Blasius represents a case of scrofulous formations in the venter of a nine or 10 year-old child: at top left and right are the liver *b* and stomach *d*; the scrofulous formations are marked *iii* below and were described as white, ashen, and almost black. He had described the same case in the *Miscellanea*, from where we learn that in his museum he had preserved a portion of the child's intestine covered with scrofulous growths of various sizes; unfortunately he does not discuss his preservation methods.[47]

Blasius was on more solid ground in representing purely structural features. In one case in February 1675 he found in a fetus that the biliary and pancreatic ducts did not join just before entering the intestine, but entered the intestine separately – this being what you expect in a dog, one may add. Blasius had seen another similar example the previous year in a woman of 40; he stated that he could offer a display at his house, so he presumably had some preparation, which would not have required great sophistication since all it had to show was the separate insertions. The figure (Illustration 10.17) shows the stomach in the center at *d*, the biliary duct *a* entering the duodenum *e* from the top, and the pancreatic duct *b* entering from the side.[48]

---

[45]Blasius 1677, 3–9; van Roonhuyse, 3–12, at 3, plate at 7.

[46]Blasius 1677, 19–24.

[47]Ibid., 24–5; and eadem., *Miscellanea*, 307–8.

[48]Blasius 1677, 46, 115. See Ragland, 632–3.

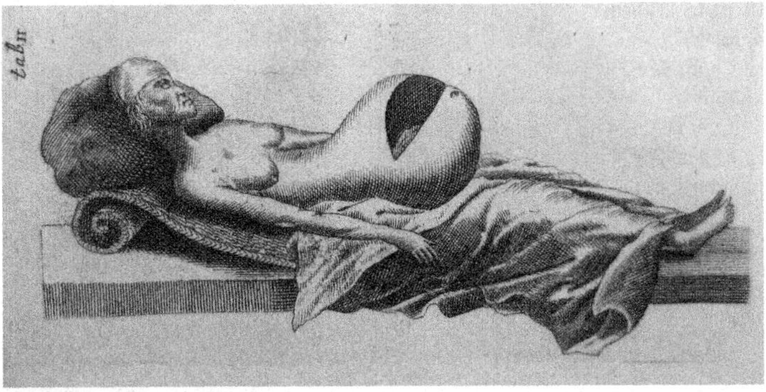

Illustration 10.14 Blasius, *Observationes*, 1677, enlarged abdomen, Plate II (Courtesy of Special Collections, Regenstein Library, University of Chicago)

10.3 Ruysch's *Observationes* and Museum

Whatever collection of specimens Blasius may have managed to assemble at his house, it would have paled in comparison to Ruysch's museum. Also Ruysch's publications documenting his collection stands out for the size and quality of their illustrations; unlike most of the works we have examined, from Tulp to Blasius, which were in small format, Ruysch's was in quarto. Unlike Kerckring's work, the only other one that was also in quarto, Ruysch's collection included many figures that occupied the whole plate, some were even folding. In 1664 Ruysch graduated MD in Leiden, where he studies with Sylvius and van Horne. In 1667 he moved to Amsterdam, where he rapidly became the leading anatomist; as physician, instructor in anatomy to surgeons and midwives, municipal obstetrician, and forensic anatomist, he had extensive access to deceased bodies. Moreover, his unparalleled skills at preserving them led to the creation of the most impressive anatomical museum of its time. Ruysch was also a prolific author whose works included many pathological plates that gained him the epithet of "founder of pathological illustration" by nineteenth-century German pathologist Hermann Lebert. Ruysch, however, did not focus on preserving and representing exclusively pathological states; rather, these were part of a larger collection including healthy human and animal specimens prepared with his sophisticated injection techniques. His last *Observatio*, for example, highlights the *vera structura* of the male reproductive organ. Ruysch wished to impress and instruct visitors to his museum by exploiting a peculiar combination of nature and art: his most elaborate constructions showed fetal skeletons in fanciful landscapes made of kidney and bladder stones and hardened vessels; in healthy specimens he often highlighted the rich vascular structure by

Illustration 10.15 Blasius, *Observationes*, 1677, ear growth, Plate 1, Figure 1, after van Roohuyse. (Courtesy of Special Collections, Regenstein Library, University of Chicago)

means of his injections, in pathological specimens he highlighted the remarkable vagaries of disease.[49]

Nonetheless, his collection of 100 *Observationes* includes many figures of pathological conditions, including one on a newborn with a tumor from the head similar to those we have seen in Meek'ren and Hildanus. Ruysch was an accomplished artist and did the drawings himself; the engraver is unknown. Stone formations found inside the body had long been a source of fascination for physicians and surgeons alike. I doubt, however, that anyone produced quite so elegant a plate as Ruysch (Illustration 10.18). Printed on a folded sheet, Ruysch's figure is the largest of those discussed in this essay. The stones are displayed on a long and narrow table, in decreasing size from right to left. A cloth hanging casually over the table edge echoes a still life touch from the Dutch golden age. From this early work one can appreciate that the size and quality of Ruysch's representations were not hampered by the material he was dealing with. Similar concerns were the hallmark of his celebrated museum, in which one could find elaborate compositions adorned with lace and ribbons in elegant jars.[50]

An especially interesting case both textually and visually concerns a condition we encountered at the outset of this essay, a lumbar tumor with spina bifida, as Tulp named it. Ruysch eschewed speculations as to its origin and discussed instead its

[49]Kooijmans, 78–82, 88–9, 97, 156, 185. See Lebert 1857–61, 1:6; I translate "iconographie pathologique" as "pathological illustrations," since the term "iconography" has a more technical meaning in English. Also Sandifort identified Ruysch as an especially significant early source in the history of the visual representation of bone diseases: see *Museum anatomicum Academiae Lugduno-Batavae*. 4 vols. (the last two were published by his son Gerard) (1793–1835), 2: 4; and eadem. 1691, 127–38.

[50]Ruysch 1691, 69 and figure 45. See Kooijmans, 169; Hansen 1996; van de Roemer 2010; Margócsy 2009; Jorink, 319–325; Knoeff 2015; Hendriksen 2015, which offers several important considerations on Ruysch.

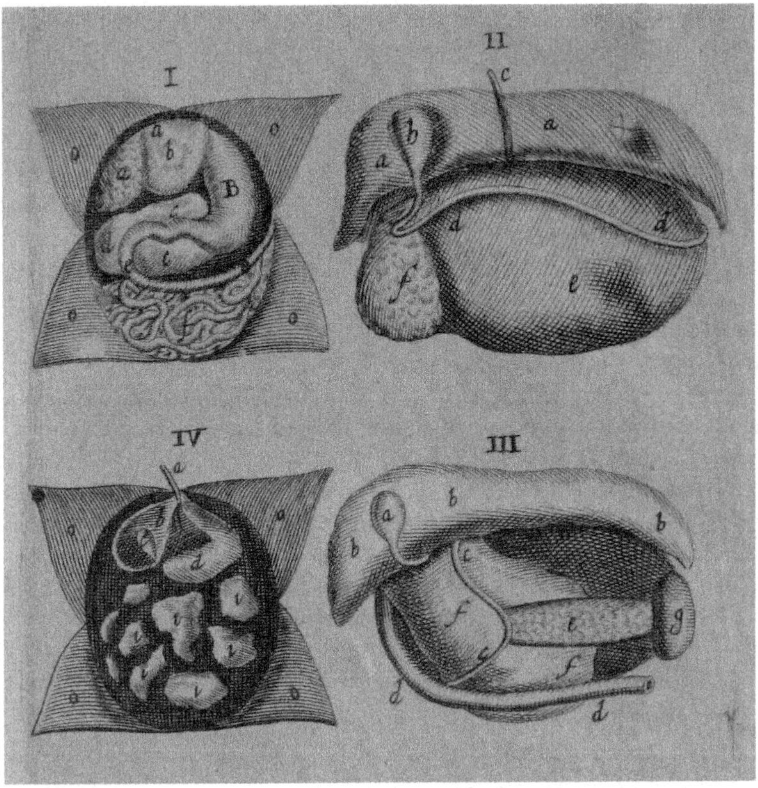

Illustration 10.16 Blasius, *Observationes*, 1677, diseased liver, Plate IV, Figures I–IV. (Courtesy of Special Collections, Regenstein Library, University of Chicago)

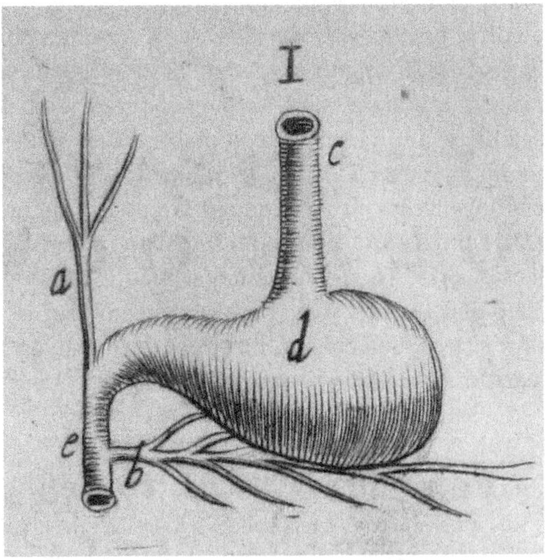

Illustration 10.17 Blasius, *Observationes*, 1677, biliary and pancreatic ducts, Plate VII Figure I. (Courtesy of Special Collections, Regenstein Library, University of Chicago)

general appearance and the occurrence of related affections with which he was familiar, having encountered ten cases. He mentioned two references in the literature: one reported by Schrader, which involved paralysis of the lower limbs, and which the author attributed to the mother's "fortem lapsus imaginationis" in falling from a ladder; the other by Tulp, "Antecessore meo piae memoriae." Ruysch agreed with him that opening the tumor hastened death. The plate (Illustration 10.19) follows Tulp's convention in relying on two figures, one of the child and the other of the spine (Illustration 10.2). Here, however, Ruysch introduced subtle and effective variations: the back of the child is shown with the tumor still attached, not excised; the figure on the left shows three vertebrae of another child affected by the same condition. The detail resembling a grasping hand is shown life size and effectively highlights the condition of the spine.[51]

Another plate in Ruysch's work involves a different representation technique with two figures showing the same specimen from different angles. The plate shows seven vertebrae of a dramatically twisted spine affected by ankylosis or bone fusion, in which the first is joined to the last, as shown at B and C in Figure 54 (Illustration 10.20). This time Ruysch had no case history and could only guess at the patient's state; of course, the subject would have had a hunched back, but Ruysch was baffled by the curvature, which he judged would have damaged the spinal cord. He also thought that if this had been a congenital condition ("morbus congenitus"), it would not have been so remarkable, but he believed the vertebrae became incurved after birth. The specimen was preserved together with related ones in his museum.[52]

10.4 Concluding Reflections

The collections of *Observationes* published in Amsterdam about the second half of the seventeenth century provide rich material for reflection about contemporary medical practices. *Observationes* were often collegial affairs: cures and especially dissections routinely included both surgeons and physicians, documenting their extensive collaboration and exchanges, and reinforcing the reliability of their accounts; at times they included medical students as well, as part of their clinical training. Works by van Meek'ren, van Roonhuyse, Blasius, and Ruysch are especially generous in mentioning witnesses. The interplay between Latin and vernacular editions and translations of several works highlights the mixed audience of these works. The professional affiliation of my protagonist is also worth reviewing: Tulp and Ruysch were both physicians and lecturers to the surgeons' guild, thus they shared medical and surgical concerns; van Meek'ren and van Roonhyuse were learned surgeons, who showed extensive familiarity with the medical and surgical

[51]Ruysch, 45–7; and Schrader, 204–5.

[52]Ruysch, 87–8. Notice that in both figures the orientation of the letters does not agree with that of the shadows, supposing the light to be coming from the left.

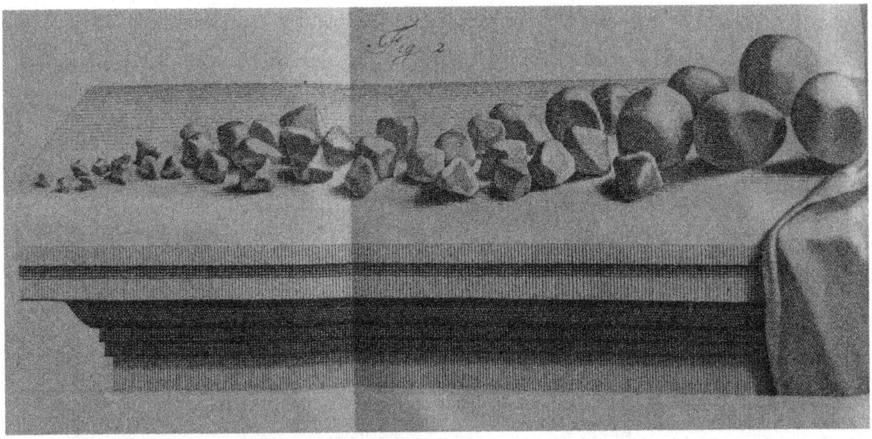

Illustration 10.18 Ruysch, *Observationes*, 1691, Figure 2, bladder stones. (Private collection)

literature in Latin; de Heide was trained as both a surgeon and a physician; Kerckring, Blasius, Schrader, and Stalpart were physician-anatomists with an interest in healthy and diseased structures.[53]

The collections of *Observationes* we have examined contain several illustrations of rare and remarkable cases. Both physician-anatomists and surgeons privileged extraordinary cases over mapping common and uncommon lesions. Although many of them concern disease, overall they cannot be characterized as "pathological" in that their contents were broader. Whereas some involved only human cases, others like Ruysch included humans and animals alike, while Tulp discussed exotic animals. Some included major malformations, often characterized as "monsters", whereas Blasius used the term "monster" more restrictively and grouped them separately. At times a malformation, such as Ingeltje Thoveling's lack of the right ovary, was seen as the cause of her complaint. Boundaries among different categorizations were rather fluid. Moreover, there was no self-conscious statement on the part of the authors of the illustrated Amsterdam *Observationes* of the birth of a new genre: this occurred at the end of the eighteenth century when Leiden professor of anatomy Eduard Sandifort stated "sic nata est anatome pathologica picta", or "thus is born illustrated pathological anatomy" in *Museum anatomicum* (Leiden, 1693).[54] Thus it would seem more accurate to consider the Amsterdam *Observationes* as akin to illustrated medical, surgical, anatomical, and even natural historical cabinet of curiosities.

There was not a single motive justifying the inclusion of illustrations, a fact underscored by the wide range of subjects. Many plates in the works of physicians

[53]Temkin 1951.

[54]Sandifort 1793–1835, vol. 1, quotation at v. See Bertoloni Meli, "The Rise of Pathological Illustrations: Baillie, Bleuland, and their Collections," forthcoming in *Bulletin for the History of Medicine*.

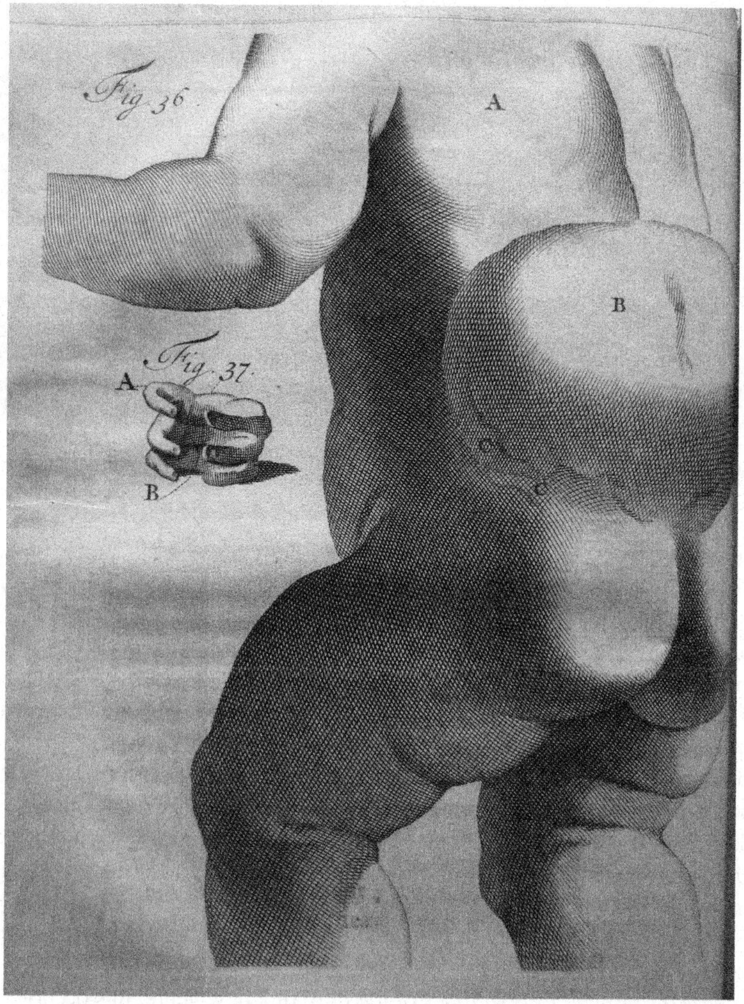

Illustration 10.19 Ruysch, *Observationes*, 1691, Figures 36–7, spina bifida (Private collection)

and especially surgeons, starting from Fabricius Hildanus, were included to advertise the author's skill and therapeutic successes, though similar motives are not always present, as for several images from postmortems – though some of these cases may offer retrospective justification for therapies. Ruysch's plates too can be seen as an advertisement, not of his therapeutic but of his preserving skills, embodied in his remarkable museum; he stated that many items discussed in the *Observationes* could be seen in his museum.[55] In many cases images provided

[55]Ruysch 1691, see for example 88, 90, 100, 108.

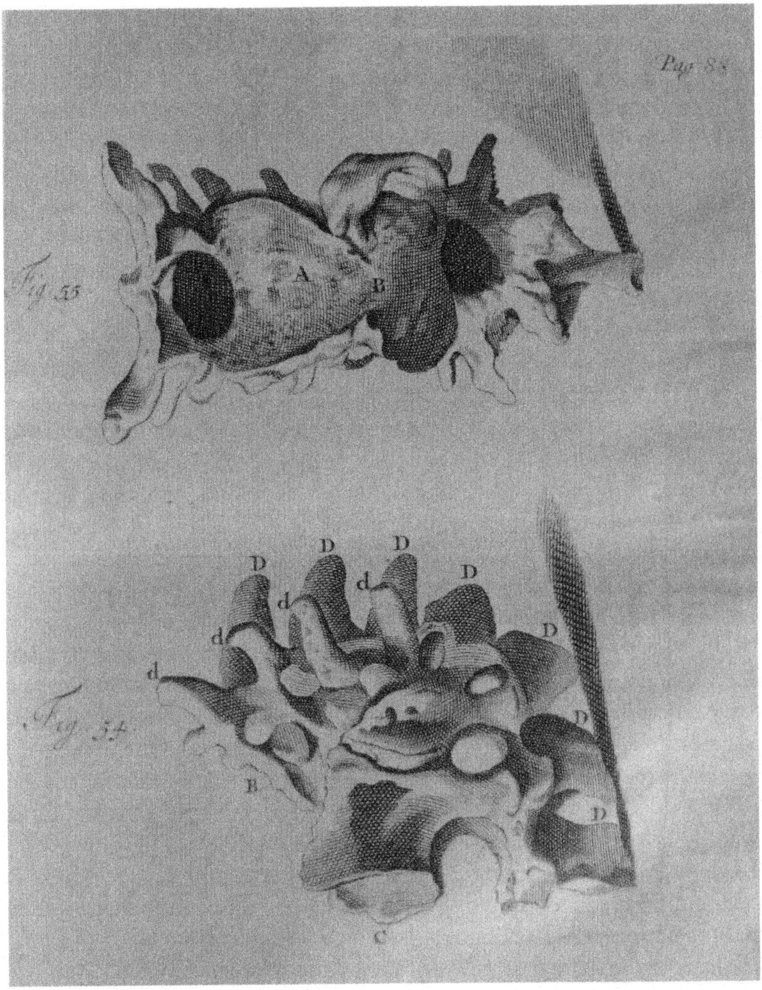

Illustration 10.20 Ruysch, *Observationes*, 1691, Figures 53–4, spine ankylosis. (Private collection)

visual documentation, almost a form of witnessing, or rare and remarkable occurrences.

The illustrations in the Amsterdam *Observationes* show notable differences not only in the topic but also in the quality of execution, ranging from the less than mediocre, as in Blasius's often diminutive figures, to the elegant quarto plates in Ruysch's treatise. Looking at the medical plates we have found a series of related dichotomies or tensions between different aspects. The first one is between external and internal conditions: external ones, especially growths, were a surgical domain and could be effectively represented; most internal conditions would not have been amenable to surgical treatment and would have been found postmortem. Closely related to this dichotomy is the one between structure and texture; artists found the former more amenable to representation, whereas the latter proved intractable in most of the plates we have seen. An effective representation of different textures in bones was produced a couple of decades later by artists working for leading London surgeon William Cheselden by combining line engraving and etching.[56] Lastly, we have seen the tension between representing conditions *in situ* or outside the body; representations *in situ* have the advantage of showing the affected body part in relation to surrounding ones, including position and relative size. However, *in situ* representations can often be cluttered and confusing, with the juxtaposition of organs and body part. By contrast, images of diseased parts seen in isolation can be hard to interpret; this was the reason why Tulp, Blasius, and Ruysch at times combined the two approaches, as with spina bifida and a liver condition. Ruysch also showed the same specimen outside the body from different angles.

With regard to the specific problem of representing disease, the plates we have seen raise several questions. In the early nineteenth century medical men produced illustrated pathology works providing comprehensive and systematic accounts of both rare and common cases; nothing of the sort occurred in the seventeenth century. The printing techniques available around 1800 were much more advanced in representing texture than those routinely employed over a century earlier and, in addition, could also involve color. This, however, was not the only or even the main issue. Rather, early modern physicians generally privileged seeking the causes of diseases; these could be ascertained from a variety of factors, including patients' narrative, some form of physical examination, and also postmortems. Overall, even with postmortems, the focus was on identifying the location of the disease in order to ascertain the organ involved, the cause of the disease, and eventually the therapy. In this framework the different specific affections of an organ were not especially significant or interesting, unless they were seen as exceptional in some form. Thus early modern physicians seeking correlations between symptoms and postmortem lesions would privilege identifying which internal organ was affected over exactly how that organ appeared. By contrast, surgeons were less interested in causes and more in the actual appearance of lesions, on which they had to act.[57]

Regardless of the initial motivation for the inclusion of illustrations, those images were later used in many ways. Images and texts did not simply represent and describe specimens, but also generated new layers of meaning. On the one hand, the works we have discussed privileged unusual and extraordinary cases; on the other, they also relied on the previous literature in order to compare case histories and images. The illustrated collections of observations by Fabricius Hildanus were frequently cited in this regard. The works we have investigated involve networks of cross references that we have been able to uncover and investigate only in part; it is clear, however, that both textual and visual objects form an integral part of those networks. Without excluding a range of other roles

[56] Kemp 1993, 107–8.

[57] See Temkin 1951; Rodari 1996; Goldschmied; Givens et al. 2006.

and purposes, both texts and images had a cognitive significance, as cases from heart polyps to spina bifida show, regardless of whether later conceptualizations of these conditions correspond to seventeenth-century ones. Thus it is not surprising if visual representations of diseased states – especially high-quality ones – could be seen retrospectively as examples in the history of pathological illustrations, as Lebert's comments on Ruysch show.

Appendix 1: Some of the Main Works Authored or Edited by Gerardus Blasius

Gerardus Blasius, *De nephritide*. Leiden: Moyaert, 1648

Johann Vesling, *Syntagma anatomicum, commentariis illustratum*. Amsterdam: Apud Joannem Janssonium, 1659.

Philipp Müller, *Miracula chymica et mysteria medica libris quinque enucleata. Ex recensione Gerardi Blasii*. Amsterdam: Apud Aegidium Jansonium Valckenier, 1659.

Jean Béguin, *Tyrocinium chymicum commentariis illustratum à Gerardo Blasio*. Amsterdam: Egidium Valkenier and Casper Commelijn, 1659.

Gerardus Blasius, *Medicina generalis, nova accurataque methodo fundamenta exhibens*. Amsterdam: Apud P. van den Berge, 1661.

Lorenzo Bellini, *Exercitatio anatomica de structura et usu renum, cui renum monstrosorum exempla, ex medicorum celebrium scriptis, addidit Gerardus Blasius*. Amsterdam: sumptibus Andreae Frisii, 1665.

Pierre Morel, *Methodus praescribendi formulas remediorum. Cum adjuncto Materiae medicae systemate. Aucta, variisque modis illustrata, nunc pro 2. ed. recensita a Gerardo Blasio*. Amsterdam: Apud Casparum Commelinum, 1665 (also 1680).

Fortunio Liceti, *De monstris, ex recensione Gerardi Blasii, qui monstra quædam nova & rariora ex recentiorum scriptis addidit*. Amsterdam: sumptibus Andreae Frisii, 1665.

Gerardus Blasius, *Medicina universa; hygieines & therapeutices fundamenta methodo nova brevissime exhibens*. Amsterdam: Petrus vanden Berge, 1665.

Johann Vesling, *Syntagma anatomicum, commentario atque appendice ex veterum, recentiorum, propriisque, observationibus, illustratum et auctum*. Amsterdam: apud Joannem Janssonium à Waesberge, 1666.

Gerardus Blasius, *Anatome medullae spinalis et nervorum inde provenientium*. Amsterdam: Apud Casparum Commelinum, 1666.

Gerardus Blasius, *Anatome contracta in gratiam discipulorum conscripta, & edita*. Amsterdam: apud Gerbrandum Schagen, 1666.

Gerardus Blasius, *Institutionum medicarum compendium, Disputationibus XII in Amstel. Athenaeo publice ventilatis, absolutum*. Amsterdam: Petrus van den Bergh, 1667.

Gerardus Blasius, *Miscellanea anatomica, hominis, brutorumque variorum, fabricam diversam magna parte exhibentia.* Amsterdam: Casper Commelijn, 1673.

Gerardus Blasius, *Observata anatomica in homine, simiâ, equo, vitulo, ove, testudine, echino, glire, serpente, ardeâ, variisque animalibus aliis. Accedunt extraordinaria in homine reperta praxin medicam aeque ac anatomen illustrantia.* Leiden: Apud Gaasbeeck, 1674.

Gerardus Blasius, *Zootomiae, seu anatomes variorum animalium pars prima.* Amsterdam: apud Abrahamum Wolfgang, 1676 [i.e. 1677].

Gerardus Blasius, *Observationes medicae rariores. Accedit monstri triplicis historia.* Amsterdam: Apud Abrahamum Wolfgang, 1677.

Gerardus Blasius, *Medicina curatoria: methodo nova in gratiam discipulorum conscripta.* Amsterdam: Apud Henricum & Theodorum Boom, 1680.

Gerardus Blasius, *Anatome animalium terrestrium variorum, volatilium, aquatilium, serpentum, insectorum, ovorumque, structuram naturalem: ex veterum recentiorum, propriisque observationibus proponens, figuris variis illustrata.* Amsterdam: Sumptibus viduæ Joannis à Someren, Henrici & viduæ Theodori Boom, 1681.

Thomas Willis, *Opera omnia, nitidius, quam unquam hactenus edita, plurimum emendata, indicibus rerum copiosissimis, ac distinctione characterum exornata.* Amsterdam: apud Henricum Wetstenium, 1682.

Appendix 2: Illustrated Collections of *Observationes*, Whether *medicae*, *chirurgicae*, or *anatomicae*, Published in Amsterdam, 1641–1691

Tulp, Nicolaas, *Observationes medicae.* Amsterdam: Apud Ludovicum Elzevirium, 1641. Second expanded edition, 1652. Reprinted in Amsterdam by Daniel Elzevier, 1672 and Henricus Wetstenius, 1685.

Roonhuyse, Hendrik van, *Heel-konstige aanmerkkingen betreffende de gebreekken der vrowven.* Amsterdam: by de Weduwe van Theunis Jacobsz. Lootsman, 1663, second edition 1672.

Collegium privatum Amstelodamensis, *Observationes anatomicae selectiores.* Amsterdam: Apud Casparum Commelinum, 1667–73.

Roonhuyse, Hendrik van, *Genees en heel-konstige aanmerkingen.* Amsterdam: By de Weduwe van Theunis Jacobsz. Lootsman, 1672.

Meek'ren, Job van, *Heel- en geneeskonstige aanmerkkingen.* Amsterdam: Casper Commelijn, 1668. Latin translation by Abraham Blasius, *Observationes medico-chirurgicae.* Amsterdam: Ex officina Henrici & Viduae Theodori Boom, 1682.

Kerckring, Theodor, *Spicilegium anatomicum, continens observationum anatomicarum rariorum centuriam unam.* Amsterdam: Andreas Frisius, 1670.

Justus Schrader, *Observationes et historiae omnes et singulae è Guiljelmi Harvei libello 'De generatione animalium' ecxerptae, & in accuratissimo ordine redactae ... Accedunt ovi faecundi singulis ab incubatione diebus factae inspectiones; ut et observationum anatomico-med[icarum] decades quatuor.* Amsterdam: Abraham Wolfgang, 1674.

Blasius, Gerardus, *Observationes medicae rariores. Accedit monstri triplicis historia.* Amsterdam: Abraham Wolfgang, 1677.

Blankaart, Steven, *Collectanea medico-physica, oft Hollands Jaar-Register der genees-en natuur-kundige Aanmerkingen van gantsch Europa.* Amsterdam: Johan ten Hoorn, 1680–6, 3 volumes.

Stalpart vander Wiel, Cornelis, *Hondert seldzame aanmerkingen so in de genees- als heel- en sny-konst.* Amsterdam: Johan ten Hoorn, 1682 (later additions, with Latin and French translations, appeared in The Hague, Leiden, and Paris).

de Heide, Antonius, *Anatome mytuli, Belgicè mossel, structuram elegantem eius motumque mirandum exponens, nec non centuria observationum medicarum.* Amsterdam: Apud Janssonio-Waesbergios, 1683 (some copies dated 1684).

Ruysch, Frederik, *Observationum anatomico-chirurgicarum centuria: accedit catalogus rariorum, quae in Museo Ruyschiano asservantur.* Amsterdam: Apud Henricum & Viduam Theodori Boom, 1691.

Bauhin, Caspar. 1621. *Theatrum anatomicum.* Frankfurt/M: Johann Theodor de Bry.

Bibliography

Acheson, Katherine. 2010. Gesner, Topsell, and the Purposes of Pictures in Early Modern Natural Histories. In *Printed Images in Early Modern Britain*, ed. Michael Hunter, 127–144. Farnham: Ashgate.

Alpers, Svetlana. 1983. *The Art of Describing: Dutch Art in the Seventeenth Century.* Chicago: University of Chicago Press.

Anderson, Nancy. 2009. Eye and Image: Looking at Visual Studies of Science. *Historical Studies in the Natural Sciences* 39: 115–125.

Baigrie, Brian S., ed. 1996. *Picturing Knowledge: Historical and Philosophical Problems Concerning the Use of Art in Science.* Toronto: University of Toronto Press.

Bartholin, Thomas. 1654, 1657, 1661. *Historiarum anatomicarum rariorum centuriae*, 3 vols. Copenhagen: Haubold.

———. 1663–1667. *Epistolarum medicinalium centuria I–IV*, 3 vols. Copenhagen: Typis Matthiæ Godicchenii, impensis Petri Haubold.

Bates, Alan W. 2005. *Emblematic Monsters: Unnatural Conceptions and Deformed Births in Early Modern Europe.* Amsterdam: Rodopi.

Bauhin, Caspar. 1605. *Theatrum Anatomicum.* Frankfurt/M: Matthaeus Becker.

Belloni, Luigi. 1956. Introduzione storica alla patologia dell'aterosclerosi. In *Società italiana di patologia, Aterosclerosi umana e sperimentale*, 25–66. Milan: Ambrosiana.

Bertoloni Meli, D. 2001. Blood, Monsters, and Necessity in Malpighi's *De polypo cordis. Medical History* 45: 511–522.

———. 2011. *Mechanism, Experiment, Disease: Marcello Malpighi and Seventeenth-Century Anatomy.* Baltimore: Johns Hopkins University Press.

———. 2013a. Early Modern Experimentation on Live Animals. *Journal of the History of Biology* 46: 199–226.

———. 2013b. Of Snails and Horsetails: Anatomical Empiricism in the Early Modern Period. *Early Science and Medicine* 18: 435–452.

———. 2015. Visual Representations of Disease: The *Philosophical Transactions* and William Cheselden's *Osteographia. Huntington Library Quarterly* 78 (2): 157–186.

Beukers, Harm. 1987–88. Clinical teaching in Leiden from Its Beginning until the End of the Eighteenth Century. *Clio Medica* 21: 139–152.

Bonet, Théophile. 1679. *Sepulchretum; sive anatomia practica, ex cadaveribus morbo denatis.* Geneva: Leonard Chouët.

———. 1700. *Sepulchretum; sive anatomia practica, ex cadaveribus morbo denatis.* Lyon and Geneva: Cramer et Perachon.

Bontius, Jacobus. 1642. *Medicina indorum.* Leiden: Franciscus Hackius.

Cole, Francis Joseph. 1944. *A History of Comparative Anatomy.* London: Macmillan.

Collegium privatum Amstelodamense. 1975. *Observationes anatomica selectiores*, facsimile edition, ed. Gerrit Arie Lindeboom. Nieuwkoop: de Graaf.

Cook, Harold J. 1992. The New Philosophy in the Low Countries. In *The Scientific Revolution in National Context*, ed. Roy Porter and Mikulás Teich, 115–149. Cambridge: Cambridge University Press.

———. 2007. *Matters of Exchange. Commerce, Medicine, and Science in the Dutch Golden Age.* New Haven: Yale University Press.

Cunningham, Andrew. 1985. Fabricius and the 'Aristotle Project' in Anatomical Teaching and Research at Padua. In *The Medical Renaissance of the Sixteenth Century*, ed. Andrew Wear, Roger K. French, and Iain M. Lonie, 195–222. Cambridge: Cambridge University Press.

———. 2010. *The Anatomist Anatomis'd.* Farnham: Ashgate.

Dackerman, Susan, ed. 2011. *Prints and the Pursuit of Knowledge in Early Modern Europe.* Cambridge, MA: Harvard Art Museum.

Daston, Lorraine, and Katharine Park. 1998. *Wonders and the Order of Nature, 1150–1750.* New York: Zone Books.

Daston, Lorraine, and Peter Galison. 2007. *Objectivity.* New York: Zone Books.

de Moulin, Daniel. 1988. *A History of Surgery with Emphasis on the Netherlands.* Dordrecht: Martinus Nijhoff.

Donato, Maria Pia. 2011. Il normale, il patologico, e la sezione cadaverica in età moderna. *Quaderni Storici* 136: 75–97.

Findlen, Paula. 1994. *Possessing Nature: Museums, Collecting, and Scientific Culture in Early Modern Italy.* Berkeley: University of California Press.

French, Roger. 2001. *Canonical Medicine: Gentile da Foligno and Scholasticism.* Leiden: Brill.

Goldschmid, Edgar. 1925. *Entwicklung und Bibliographie der pathologisch-anatomischen Abbildung.* Leipzig: Hiersemann.

Guerrini, Anita. 2010. The King's Animals and the King's Books: The Illustrations for the Paris Academy's 'Histoire des animaux'. *Annals of Science* 67: 383–404.

Hansen, Julie V. 1996. Resurrecting Death: Anatomical Art in the Cabinet of Dr. Frederik Ruysch. *Art Bulletin* 78: 663–679.

Hendriksen, Marieke M.A. 2015. *Elegant Anatomy: The Eighteenth-Century Leiden Anatomical Collections.* Leiden: Brill.

Hildanus, Guilhelmus Fabricius. 1646. *Opera.* Frankfurt: Sumptibus J. Beyeri.

Huisman, Tim. 2009. *The Finger of God. Anatomical Practice in Seventeenth-Century Leiden.* Leiden: Primavera.

Impey, Oliver, and Arthur Macgregor, eds. 1985. *The Origins of Museums: The Cabinet of Curiosities in Sixteenth- and Seventeenth-Century Europe.* Oxford: Clarendon Press.

Jones, Ellis W. P. 1960. The Life and Works of Guilhelmus Fabricius Hildanus (1560–1634). *Medical History* 4: 112–134; 196–209.

Jorink, Erik. 2010. *Reading the Book of Nature in the Dutch Golden Age, 1575–1715.* Leiden: Brill.

Kemp, Martin. 1993. 'The Mark of Truth': Looking and Learning in Some Anatomical Illustrations in the Renaissance and Eighteenth Century. In *Medicine and the Five Senses*, ed. William F. Bynum and Roy Porter, 85–121. Cambridge: Cambridge University Press.

———. 2010. Style and Non-Style in Anatomical Illustration: From Renaissance Humanism to Henry Gray. *Journal of Anatomy* 216: 192–208.

Knoeff, Rita. 2015. Touching Anatomy: On the Handling of Preparations in the Anatomical Cabinets of Frederik Ruysch (1638–1731). *Studies in History and Philosophy of Science* 49: 32–44.

Kooijmans, Luuc. 2010. *Death Defied: The Anatomy Lessons of Frederik Ruysch*. Leiden: Brill.

Kusukawa, Sachiko. 2012. *Picturing the Book of Nature: Image, Text, and Argument in Sixteenth-Century Human Anatomy and Medical Botany*. Chicago: University of Chicago Press.

Lebert, Hermann. 1857–61. *Traité d'anatomie pathologique générale et speciale*, 4 vols. Paris: Ballière.

Liedtke, Alter. 2007. *Dutch Paintings in the Metropolitan Museum of Art*, 2 vols. New Haven: Yale University Press.

Lindeboom, Gerrit A. 1984. *Dutch Medical Biography*. Amsterdam: Rodopi.

Long, Pamela. 2011. *Artisan/Practitioners and the Rise of the New Science, 1400–1600*. Corvallis: Oregon State University Press.

Malpighi, Marcello. 1995. *De polypo cordis*. An Annotated Translation, trans. and ed. John M. Forrester, *Medical History* 39: 477–492.

Margócsy, Daniél. 2009. Advertising Cadavers in the Republic of Letters: Anatomical Publications in Early Modern Netherlands. *British Journal of the History of Science* 42: 187–210.

———. 2011. A Museum of Wonders or a Cemetery of Corpses? The Commercial Exchange of Anatomical Collections in Early Modern Netherlands. In *Silent Messengers: The Circulation of Material Objects of Knowledge in the Early Modern Low Countries*, ed. Sven Dupré and Christoph Lüthy, 185–215. Berlin: Lit.

Mooij, Annet. 1999. *Doctors of Amsterdam. Patient Care, Medical Training and Research (1650–2000)*. Amsterdam: Amsterdam University Press.

Ogilvie, Brian. 2003. Image and Text in Natural History, 1500–1700. In *The Power of Images in Early Modern Science*, ed. Wolfgang Lefèvre, Jürgen Renn, and Urs Schoepflin, 144–166. Basel: Birkhäuser.

———. 2006. *The Science of Describing: Natural History in Renaissance Europe*. Chicago: University of Chicago Press.

Olmi, Giuseppe. 1992. *L'inventario del mondo*. Bologna: il Mulino.

Parapia, Liakat A., and Carolyn Jackson. 2008. Ehlers-Danlos Syndrome–A Historical Review. *British Journal of Hœmatology* 141: 32–35.

Park, Katharine. 1998. Impressed Images: Reproducing Wonders. In *Picturing Science, Producing Art*, ed. Caroline A. Jones and Peter Galison, 254–271. New York: Routledge.

———. 2006. *Secrets of Women: Gender, Generation, and the Origins of Human Dissection*. New York: Zone Books.

———. 2011. Observation in the Margins, 500–1500. In *Histories of Scientific Observation*, ed. Lorraine Daston and Elizabeth Lunbeck, 15–44. Chicago: University of Chicago Press.

Pomata, Gianna. 2005. Praxis historialis: The Uses of *Historia* in Early Modern Medicine. In *"Historia": Empiricism and Erudition in Early Modern Europe*, ed. Nancy G. Siraisi and Gianna Pomata, 105–146. Cambridge: MIT Press.

———. 2010. Sharing Cases: The *Observationes* in Early Modern Medicine. *Early Science and Medicine* 15: 193–236.

———. 2011a. A Word of the Empirics: The Ancient Concept of Observation and Its Recovery in Early Modern Medicine. *Annals of Science* 68: 1–25.

———. 2011b. Observation Rising: Birth of an Epistemic Genre, ca. 1500–1650. In *Histories of Scientific Observation*, ed. Lorraine Daston and Elizabeth Lunbeck, 45–80. Chicago: University of Chicago Press.

Ragland, Evan. 2008. Experimenting with Chemical Bodies: Reinier de Graaf's Investigations of the Pancreas. *Early Science and Medicine* 13: 615–664.

Ricci, James V. 1949. *The Development of Gynaecological Surgery and Instruments*. Philadelphia: Blakiston.

Roberts, K.B., and J.D.W. Tomlinson. 1992. *The Fabric of the Body. European Traditions of Anatomical Illustration*. Oxford: Clarendon Press.

Rodari, Florian, ed. 1996. *Anatomie de la couleur. L'invention de l'estampe en couleur*. Paris/ Lausanne: Bibliothèque Nationale de France/Musée Olympique.

Sandifort, Eduard. 1793–1835. *Museum anatomicum Academiae Lugduno-Batavae*, 4 vols. Leiden: Luchtmans.

Sappol, Michael. 2006. *Dream Anatomy*. Bethesda: U.S. Department of Health and Human Services.

Schouten, Jan. 1974. Johannes Walaeus (1604–1649) and His Experiments on the Circulation of the Blood. *Journal of the History of Medicine* 29: 259–279.

Silver, Larry. 2006. *Peasant Scenes and Landscapes: The Rise of Pictorial Genres in the Antwerp Art Market*. Philadelphia: University of Pennsylvania Press.

Siraisi, Nancy. 2001a. Segni evidenti, teoria e testimonianza nelle narrazioni di autopsie del Rinascimento. *Quaderni Storici* 108: 719–744.

———. 2001b. Signs and Evidence: Autopsy and Sanctity in Late-Sixteenth Century Italy. In *Medicine and the Italian Universities, 1250–1600*, 356–380. Leiden: Brill.

———. 2001c. 'Remarkable' Diseases, 'Remarkable' Cures, and Personal Experience in Renaissance Medical Texts. In *Medicine and the Italian Universities*, 226–252. Leiden: Brill.

———. 2001d. Vesalius and Human Diversity. In *Medicine and the Italian Universities*, 287–327. Leiden: Brill.

———. 2013. *Communities of Learned Experience: Epistolary Medicine in the Renaissance*. Baltimore: Johns Hopkins University Press.

Soojung-Kim Pang, Alex. 1997. Visual Representation and Post-Constructivist History of Science. *Historical Studies in the Physical and Biological Sciences* 28: 139–171.

Stalpart vander Wiel, Cornelius. 1682. Hondert seldzame aanmerkingen. Amsterdam: n.p.

Steno, Nicolaus. 1986. *Opere scientifiche*, ed. Enrico Coturri, 2 vols. Prato: Cassa di Risparmi e Depositi.

Swan, Claudia. 2006. The Uses of Realism in Early Modern Illustrated Botany. In *Visualizing Medieval Medicine and Natural History, 1200–1550*, ed. Jean A. Givens, Karen M. Reeds, and Alain Touwaide, 239–249. Aldershot: Ashgate.

———. 2008a. Making Sense of Medical Collections in Early Modern Holland: The Uses of Wonder. In *Making Knowledge in Early Modern Europe: Practices, Objects, and Texts, 1400–1800*, ed. Pamela H. Smith and Benjamin Schmidt, 199–213. Chicago: University of Chicago Press.

———. 2008b. The Uses of Botanical Treatises in the Netherlands ca. 1600. In *The Art of Natural History: Illustrated Treatises and Botanical Paintings, 1400–1850*, ed. Therese O'Malley and Amy R.W. Meyers, 63–81. Washington, DC: National Gallery of Art.

———. 2011. Illustrated Natural History. In *Prints and the Pursuit of Knowledge in Early Modern Europe*, ed. Susan Dackerman, 186–190. Cambridge, MA: Harvard Art Museum.

Temkin, Owsei. 1951. The Role of Surgery in the Rise of Modern Medical Thought. *Bulletin of the History of Medicine* 25: 248–259.

Trioen, Cornelis. 1743a. *Observationum medico-chirurgicarum fasciculus*. Leiden: Petrum van der Eyk et Jacobum van der Kluis.

———. 1743b. *Genees- en heelkundige waarneemingen*. Leiden: Petrum van der Eyk et Jacobum van der Kluis.

Tyson, Edward. 1678–79. Anatomical Observations. *Philosophical Transactions* 12: 1035–1039.

van de Roemer, G.M. (Bert). 2010. From *vanitas* to Veneration. The Embellishments in the Anatomical Cabinet of Frederik Ruysch. *Journal of the History of Collections* 22: 169–186.

van Heel, Sebastien A.C. Dudok. 1998. *Nicolaes Tulp: The Life and Work of an Amsterdam Physician and Magistrate in the Seventeenth Century*. Amsterdam: Six Art Promotion.
van Meekeren, Job. 1979. *Heel- en geneeskonstike aenmerkingen*. Rotterdam: Alphen aan den Rijn.
van Roonhuyse, Hendrik. 1672. *Genees en heel-konstige aanmerkingen*. Amsterdam: de Weduwe van Theunis Jacobsz.
van Miert, Dirk. 2009. *Humanism in an Age of Science: the Amsterdam Athenaeum in the Golden Age, 1632–1704*. Trans. Michiel Wielema, with Anthony Ossa-Richardson. Leiden: Brill.
Vesalius, Andreas. 1543. *De humani corporis fabrica*. Basel: Oporinus.
von Gersdorff, Hans. 1517. *Feldtbüch der Wundtartzney*. Strassburg: Schott.
Weber, Giorgio. 2006. *Sensata veritas: l'affiorare dell'anatomia patologica, ancora innominata, in scritti di anatomisti del '500*. Florence: Olschki.
Wilkin, Rebecca M. 2008. Essaying the Mechanical Hypothesis: Descartes, La Forge, and Malebranche on the Formation of Birthmarks. *Early Science and Medicine* 13: 533–567.

Publications of Nancy Siraisi

Monographs

Arts and Sciences at Padua: The Studium of Padua Before 1350. Toronto: Pontifical Institute of Mediaeval Studies, 1973.

Taddeo Alderotti and His Pupils: Two Generations of Italian Medical Learning. Princeton: Princeton University Press, 1981.

Avicenna in Renaissance Italy: The Canon and Medical Teaching in Italian Universities After 1500. Princeton: Princeton University Press, 1987.

Medieval and Early Renaissance Medicine: An Introduction to Knowledge and Practice. Chicago: University of Chicago Press, 1990.

The Clock and the Mirror: Girolamo Cardano and Renaissance Medicine. Princeton: Princeton University Press, 1997.

Medicine and the Italian Universities, 1250–1600. Leiden: Brill, 2001. (A collection of my previously published articles with a new introduction.)

History, Medicine, and the Traditions of Renaissance Learning. Ann Arbor: University of Michigan Press, 2007.

Communities of Learned Experience: Epistolary Medicine in the Renaissance. Baltimore: Johns Hopkins University Press, 2013.

Edited Collections

Luke Demaitre and Nancy G. Siraisi, eds., *Science, Medicine and the University: Essays in Honor of Pearl Kibre*, published as two issues of *Manuscripta*, vol. 20, July and November, 1976.

A. C. Crombie and Nancy G. Siraisi, eds., *The Rational Arts of Living*. Smith College Studies in History, vol. 50. Northampton, MA, 1987.

G. Manning, C. Klestinec (eds.), *Professors, Physicians and Practices in the History of Medicine*, Archimedes 50, DOI 10.1007/978-3-319-56514-9

Michael R. McVaugh and Nancy G. Siraisi, eds., *Renaissance Medical Learning: Evolution of a Tradition*, *Osiris* 6 (1990).
Anthony Grafton and Nancy G. Siraisi, eds., *Natural Particulars: Nature and the Disciplines in Renaissance Europe*. Cambridge, MA: MIT Press, 1999.
Gianna Pomata and Nancy G. Siraisi, eds., *Historia: Empiricism and Erudition in Early Modern Europe*. Cambridge, MA: MIT Press, 2005.

Articles in Refereed Journals

"The Expositio Problematum Aristotelis of Peter of Abano," *Isis* 61 (1970): 321–39.
"Humanism and Evangelism: Italian Religious Thought and Expression in the Fourteenth and Fifteenth Centuries," *Montclair Journal of Social Sciences and Humanities* 2 (1973): 131–37 (essay review).
"The Music of Pulse in the Writings of Some Italian Physicians of the Fourteenth and Fifteenth Centuries," *Speculum* 50 (1975): 689–710.
"Matheolus of Perugia's Commentary on the Preface to the Aphorisms of Hippocrates," *Bulletin of the History of Medicine* 49 (1975): 405–28. With Pearl Kibre.
"The libri morales in the Faculty of Arts and Medicine at Bologna: Bartolomeo da Varignana and the Pseudo-Aristotelian Economica," *Manuscripta* 20 (1976): 105–17.
"Concerning Caruncles or Callosities at the Neck of the Bladder: Translation of Surgical Text of Francesco Peccetti (1616)," *Urology* 10 (1977): 609–14. With Adrian Zorgniotti, M.D.
"Taddeo Alderotti and Bartolomeo da Varignana on the Nature of Medical Learning," *Isis* 68 (1977): 27–39. Reprinted in *The Scientific Enterprise in Antiquity and the Middle Ages*, Michael H. Shank, ed. Chicago: University of Chicago Press, 2001.
"The De monarchia attributed to Apuleius," *Medievalia* 7 (1981): 1–39. With Benjamin G. Kohl.
"Some Recent Work on Western European Medical Learning, ca. 1200–ca.1500," *History of Universities* 2 (1982): 225–38 (essay review).
"Some Current Trends in the Study of Renaissance Medicine," *Renaissance Quarterly* 37 (1984): 585–600.
"Renaissance Commentaries on Avicenna's Canon, Book 1, Part 1, and the Teaching of Medical theoria in the Italian Universities," *History of Universities* 4 (1984): 47–97.
"Pietro d'Abano and Taddeo Alderotti: Two Models of Medical Culture," *Medioevo* 11 (1985): 139–62.
"Medieval and Renaissance Medicine: Continuity and Diversity." Editorial, *Journal of the History of Medicine and Allied Sciences* 41 (1986), 391–94.
"Sixteenth-Century Medical Innovation Between Princely Patronage and Academic Controversy: The Case of Giovanni Argenterio." *Osiris* 6 (1990): 162–80.

"Girolamo Cardano and the Art of Medical Narrative," *Journal of the History of Ideas*, 52 (1991), 581–602.

"The New University History and Sapere e/è potere," *Universitas: Newsletter of the International Center for the History of Universities and Science*, No. 2 (April, 1992): 1–4.

"Vesalius and Human Diversity in De humani corporis fabrica," *Journal of the Warburg and Courtauld Institutes* 57 (1994): 60–88.

"Early Anatomy in Comparative Perspective: Introduction," in *Body and Culture: Early Anatomy in Comparative Perspective*, special issue of the *Journal of the History of Medicine and Allied Sciences* 50 (1995): 1–10.

"Vesalius and the Reading of Galen's Teleology," *Renaissance Quarterly* 50 (1997): 1–38.

"Anatomizing the Past: Physicians and History in Renaissance Culture." *Renaissance Quarterly* 53 (2000): 1–31.

"Segni evidenti, teoria e testimonianza nelle narrazioni di autopsie del Rinascimento." In *Fatti: storie dell'evidenza empirica*. Simona Cerutti and Gianna Pomata, eds. Special issue, *Quaderni Storici* 108 (2001): 719–44.

Introduction to *Science and Universities of Early Modern Europe: Teaching, Specialization, Professionalization*. David A. Lines, ed. Special issue, *Early Science and Medicine* 6 (2001): 259–66.

"History, Antiquarianism and Medicine: The Case of Girolamo Mercuriale." *Journal of the History of Ideas* 64 (2003): 231–51.

"Medicine and the Renaissance World of Learning," *Bulletin of the History of Medicine* 78 (2004): 1–36.

"Oratory and Rhetoric in Renaissance Medicine," *Journal of the History of Ideas* 65 (2004): 191–211.

"Medicine 1450–1620 and the History of Science," *Isis* 103 (2012): 1–24 (revised version of the 2010 History of Science Society Distinguished Lecture, presented at the annual meeting, Montreal, Quebec).

Chapters in Edited Volumes, Essays

"The Institutional Setting: The Universities," in *Science in the Middle Ages*. David C. Lindberg, ed. Chicago: University of Chicago Press, 1978. With Pearl Kibre. 120–44.

"The Medical Learning of Albertus Magnus," in *Albertus Magnus and the Sciences*. James A. Weisheipl, ed. Toronto: Pontifical Institute of Mediaeval Studies, 1980. 379–404.

Introduction to *The Black Death*. Daniel Williman, ed. Binghamton: SUNY Press, 1982.

"Reflections on Italian Medical Writings of the Fourteenth and Fifteenth Centuries," *Annals of the New York Academy of Sciences* 412 (1983): 155–68.

"The Changing Fortunes of a Traditional Text: Goals and Strategies in Sixteenth-Century Latin Editions of the Canon of Avicenna," in *The Medical Renaissance of the Sixteenth Century*. A. Wear, R.K. French and I.M. Lonie, eds. Cambridge: Cambridge University Press, 1985. 16–41.

"Dante and the Art and Science of Medicine Reconsidered," in *The Divine Comedy and the Encyclopedia of Arts and Sciences*. Giuseppe di Scipio and Aldo Scaglione, eds. Amsterdam/ Philadelphia: John Benjamin Publishing Company, 1987. 224–45.

"Changing Concepts of the Organization of Medical Knowledge in the Italian Universities, 14th to 16th Centuries," in *La diffusione delle scienze islamiche nel Medio Evo europeo*. Rome: Accademia Nazionale dei Lincei, 1987. 291–321.

"Medicine, Physiology, and Anatomy in Early Sixteenth-Century Critiques of the Arts and Sciences," in *New Perspectives in Renaissance Thought*. John Henry and Sarah Hutton, eds. London: Duckworth, 1990. 214–29.

"L'insegnamento della medicina ippocratica di Girolamo Cardano a Bologna," in *Sapere e\è potere: Discipline, Dispute e Professioni nell'Università Medievale e Moderna, Il caso bolognese a confronto*. Atti del 4o Convegno, Bologna, 13–15 aprile 1989. Vol II. Verso un nuovo sistema del sapere. Andrea Cristiani, ed. Bologna: Istituto per la Storia di Bologna, 1990. 153–71.

"The Faculty of Medicine." *A History of the University in Europe. Vol. 1. Universities in the Middle Ages*. Hilde De Ridder-Symoens, ed. Cambridge: Cambridge University Press, 1991. 360–387. Editions also in German, Spanish, and Portuguese.

"New drugs and 'new' diseases," in *New Worlds, Ancient Texts: The Power of Tradition and the Shock of Discovery*. Anthony Grafton with April E. Shelford and Nancy G. Siraisi, eds. Cambridge, MA: Harvard University Press and the New York Public Library, 1992. 176–93.

"Life Sciences and Medicine in the Renaissance World," in *Rome Reborn: The Vatican Library and Renaissance Culture*. Anthony Grafton, ed. Washington, DC: Library of Congress, 1993. 169–97 (volume accompanying exhibition at the Library of Congress).

"How to Write A Latin Book on Surgery: Organizing Principles and Authorial Devices in Guglielmo da Saliceto and Dino del Garbo," in *Practical Medicine from Salerno to the Black Death*. Luis Garcia Ballester, et al., eds. Cambridge: Cambridge University Press, 1994. 88–109.

"Il Canone di Avicenna e l'insegnamento della medicina pratica in Europa," in *L'insegnamento della Medicina in Europa (secoli XIV–XIX)*. Atti del Convegno tenutosi a Siena in occasione delle celebrazioni dei 750 anni dalla fondazione dell'Università di Siena, ed. Francesca Vannozzi. Monografie di Quaderni Internazionali di Storia della Medicina e della Sanità, No. 6. Siena: Typografia Senese, 1994. 9–21.

"Cardano, Hippocrates, and Criticism of Galen," in *Girolamo Cardano: Philosoph, Naturforscher, Arzt*. Eckhard Kessler, ed. Wiesbaden: Harrassowitz Verlag, 1994. 131–55.

"'L'individuale' nella medicina tra Medioevo e Umanesimo: I casi clinici," in *Umanesimo e medicina: il problema dell''individuale'*. Roberto Cardini and Mariangela Regoliosi, eds. Rome: Bulzoni Editore, 1996. 33–62.

"'Remarkable' Diseases and 'Remarkable' Cures: The Natural, the Supernatural, and the Extraordinary in Renaissance Medical Texts," in *Disease and Society*. Yasuo Otsuka and Shizu Sakai, eds. Tokyo: Ishiyaku EuroAmerica, Inc., 1997. 31–58.

"La comunicazione del sapere anatomico ai confini tra diritto e agiografia: due casi del secolo XVIe" in *Le forme della comunicazione scientifica*. Massimo Galuzzi, Gianni Micheli and Maria Teresa Monti, eds. Milan: Franco Angeli, 1998. 419–38.

"Cardano and the History of Medicine," in *Girolamo Cardano. Le opere, le fonti, la vita*. Marialuisa Baldi and Guido Canziani, eds. Milan: FrancoAngeli, 1998. 341–62.

Anthony Grafton and Nancy G. Siraisi, "Introduction," in *Natural Particulars: Nature and the Disciplines in Renaissance Europe*. Anthony Grafton and Nancy G. Siraisi, eds. Cambridge, MA: MIT Press, 1999. 1–21.

"In Search of the Origins of Medicine: Egyptian Medicine and Paduan Physicians," in *Generation and Degeneration*. Valeria Finucci and Kevin Brownlee, ed. Durham, NC: Duke University Press, 2001. 235–61.

"L'insegnamento della medicina," in *Storia della scienza*. Vol. 4, section 2, chapter 26. 1. Rome: Istituto dell' Enciclopedia Italiana, 2001. 443–47.

"La scienza medica," in *Storia della scienza*. Vol. 4, section 3, chapter 14. 1. Rome: Istituto dell' Enciclopedia Italiana, 2001. 852–65.

Anthony Grafton and Nancy G. Siraisi, "Between the Election and My Hopes: Girolamo Cardano and Medical Astrology," in *Secrets of Nature: Astrology and Alchemy in Early Modern Europe*. William R. Newman and Anthony Grafton, eds. Cambridge, MA: MIT Press 2001. 69–131.

"Hermes Among the Physicians," in *Das Ende des Hermetismus: Historische Kritik und neue Naturphilosophie in der Spätrenaissance*. Martin Mulsow, ed. Tübingen: Mohr Siebeck, 2002. 189–212.

"Disease and Symptom as Problematic Categories in Renaissance Medicine," in *Res et Verba in der Renaissance*. Ian Maclean and Eckhard Kessler, eds. Wiesbaden: Harrassowitz Verlag, 2002. 217–40.

"Historia, actio, utilitas: Fabrici e le scienze della vita nel Cinquecento," in *Il teatro dei Corpi. Le pitture colorate d'anatomia di Girolamo Fabrici d'Acquapendente*. Maurizio Rippa Bonati and José Pardo-Tomás, eds. Catalogue of an exhibition at the Biblioteca Nazionale Marciana, Venice, 17 December 2004–8 May 2005. Milan: Mediamed 2004. 63–73.

"Historiae, Natural history, Roman Antiquity, and Some Roman Physicians," in *Historia: Empiricism and Erudition in Early Modern Europe*. Gianna Pomata and Nancy G. Siraisi, eds. Cambridge, MA: MIT Press, 2005. 325–54.

"Medicina, scienza e vita intelletuale," in *Harvey e Padova: Atti del convegno celebrativo del quarto centenario della laurea di William Harvey, Padova,*

21–22 novembre 2002. Giuseppe Ongaro, Maurizio Rippa Bonati, and Gaetano Thiene, eds. Maria Galetta, trans. Treviso: Edizioni Antilia, 2005. 405–15.

"Medicina Practica: Girolamo Mercuriale as Teacher and Textbook Author," in *Scholarly Knowledge: Textbooks in Early Modern Europe*. Emidio Campi, Simone De Angelis, Anja-Silvia Goeing and Anthony T. Grafton, eds. Geneva: Droz, 2008. 287–305.

"Mercuriale's Letters to Zwinger and Humanist Medicine," in *Girolamo Mercuriale. Medicina e cultura nell'Europa del Cinquecento*. Atti del Convegno «Girolamo Mercuriale e lo spazio scientifico e culturale del Cinquecento» (Forlì, 8–11 novembre 2006). Alessandro Arcangeli and Vivian Nutton, eds. Florence: Leo S. Olschki, 2008. 77–95.

"Hippocrates in the Eyes of Some Sixteenth-Century Medical Commentators," in *Geschichte der Medizingeschichtsschreibung: Historiographie unter dem Diktat literarischer Gattungen von der Antike bis zur Aufklärung*. Thomas Rütten, ed. Remscheid: Gardez!, 2009. 233–63.

"The Life of Learning," *Charles Homer Haskins Lecture*, occasional papers American Council of Learned Societies, 2010. Reprinted in *Professors, Physicians and Practices in the History of Medicine: Essays in Honor of Nancy Siraisi*. Cynthia Klestinec and Gideon Manning, eds. New York: Springer, 2017.

"Theory, Experience, and Customary Practice in the Medical Writings of Francisco Sanches," in *Between Text and Patient: The Medical Enterprise in Medieval and Early Modern Europe*. Brian Nance and Florence Eliza Glaze, eds. Micrologus Library 39. Florence: SISMEL, 2011. 441–64.

"Styles of Medical Antiquarianism," in *Antiquarianism and Intellectual Life in Europe and China, 1500–1800*. Peter N. Miller and François Louis, eds. Ann Arbor, MI: University of Michigan Press, 2012. 207–21.

"Psychology in Some Sixteenth- and Early Seventeenth-Century General Works on Medicine," in *Psychology and the Other Disciplines: A Case of Cross-Disciplinary Interaction (1250–1750)*. Paul J. J. M. Bakker, Sander W. de Boer, and Cees Leijenhorst, eds. Leiden: Brill 2012. 325–43.

"Il corso della vita umana, la durata della vita, e le possibilità della medicina: dibattiti su medicina e filosofia naturale tra il XIV e il XVII secolo," in *Interpretare e curare: medicina e salute nel Rinascimento*. Andrea Carlino, Antonio Clericuzio, and Maria Conforti, eds. Andrea Carlino, trans. 232–248. Rome: Carocci editore 2013.

In Preparation

"Medical Humanism," a collection of texts and translations to be published as a volume of the *I Tatti Renaissance Library* (James Hankins, general editor). Introduction and selection of texts by Nancy G. Siraisi, translations by Craig Martin.

Book Reviews

F. Borchardt, *German Antiquity in Renaissance Myth* in *JEGP* 72 (1973): 70–71.

N.H. Steneck, *Science and Creation in the Middle Ages* in *International Studies in Philosophy* 9 (1977): 211–12.

Ian Maclean, *The Renaissance Notion of Woman* in *Bulletin of the History of Medicine* 54 (1980): 616–18.

Manlio Bellomo, *Saggio sull'università nel età del diritto comune* in *The American Historical Review* 85 (1980): 615.

Brian Lawn, *The Prose Salernitan Questions* in *Clio Medica* 16 (1981): 170.

Giovanni Santini, *Università e società nel XII secolo: Pillio da Medicina e lo studio di Modena* in *Speculum* 56 (1981): 285–86.

Tiziana Pesenti Marangon, *La biblioteca universitaria di Padova* in *The Journal of Modern History* 55 (1981): 149–50.

Luke Demaitre, *Doctor Bernard de Gordon, Professor and Practitioner* in *Bulletin of the History of Medicine* 55 (1981): 285–86.

Peter Kopp, *Vademecum eines frühmittelalterlichen Ärztes* and Wolfgang Schneider, *Ein neuer Textbeleg vom Liber praeparationum* in *Archives internationales d'histoire des sciences* 32 (1982): 332–33.

Edward J. Kealey, *Medieval Medicus: A Social History of Anglo-Norman Medicine* in *Journal of the History of Medicine and Allied Sciences* 37 (1982): 343–44.

Danielle Jacquart, *Le milieu médical en France du XIIe au XVe siècle* in *Journal of the History of Medicine and Allied Sciences* 37 (1982): 463.

Beryl Rowland, ed. and tr., *Medieval Woman's Guide to Health* in *The American Historical Review* 87 (1982): 435–36.

Sandra Tugnoli Pattaro, *Metodo e sistema delle scienze nel pensiero di Ulisse Aldrovandi* in *Isis* 3 (1982): 313.

Arnaldi de Villanova, *Opera medica omnia*, vol. 16, Michael R. McVaugh, ed., in *Isis* 74 (1983): 127–28.

Richard Palmer, *The Studio of Venice and Its Graduates in the Sixteenth Century* in *Bulletin of the History of Medicine* 58 (1984): 115–17.

Bernardo Torni, *Opuscoli filosofici e medici* in *Renaissance Quarterly* 37 (1984): 59–61.

Markus Fierz, *Girolamo Cardano, 1501–1576: Physician, Natural Philosopher, Mathematician, Astrologer and Interpreter of Dreams* in *Isis* 75 (1984): 417.

Robert S. Gottfried, *The Black Death* in *Isis* 75 (1984): 753–54.

G. Zanier, *Medicina e filosofia tra '500 e '600* in *Annals of Science* 41 (1984): 413–15.

W. Steffen, *Die studentische Autonomie im mittelalterlichen Bologna* in *History of Universities* 4 (1984): 191–93.

Stefano De Rosa, *Una biblioteca universitaria del secondo '600: La libreria di Sapienza dello studio Pisano* in *History of Universities* 4 (1984): 213–14.

Constantini Liber de coitu in *Journal of the History of Medicine and Allied Sciences* 40 (1985): 110–11.

Antonio Poppi, ed., *Scienza e filosofia all'Università di Padova nel Quattrocento* in *Renaissance Quarterly* 38 (1985): 309–12.

T. Pesenti, *Professori e promotori di medicina nello Studio di Padova dal 1404 al 1509* in *Bulletin of the History of Medicine* 59 (1985): 261.

G. Federici Vescovini, *'Arti' e filosofia nel secolo XIV* in *Speculum* 60 (1985): 493–94.

Linda E. Voigts and Michael R. McVaugh, eds., *A Latin Technical Phlebotomy and Its Middle English Translation* in *Annals of Science* 42 (1985): 184.

John E. Murdoch, *Album of Science: Antiquity and the Middle Ages* in *Bulletin of the History of Medicine*, 60 (1986): 245–46

Charles B. Schmitt, *The Aristotelian Tradition and Renaissance Universities*, in *Journal of the History of Philosophy*, 24 (1986): 408–10.

Arnaldi de Villanova *Opera medica omnia*, vol. 15, Luis Garcia Ballester, Eustaquio Sanchez Salor and Richard J. Durling, ed., in *Isis* 77 (1986): 367–68.

John W. Shirley and F. David Hoeniger, eds., *Science and the Arts in the Renaissance* in *Bulletin of the History of Medicine* 61 (1987): 121–22.

Ann Carmichael, *Plague and the Poor in Renaissance Florence* in *Journal of the History of Medicine and Allied Sciences* 42 (1987): 91–93.

Sanità e Società. Vol. 1. Friuli-Venezia-Giulia, Secoli XVI–XX in *Journal of the History of Medicine and Allied Sciences* 43 (1988): 213–15.

Vivian Nutton, *John Caius and the Manuscripts of Galen* in *Bulletin of the History of Medicine* 62 (1988): 299–300.

Jole Agrimi and Chiara Crisciani, *Edocere medicos: medicina scolastica nei secoli XIII–XV* in *Bulletin of the History of Medicine* 63 (1989): 666–67.

Guy Sabbah, et al., eds. *Bibliographie des textes médicaux latins: Antiquité et haut moyen âge* in *Bulletin of the History of Medicine* 63 (1989): 693–94.

Charles B. Schmitt, et al., eds., *The Cambridge History of Renaissance Philosophy* in *The English Historical Review* (1989): 409–11.

Arnald of Villanova, *Arnaldi de Villanova Opera medica omnia IV*, Luke Demaitre, et al., eds., in *Bulletin of the History of Medicine* 64 (1990): 301–2.

G. R. Dunstan, ed., *The Human Embryo: Aristotle and the Arabic and European Traditions* in *Bulletin of the History of Medicine* 65 (1991): 266–67.

Charles B. Schmitt, *Reappraisals in Renaissance Thought* in *Isis* 82 (1991): 554–55.

Berengario da Carpi, *On Fracture of the Skull or Cranium*, L. R. Lind, tr., in *Bulletin of the History of Medicine* 65 (1991): 583–85.

Alfredo Serrai, Conrad Gesner, ed. *Maria Cochetti*, in *History of Universities* 12 (1993): 415–16.

Micrologus: Natura, scienze e società medievali. Vol. 1, 1993: I discorsi dei corpi in *Bulletin of the History of Medicine* 68 (1994): 707–8.

W. F. Bynum and Roy Porter, eds., *Medicine and the Five Senses* in *Social History of Medicine* (1994): 321–22.

Stefano Caroti, *I codici di Bernardo Campagna: Filosofia e medicina alla fine del sec. XIV* in *Isis* 86 (1995): 100–1.

Mirko D. Grmek et al., Storia *del pensiero medico occidentale, vol. 1, Antichità e Medioevo*, in *Physis* 33 (1996): 335–36.

William Newman, *Gehennical Fire: The Lives of George Starkey, an American Alchemist in the Scientific Revolution* in *Renaissance Quarterly* 50 (1997): 331–32.

Jon Arrizabalaga, et al., *The Great Pox* in *Nuncius: Annali di Storia della Scienza* 12 (1997): 554–58.

Ana Isabel Martín Ferreira, *El humanismo médico en la Universidad de Alcalá (siglo XVI)* in *Bulletin of the History of Medicine* 72 (1998): 324–25.

Giovanna Ferrari, *L'esperienza del passato: Alessandro Benedetti filologo e medico umanista* in *Renaissance Quarterly* 51 (1998): 1339–40.

John M. Headley, *Tommaso Campanella and the Transformation of the World* and Ronnie H. Terpening, *Lodovico Dolce, Renaissance Man of Letters* in *Renaissance Quarterly* 52 (1999): 496–98.

Agostino Paravicini Bagliani and Francesco Santi, eds., *The Regulation of Evil: Social and Cultural Attitudes to Epidemics in the Late Middle Ages* in *Bulletin of the History of Medicine* 73 (1999): 488–89.

Joseph Ziegler, *Medicine and Religion c. 1300: The case of Arnau de Vilanova* in *American Historical Review* (April 2000): 603.

Monica H. Green, ed. and tr., *The Trotula: A Medieval Compendium of Women's Medicine* in *Bulletin of the History of Medicine* 76 (2002): 121–22.

Giorgio Weber, *Sensata veritas. L'affiorare dell'anatomia patologica, ancora innominata, in scritti di anatomisti del '500, in appendice il Liber introductorius anatomiae (1536) di Niccolò Massa* in *Renaissance Quarterly* 61 (2008): 255–56.

John Henderson, *The Renaissance Hospital: Healing the Body and Healing the Soul* in *Renaissance Quarterly* 60 (2007): 637–39.

David Gentilcore, *Medical Charlatanism in Early Modern Italy* in *Social History of Medicine* 20 (2007): 393–394.

Pieter De Leemans and Michèle Goyens, eds., *Aristotle's Problemata in Different Times and Tongues* in *Early Science and Medicine* 13 (2008): 199–201.

Jorge Canizares-Esguerra, *Nature, Empire, and Nation: Explorations of the History of Science in the Iberian World* in *Journal of Interdisciplinary History* 39 (2008): 423–24.

CPSIA information can be obtained
at www.ICGtesting.com
Printed in the USA
BVOW06*0226240517
485027BV00004B/7/P